Enzymes as Catalysts in Organic Synthesis

NATO ASI Series

Advanced Science Institutes Series

A series presenting the results of activities sponsored by the NATO Science Committee, which aims at the dissemination of advanced scientific and technological knowledge, with a view to strengthening links between scientific communities.

The series is published by an international board of publishers in conjunction with the NATO Scientific Affairs Division

A Life Sciences B Physics	Plenum Publishing Corporation London and New York
C Mathematical and Physical Sciences	D. Reidel Publishing Company Dordrecht, Boston, Lancaster and Tokyo
D Behavioural and Social Sciences E Engineering and Materials Sciences	Martinus Nijhoff Publishers The Hague, Boston and Lancaster
F Computer and Systems Sciences G Ecological Sciences	Springer-Verlag Berlin, Heidelberg, New York and Tokyo

Series C: Mathematical and Physical Sciences Vol. 178

Enzymes as Catalysts in Organic Synthesis

edited by

M. P. Schneider

University of Wuppertal, F.R.G.

D. Reidel Publishing Company

Dordrecht / Boston / Lancaster / Tokyo

Published in cooperation with NATO Scientific Affairs Division

Proceedings of the NATO Advanced Research Workshop on
Enzymes as Catalysts in Organic Synthesis
Reisensburg/Ulm (Donau), F.R.G.
June 16-22, 1985

Library of Congress Cataloging in Publication Data

NATO Advanced Research Workshop on Enzymes as Catalysts in Organic Synthesis (1985 : Reisensburg, Germany)
Enzymes as catalysts in organic synthesis.

(NATO ASI series. Series C, Mathematical and physical sciences; vol. 178)
"Proceedings of the NATO Advanced Research Workshop on Enzymes as Catalysts in Organic Synthesis, Reisensburg / Ulm (Donau), F.R.G., June 16–22, 1985"—T.p. verso.
Includes index.
1. Enzymes—Industrial applications—Congresses. 2. Chemistry, Organic—Synthesis—Congresses. 3. Catalysts—Congresses. I. Schneider, M. P. II. Title. III. Series: NATO ASI series. Series C, Mathematical and physical sciences; vol. 178.
TP248.E5N37 1985 660.2'995 86–10222
ISBN 90–277–2267–6

Published by D. Reidel Publishing Company
P.O. Box 17, 3300 AA Dordrecht, Holland

Sold and distributed in the U.S.A. and Canada
by Kluwer Academic Publishers,
101 Philip Drive, Assinippi Park, Norwell, MA 02061, U.S.A.

In all other countries, sold and distributed
by Kluwer Academic Publishers Group,
P.O. Box 322, 3300 AH Dordrecht, Holland

D. Reidel Publishing Company is a member of the Kluwer Academic Publishers Group

Printed in The Netherlands

CONTENTS

Posters

LIST OF PARTICIPANTS

Benner, S. A., Department of Chemistry, Harvard University, 12 Oxford Street, Cambridge Massachusetts 02138, USA

Berti, G., Institute of Organic Chemistry, Faculty of Pharmacy, University of Pisa, Via Bonanno 6, I-56100 Pisa, Italy

de Bont, J. A. M., Department of Microbiology, Agricultural University, De Dreijen 12, NL-6703 BC Wageningen, The Netherlands

Brooks, D. W. Abbott Laboratories, Pharmaceutical Products Division, Abbott Park, Illinois 60064, USA

Collins, S. H., Biological Products Research, ICI Agricultural Division, P.O. Box 1, Billingham, Cleveland, U. K.

Crout, D. H. G., Department of Chemistry, University of Warwick, Coventry CV4 7AL, U. K.

Duarte, J. C., Department of Chemical Industries, LNETI, 2745 Queluz, Portugal

van der Eycken, J., Laboratorium voor Organische Chemie, Rijksuniversiteit Gent, Krijgslaan 281, B-9000 Gent, Belgium

Fuganti, C., Dipartimento di Chimica, Politecnico di Milano, Piazza Leonardo da Vinci 32, I-20133 Milano, Italy

Furstoss, R. Laboratory of Organic and Bioorganic Chemistry, Department of Chemistry, Université d'Aix Marseille, 70 route Léon Lachamp, F-13288 Marseille Cedex 9, France

Gais, H. - J., Institut für Organische Chemie und Biochemie, TH Darmstadt, D-6100 Darmstadt, FGR

Ghisalba, O., Central Research Laboratories, Ciba-Geigy Ltd., CH-4002 Basel, Switzerland

Godtfredsen, S. E., Novo Industri A/S, Novo Alle, DK-2880 Bagsvaerd, Denmark

Goodhue, C. T., Eastman Kodak, Biosciences Division, Research Laboratories, 1669 Lake Ave., Rochester, N.Y. 14650, USA

Guanti, G., Institute of Organic Chemistry, University of Genova, Centro C.N.R., Palazzo delle Scienze, Corso Europa, I-16132 Genova, Italy

Hilvert, D. M., Laboratory of Bioorganic Chemistry and Biochemistry, The Rockefeller University, 1230 York Avenue, New York, N. Y. 10021, USA

Holm, K., Kjemisk Institutt, Universitetet i Oslo, P.B. 1033, Blindern, Oslo 3, Norway

Hoppe, D., Department of Organic Chemistry, University of Göttingen, Tammannstr. 2, D-3400 Göttingen, FRG

Krämer, D., Röhm GmbH, Postfach 4242, D-6100 Darmstadt, FRG

Laumen, K., FB9-Organische Chemie, Bergische Universtität-GH-Wuppertal, D-5600 Wuppertal 1, FRG

Lemière, G. L., Laboratory of Organic Chemistry, University of Antwerp (R.U.C.A.), Groenenborgerlaan 171, B-2020 Antwerp, Belgium

Leatherbarrow, R. J., Department of Chemistry, Imperial College of Science & Technology, London SW7 2AY, U.K.

Mitchell, M.B., Smith, Kline and French, The Frythe, Welwyn AC6 9AR, Herts, U.K.

Ohno, M., Faculty of Pharmaceutical Sciences, University of Tokyo, 7-3-1-Hongo, Bunkyo-ku, Tokyo 113, Japan

Procter, G., Department of Chemistry, University of Salford, Salford, U.K.

Pugh, S.Y.R., Biotechnology Centre, Cranfield Institute of Technology, Cranfield, Bedfordshire, MK43 0AL, U.K.

Reimerdes, E. H., FB9-Lebensmittelchemie, Bergische Universität-GH-Wuppertal, D-5600 Wuppertal 1, FRG

Riefling, B., Merck Darmstadt, Pharmazeutische Chemie, Frankfurter Str. 250, D-6100 Darmstadt, FRG

Roberts, S. M., Department of Microbiological Chemistry,
Glaxo Group Research Ltd., Greenford, Middlesex UB6 OHE, U.K.

Schneider, M., FB9-Organische Chemie, Bergische Universität-GH-Wuppertal,
D-5600 Wuppertal 1, FRG

Schoemaker, H. E., DSM, Research and Patents, P.O. Box 18,
NL-6160 MD Geleen, The Netherlands

Schröder, T., Bayer AG, Verfahrensentwicklung Biochemie,
D-5600 Wuppertal 1, FRG

Sinay, P., Laboratoire de Biochimie Structurale, U.E.R. de Sciences
Fondamentales et Appliquées, Université d'Orléans, F-45046 Orléans-Cedex,
France

Stegelmeyer, H., Bayer AG, Pharmaforschung, D-5600 Wuppertal 1, FRG

Simon, H., Institute for Organic Chemistry, Technical University Munich,
Lichtenbergstr. 4, D-8046 Garching, FRG

Tramper, J., Department of Food Science, Agricultural University,
De Dreijen 12, NL-6703 BC Wageningen, The Netherlands

Wandrey, C., Nuclear Research Center, Institute of Biotechnology,
P.O. Box 1913, D-5170 Jülich, FRG

Winterfeldt, E., Institut für Organische Chemie, Universität Hannover,
D-3000 Hannover, FRG

Wong, C.-H., Department of Chemistry, Texas A & M University, College
Station, Texas 77843-3255, USA

Wullbrandt, D., Hoechst AG, Hauptlaboratorium G830, D-6230 Frankfurt
am Main 80, FRG

Young, D. W., School of Chemistry and Molecular Sciences, University of
Sussex, Falmer, Brighton, BN1 9QJ, U.K.

Ziffer, H., National Institutes of Health, Bethesda, Maryland 20205,
USA

PREFACE

Enzymatic reactions can frequently be employed to effect transformations in organic syntheses that otherwise would be difficult to carry out. However, it is only recently that systematic attempts have been made by organic chemists to explore the scope and utility of such methods.

It ist now becoming widely appreciated that, compared with chemical methods, enzyme catalyzed processes often offer significant advantages including those of efficiency, chemoselectivity, regioselectivity, diastereo- and enantioselectivity. It is not surprising therefore that these reactions are coming into ever increasing prominence both on a laboratory and industrial scale for the preparation of chiral building blocks and the production of pharmaceuticals, agro- and fine chemicals. However, the ground rules for such applications are still sketchy and in this respect interdisciplinary, fundamental, research in academic laboratories of the scope and applicability of enzymatic methods is urgently needed.

In order to bring together scientists from various different backgrounds with an interest in this area, the present NATO Advanced Research Workshop was organized. It was attended by synthetic organic chemists, bioorganic chemists, biochemists and microbiologists from academic, government and industrial laboratories. A particular feature of this meeting was the main focus on the requirements of synthetic organic chemistry. The participating organic chemists were giving a broad view of the current state of organic synthesis against which the putative advantages of enzymatic methods would have to be measured.

The activities of the workshop consisted of lectures, poster sessions and round-table discussions. Important themes illustrated in the lectures included applications of hydrolytic enzymes and oxido-reductases (isolated and whole cell systems) for the preparation of enantiomerically pure chiral building blocks, the use of aldolases for enzymatic carbon-carbon bond formation, the preparation of chirally labelled compounds, co-factor recycling and the application of artificial co-factors, the application of microbial transformations in the synthesis of natural products, the development of bioreactors using immobilized enzymes for the industrial production of chiral organic compounds in optically pure form (e. g. L-aminoacids) and for the treatment of byproducts from the food industry and chemical plants, and the industrial production of enzymes.

Recent achievements in enzyme engineering, the preparation and properties of semisynthetic enzymes and the possibilities of genetic engineering in producing enzymes tailored to specific purposes were reported.

The conclusion emerging from the workshop was clear: enzyme-catalyzed reactions play an important role both in academic and industrial laboratories, particularly for the preparation of enantiomerically pure building blocks needed in the syntheses of biologically active compounds. In the discussions frequently the different viewpoints of academic and industrial participants emerged. A particularly useful outcome of the meeting was therefore the clear underlining of the need for more basic research in academic laboratories and increased collaborations with industrial groups.

We are grateful to the NATO Scientific Affairs Division for funding a meeting which was intended

- to close a certain gap between a wealth of biochemical knowledge and its practical use in organic synthesis; and

- to amalgamate the diverse, yet complementary, viewpoints of different scientific disciplines in order to serve its main purpose: to promote an increased use of

Enzymes as Catalysts in Organic Synthesis.

M. Schneider

Lectures

BAKER'S YEAST MEDIATED PREPARATION OF CARBOHYDRATE-LIKE CHIRAL SYNTHONS

Claudio Fuganti
Dipartimento di Chimica del Politecnico, Centro del CNR per la Chimica delle Sostanze Organiche Naturali, 20133 Milano, Italy

ABSTRACT: Baker's yeast fermenting on D-glucose converts C_6-C_3 α,β-unsaturated aromatic aldehydes into C_6-C_5, carbohydrate-like, (2S,3R) methyl diols (4), with yield of about 25%, as well as into products (2) and (3). Products (4) and homologous diols prepared by enantio- and stereoselective reduction of α-acetoxy ketones such as (75)-(77), can be used instead of natural carbohydrates as starting materials for the synthesis of optically active forms of natural products belonging to quite different structural classes.

A current approach[1] to the synthesis of enantiomerically pure forms of natural products and drugs relies on the use as starting materials of components of the collection of inexpensive, readily available optically active compounds produced by nature, called the 'pool of chirality', which includes, among others, carbohydrates, amino acids, hydroxyacids like tartaric, malic, lactic, and citramalic, alcohols like D-mannitol, and few terpenes. However, the present composition of the 'pool of chirality' is far from being satisfactory, the major drawback arising from the fact that most of the components are really available in only one enantiomeric form, a circumstance which dictates, when the absolute configuration of the target molecule is opposite to the one of the chosen starting material, chemical manipulation of the chiral center(s), usually through multistep, low-yield sequences.

This is the case of the synthesis of the 2,3,6-trideoxy-3-amino hexoses of the L series present in the therapeutically important anthracycline glycosides adriamycin[2] and its 4'-epimer,[3] which can be realized either from inexpensive hexoses of the D series,[4] but with a critical inversion of configuration at position 5 at some stage of the sequence, or from the rather rare 6-deoxy sugar L-rhamnose.[5]

Furthermore, the choice of types of chirality is rather poor, those

M. P. Schneider (ed.), Enzymes as Catalysts in Organic Synthesis, 3–17.

of the type $R,R^{1}CHX$, where X= oxygen or nitrogen functions, being particularly abundant, whereas those of the type $R,R^{1},R^{2}CH$ and $R,R^{1},R^{2}C(OH)$, quite frequent, for example, amongst the insect pheromones, occur rather rarely.

Consequently, there is at present an interest in expanding the composition of the 'pool of chirality' and new chiral products are expected to be accessible from the microbial transformations of non-conventional substrates. Microbes are indeed capable of performing a variety of transformations of non-conventional substrates using enzymes either of the primary or of the secondary metabolism.[6] From a practical point of view, we would expect to be synthetically useful those transformations of non-conventional substrates leading to optically active products performed by microorganisms commercially available at low cost, possessing large quantities of enzymes (usually of the primary metabolism), and showing a large substrate specificity, while mantaining a precise reaction stereospecificity.

Baker's yeast meets most of the above requirements, and its ability to transform stereoselectively non-conventional substrates into chiral products using its constitutive enzymes has long been recognized.[7] However, apart from (R) phenylacetylcarbinol, obtained in an acyloin-type condensation from benzaldehyde and 'statu nascendi' acetaldehyde,[8] most of the optically active products obtained up to now by this means contain chiral centres of the type $R,R^{1},R^{2}CH$ and/or $R,R^{1}CHOH$, which arise by formal addition of hydrogen onto (activated) double bonds and/or carbonyl carbon of the precursors. These processes occur without modification of the original carbon framework, and are now being mimicked in non-enzymic asymmetric synthesis.[9]

In the reawakening of interest for the microbially aided preparation of synthetically useful chiral products, our observation[10] that baker's yeast fermenting on D-glucose converts aromatic α,β-unsaturated aldehydes (1) into C_6-C_5 (2S,3R) 2,3-methyldiols (4), with yields of 25%, as well as into products (2) and (3) (equation 1) is particularly fruitful from a synthetic point of view.

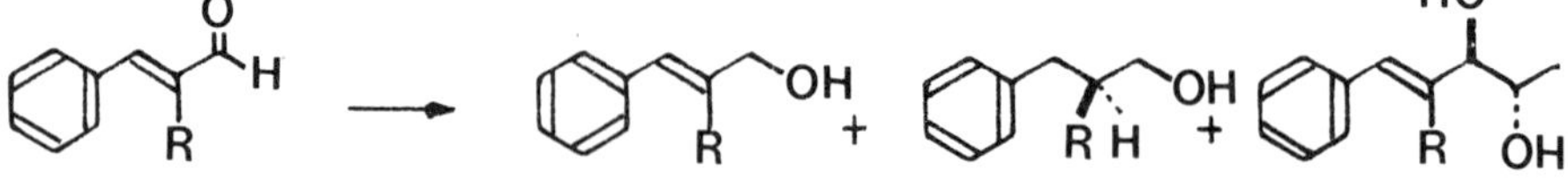

(1) (2) (3) (4)

R= H, Me, Br

equation 1

Indeed, whereas the production from (1) of (2) and (3) falls amongst the known baker's yeast capacities,[11] the formation of the methyldiols (4), which contain two more carbon atoms than the precursor aldehydes and two adjacent chiral centres of the type R,R^1CHOH, is new and, to our knowledge, without equivalent in non-enzymic asymmetric synthesis.

The formation of the (2*S*,3*R*) diols from α,β-unsaturated aromatic aldehydes by fermenting baker's yeast can be viewed as the overall consequence of a complex aldehyde condensation-reduction involving two distinct chemical operations. (i) Addition of a C_2 unit equivalent of acetaldehyde onto the *si* face of the carbonyl carbon of the α-position unsaturated aldehydes forms (*R*) α-hydroxyketones, and (ii) reduction of the latter intermediates on the *re* face of the carbonyl gives rise to the diols actually isolated (equation 2).

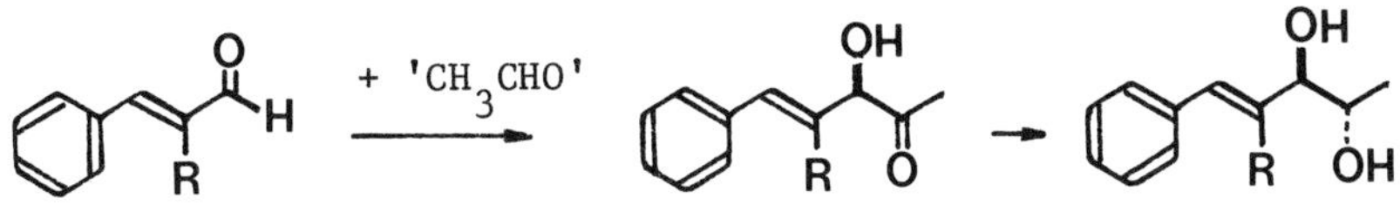

equation 2

Under suitable experimental conditions,[12] (*R*) hydroxyketones can be obtained as sole transformation products of the aldehydes by actively fermenting baker's yeast.

Experiments[13] designed to elucidate the substrate specificity of the sequence of equation 2 indicated that there is some tolerance by the enzymic system(s) involved as far as the structure of the aromatic aldehydes and the substituents in the α-position are concerned. However, α-ethyl and α-propyl cinnamaldehydes are not converted into the corresponding methyl diols. Furthermore, acetaldehyde is the only aldehyde accepted as second terminus of the reaction; when cinnamaldehyde was incubated with yeast in the presence of propionaldehyde or butyraldehyde, ((2*S*,3*R*) (7) was isolated as the sole transformation product. The failure of yeast to convert α-ethylcinnamaldehyde and propionaldehyde into the type of diols of equation 2 is probably due to the inability of these materials to be accepted as substrates by the condensing enzyme(s) (first part of equation 2). The substrate specificity of the second part of equation 2 is not as restricted, since the synthetic α-hydroxyketones derived from these aldehydes, or from other carbonyl compounds, are stereospecifically reduced by yeast (Figure 1).[14]

The (2*S*,3*R*) diols (7) and (8), obtained from cinnamaldehyde and α-methylcinnamaldehyde, respectively, have an absolute configuration matching that at positions 5 and 4 of 6-deoxy-L-sugars and they have been converted, in the work designed to establish their absolute configuration,[15] into L-amicetose (23) and L-olivomycose (24).

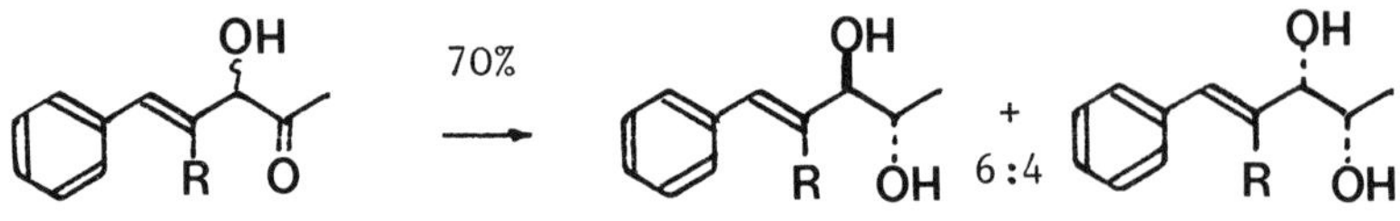

(5) R= H (7) R= H (9) R= H
(6) R= Me (8) R= Me (10) R= Me

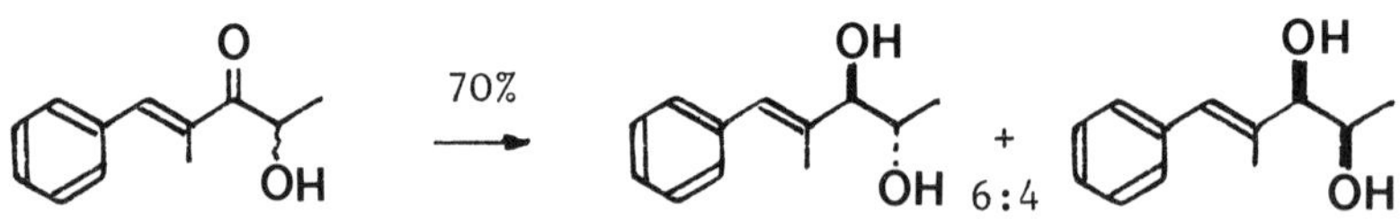

(11) (8) (12)

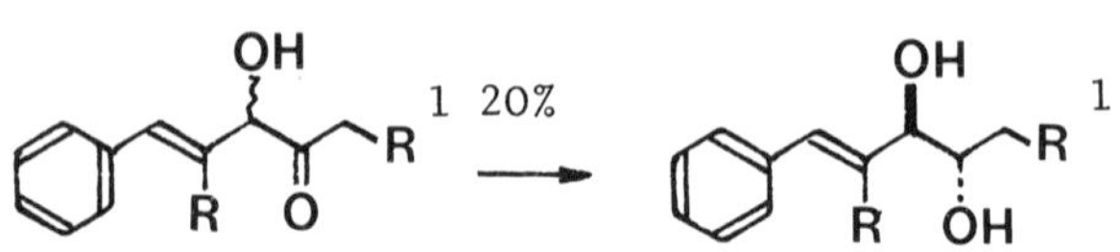

+ THREO ISOMER
9:1 (10%)

(13) R= Et; R^1 = H (15) R= Et; R^1 = H
(14) R= H; R^1 = Me (16) R= H; R^1 = Me

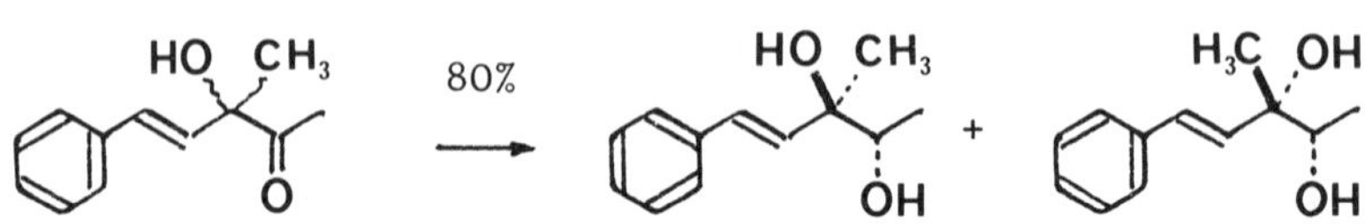

(17) (18) 1:1 (19)

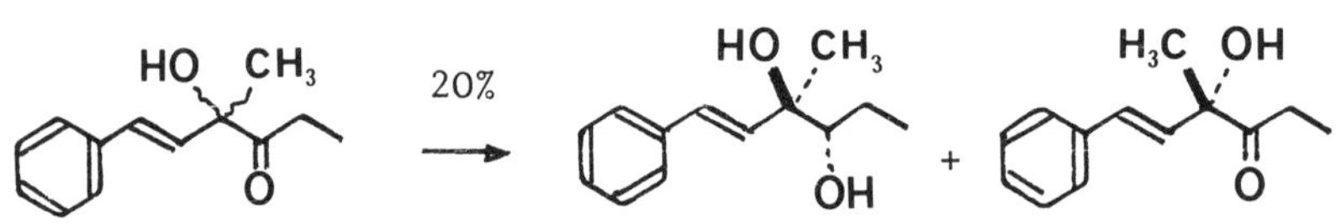

(20) (21) (22)

Figure 1: Conversion of racemic α-hydroxyketones to diols by baker's yeast

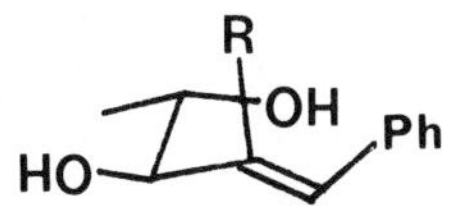

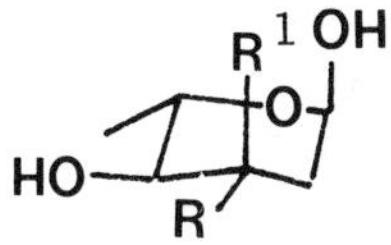

(7) R= H (23) R= R^1 = H

(8) R= Me (24) R= Me; R^1 = OH

Thus, (7) and (8) can be considered as carbohydrate-like, non-carbohydrate derived chiral synthons, the value in synthesis of which is associated with the fact that it is possible to obtain, from protected forms of (7) and (8), the C_4 and C_5 chiral carbonyl compounds (25) and (27), which can be converted by treatment with base into their α-epimers (26) and (28). Similarly, from the diols (9), (10), (12), (15), (16) and (21), prepared in fermenting baker's yeast from the synthetic α-hydroxyketones (5), (6), (11), (13), (14) and (20), the chiral <u>erythro</u> and <u>threo</u> carbonyl compounds (29)-(35) have been prepared (Figure 2).

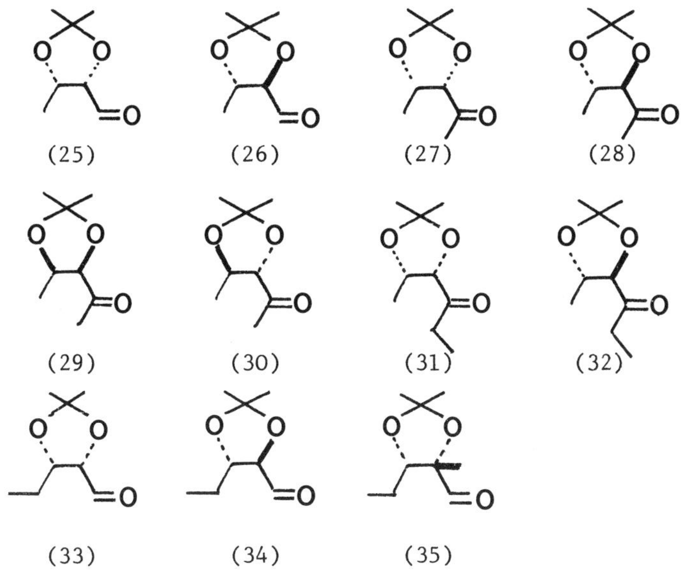

Figure 2: Chiral carbonyl compounds obtained from the diols of Figure 1

Thus, these synthons seem particularly useful for the preparation of molecules containing relatively few carbon atoms that are chiral due to oxygen substitution, and their value is enhanced because some of them are available in both the enantiomeric forms. Furthermore, as a result of the increasing knowledge of methods for stereocontrolled chain elongation through nucleophilic addition onto sp^2 carbon, the number of derived synthons available has been increased considerably.

An application of the above chiral synthons as starting materials alternative to natural carbohydrates in the synthesis of carbohydrates is illustrated by the straightforward obtainment from (25) and (26) of N-trifluoroacetyl-L-acosamine (39) and L-daunosamine (43). Thus, the <u>erythro</u> C_4 aldehyde (25) was converted into the C_6 α,β-unsaturated ester (36), which added ammonia stereoselectively to give the (3S,4S,5S) C_6-N product (37), yielding, in turn, upon hydrolysis to (38), N-protection and DIBAH reduction, the required aminosugar (39).[16] (Figure 3)

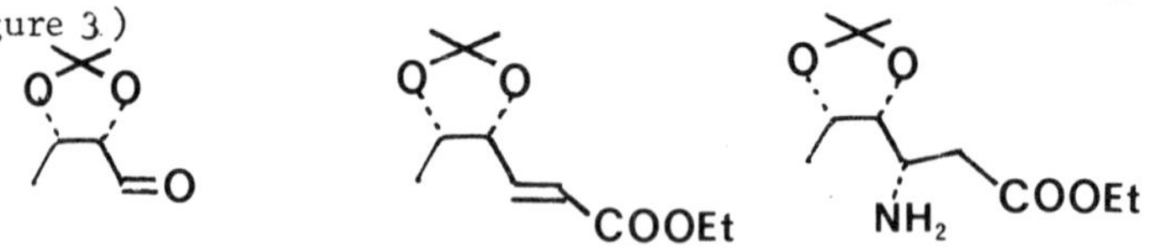

(25) (36) (37)

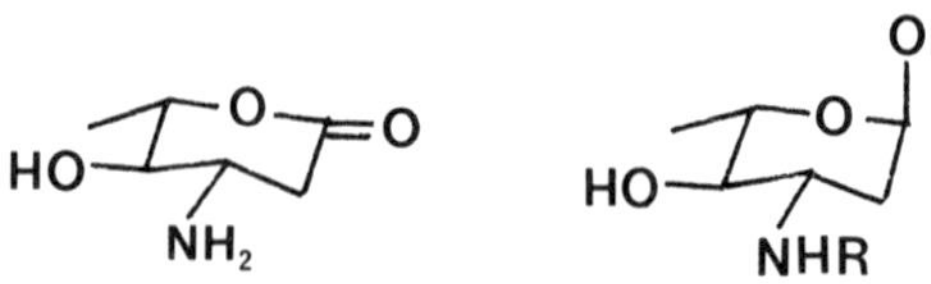

(38) (39) R= $COCF_3$

Figure 3: Steps in the synthesis of trifluoroacetyl-L-acosamine (39)

The synthesis of N-trifluoroacetyl-L-daunosamine (43) proceeded from the <u>threo</u> aldehyde (26), the C_4-N phenylsulfenimine (40), addition of diallylzinc to give stereoselectively the C_7-N eductu (42), converted upon N-protection and ozonolysis, into (43).[17] (Figure 4)

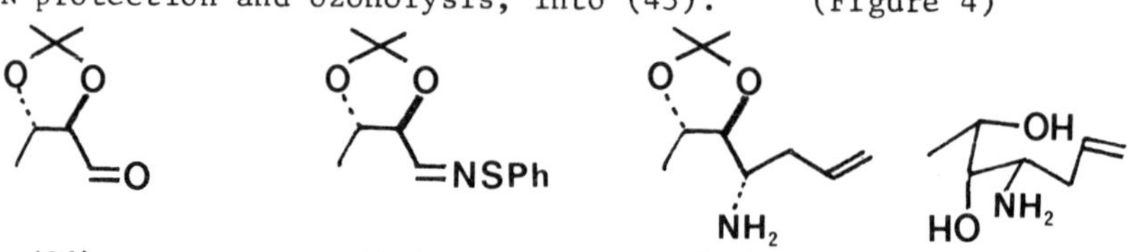

(26) (40) (41) (42)

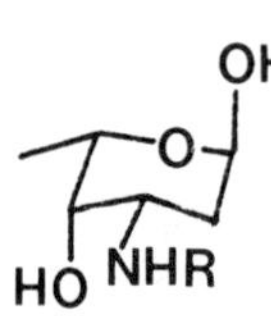

(43) R= $COCF_3$

Figure 4: Steps in the synthesis of trifluoroacetyl-L-daunosamine

Similar strategy applied onto the phenylsulfenimine (44), prepared from the *threo* methylketone (28), yielded N-protected vancosamine (48). However, using allylmagnesiumbromide as carbon nucleophile, from (44) the adduct (45) was prepared, yielding, as above, the *xylo* isomer (47).[18] (Figure 5)

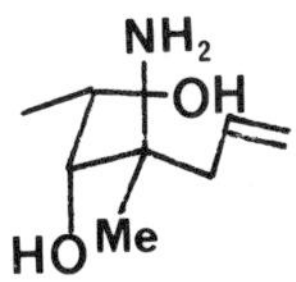

(45)

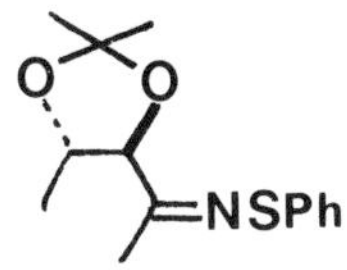

(44)

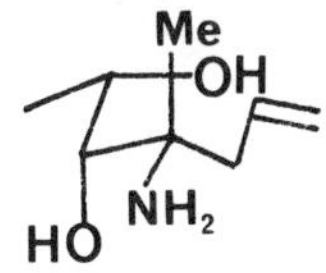

(46)

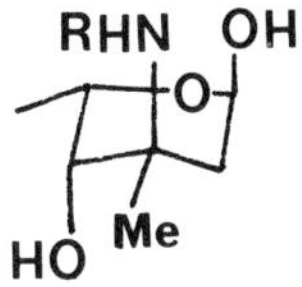

(47)

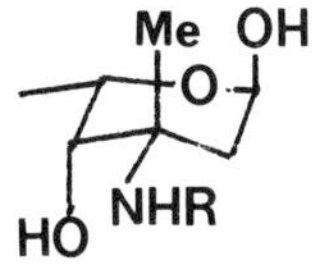

(48)

Figure 5: Steps in the synthesis of N-protected vancosamine (48) and of the *xylo* isomer (47).

The synthesis of the aminodeoxysugar derivatives (39), (43), (47) and (48) from carbohydrate-like (7) and (8) can be compared favourably with those based on natural carbohydrates as starting materials in terms of number of steps and lack of manipulations of the chiral centres once the carbon framework has been assessed.

The aldehyde (25) resulted also starting material suitable for the preparation of the aldehydes (51) and (52), key intermediates in the synthesis of LTB$_4$ (53). Previously, these materials had been prepared from 2-deoxy-D-ribose and L-arabinose, respectively, in 9 steps each. Again, upon suitable choice of the carbon nucleophiles, from (25), the two adducts (49) and (50) were preferencially or exclusively obtained, subsequently converted into (51) and (52) through unexceptional steps.[19] (Figure 6)

Furthermore, addition of diallylzinc[20] onto the epimeric aldehydes (25) and (26) afforded the carbohydrate-like C$_7$ adducts (54) and (55).

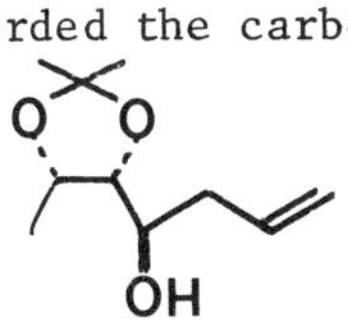

(54)

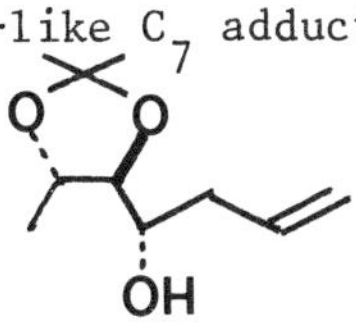

(55)

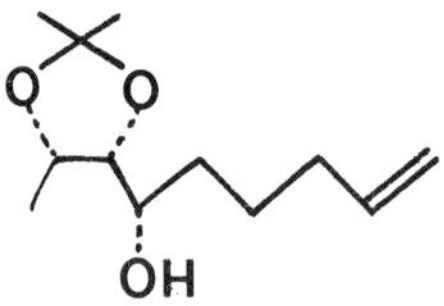

(49)

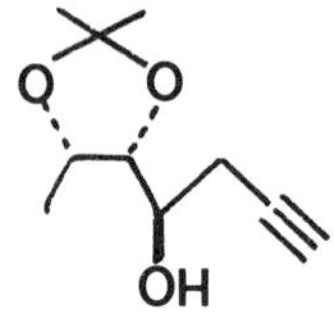

(50)

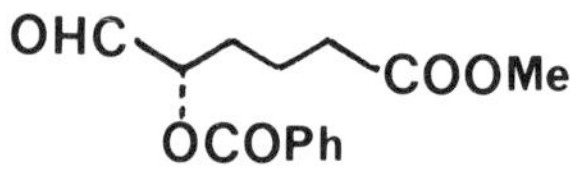

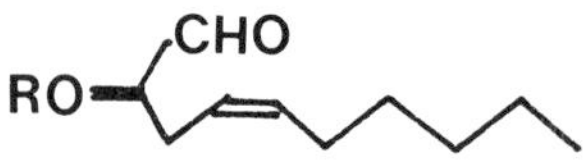

(51) (52) R= $SiPh_2Bu^t$

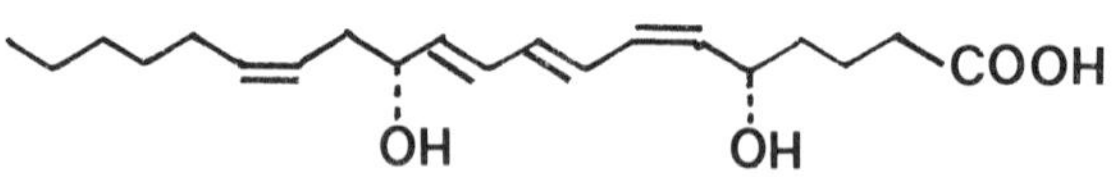

(53)

Figure 6: Steps in the synthesis of LTB_4(53) from (25)

The synthetic value of such products as (54) and (55) is not limited to the preparation of sugar derivatives, since they contain, in masked form, (R) and (S)-malaldehyde, the two carbonyl functions of which can be generated regioselectively by the appropriate choice of reagent. Moreover, in (54) and (55), the three adjacent chiral centres can be manipulated easily by the appropriate use of protecting groups. Accordingly, (54) has been used, as an alternative to natural carbohydrates, as a chiral starting material for the synthesis of (+) exo-brevicomin (56), (R)[21] hexanolide (57) and (5R,6S,7S) 6,7-isopropylidenedioxy octanolide (58).

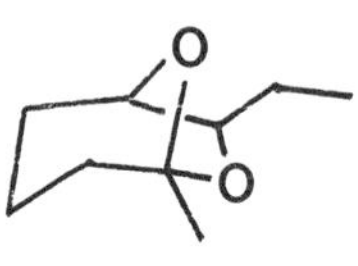

(56)

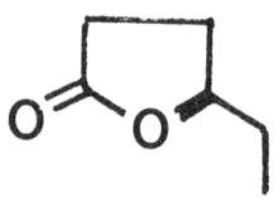

(57)

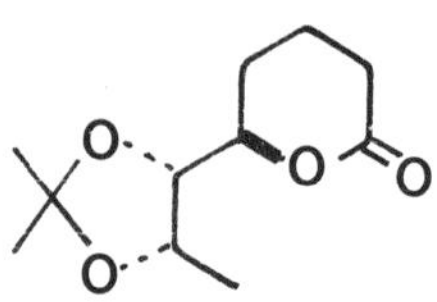

(58)

Moreover, from (54), through the key intermediacy of the epoxy alcohols (59), (60) and (61), the isomeric 2,3,6-trideoxy-4-amino-L-hexose derivatives (62)-(64) were prepared.[22] (Figure 7)

Taking advantage of the control in the addition of carbon nucleophiles onto the carbonyl carbon of the C_5 methyl ketone (27) exerted by the oxygen functions in α and β-positions, the pheromone (-)-frontalin (65)[23] and the C_{14} chiral chromanyl moiety of natural α-tocopherol

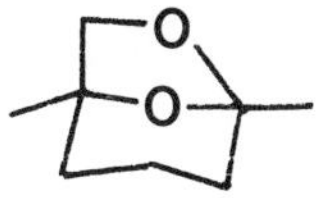

(65)

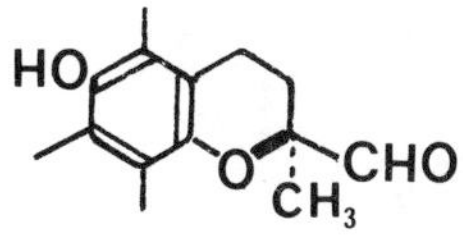

(66)

(66)[24] were prepared.

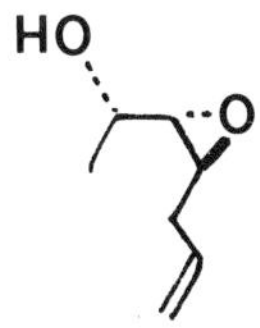

(59)

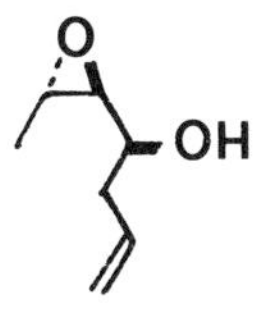

(60)

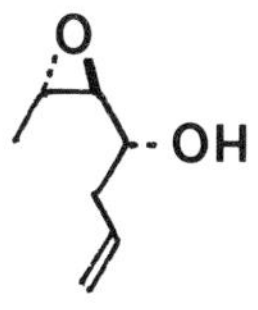

(61)

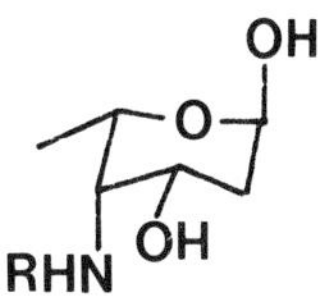

(62)

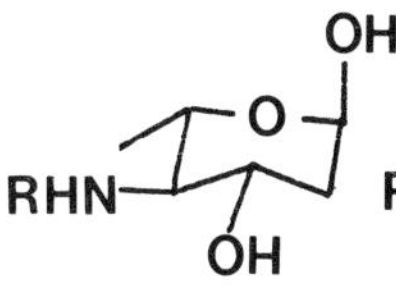

(63)

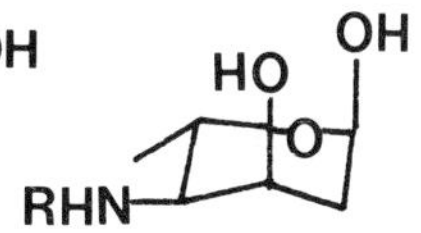

(64)

R= $COCF_3$

Figure 7: Steps in the synthesis of the isomeric aminodeoxy sugar derivatives (62)-(64) from (54)

From (7) and (8) we have prepared a variety of relatively small chiral synthons in the two enantiomeric forms, including products (67)-(74),[25,26] by means of suitable manipulation of the protecting groups.

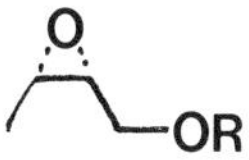

(67)

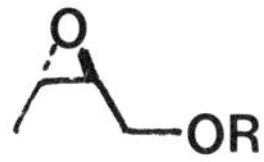

(68)

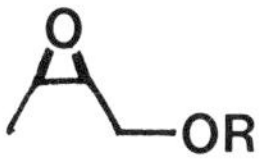

(69)

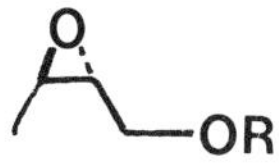

(70)

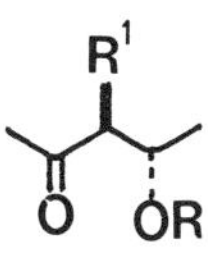

(71)

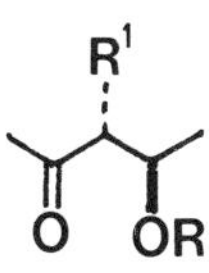

(72)

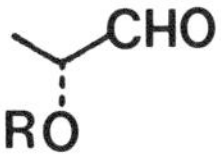

(73)

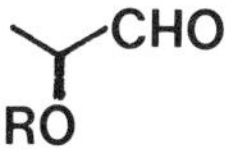

(74)

R= CH_2Ph; R^1 = H or Me

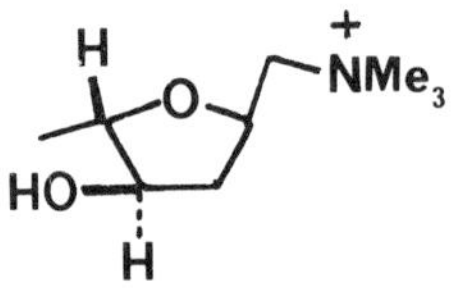

D(-)-allomuscarine[a]

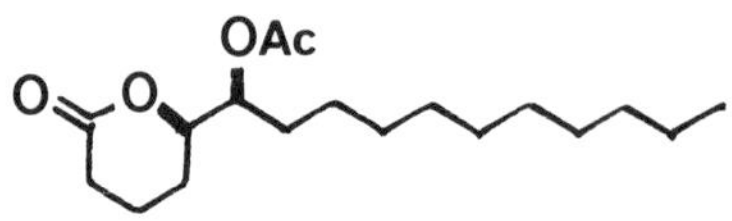

(5R,6S) 6-acetoxy-5-hexadecanolide[b]

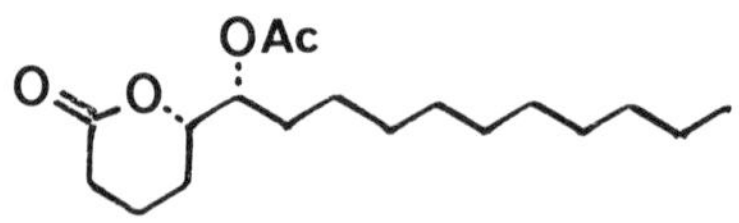

(5S,6R) 6-acetoxy-5-hexadecanolide[b]

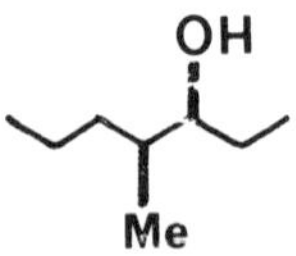

(3S,4S) 4-methylheptan-3-ol[c]

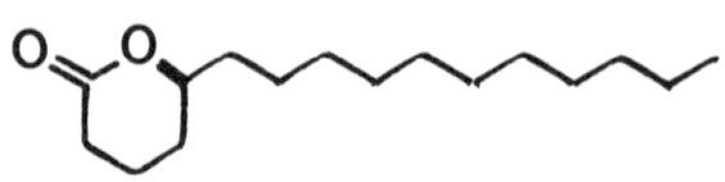

(S) 5-hexadecanolide[d]

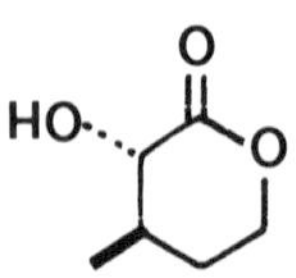

(-) verrucarino lactone[e]

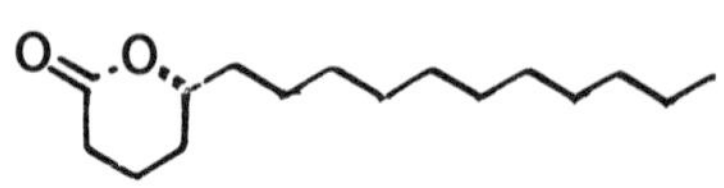

(R) 5-hexadecanolide[d]

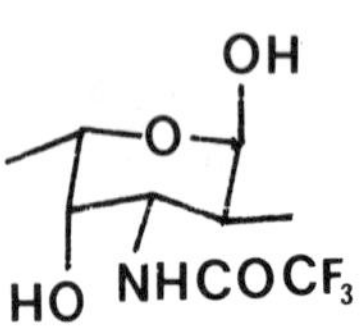

2-C-methyl-N-trifluoroacetyl L-daunosamine[f]

a) G.Fronza, C.Fuganti and P.Grasselli, Tetrahedron Letters,1978,3941
b) C.Fuganti, P.Grasselli and S.Servi, J.C.S.Chem.Commun.,1982,1285
c) ref. 25 d) S.Servi, Tetrahedron Letters,1983,2023 e) C.Fuganti and S.Servi, to be published f) C.Fuganti and G.Pedrocchi-Fantoni, to be published

Figure 8: Chiral products prepared from the (2S,3R) diol (7) prepared in fermenting baker's yeast from cinnamaldehyde

However, the present approach to the synthesis of enantiomerically pure forms of natural products based on the use as starting materials of (7) and (8), considered as carbohydrate-like chiral synthons, suffers of the drawback that only one enantiomeric form (namely the L) of the precursors are really available. We thought to overcome this problem by submitting to the action of baker's yeast the α-acetoxyketones (75)-(77), bearing two masked carbonyl functions (being the two carbonyl moieties revealable regioselectively with different reagents) in 1,6 and 1,5 relationships, respectively, in the expectation that baker's yeast will act on these substrates with the enantio- and stereoselectivity shown towards the α-hydroxyketones homologous of (5) and (6) (see Figure 1).

(75) (78)

(76) (79)

(77) (80)

If this had to be the case, from the reduction products of (75) and (77) we should have access to 2,3-dideoxyhexose and 2-deoxypentose of the D series, and from the reduction material of (76) to 2,3,6-trideoxy hexoses of the D and/or L series, depending upon the steric course of the chemical reduction of the methylketone liberated by ozonolysis at some stage of the sequence. Indeed, baker's yeast afforded from (75) optically pure (78), soon converted *via* (81) into 2,3-dideoxy-D-glucose (82). The lower homologue (76), on yeast treatment, afforded a reduction material containing *ca*. 75% of the (3*S*,4*R*) enantiomer (79). The diol obtained from (79) was crystalline, and afforded, on repeated crystallization, an optically pure material, from which carbohydrate-like (83) and (84) (cfr (54)) were obtained. Finally, product (77), under identical conditions, afforded a reduction material containing 65-70% of the (3_R_,4_S_) enantiomer (80). Again, the corresponding diol, on crystallization, gave a nearly optically pure material, from which the masked 2-deoxy-L-ribose (85) was obtained.

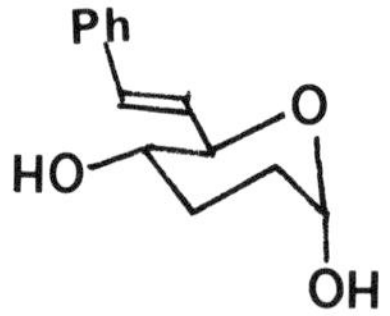

(81)

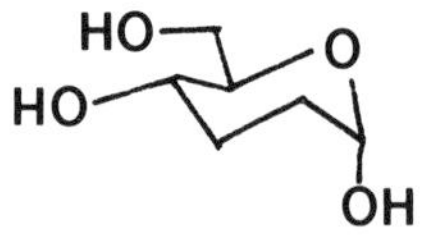

(82)

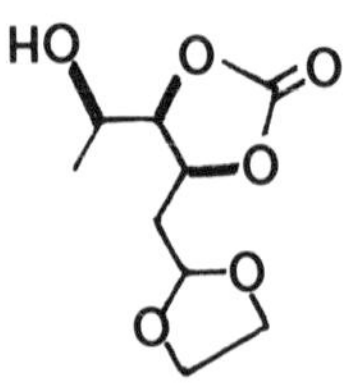

(83)

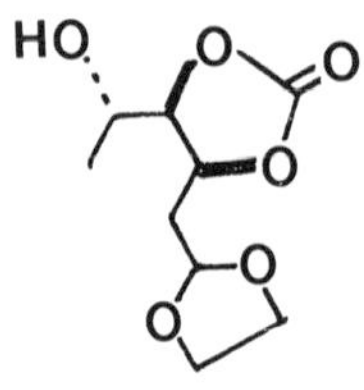

(84)

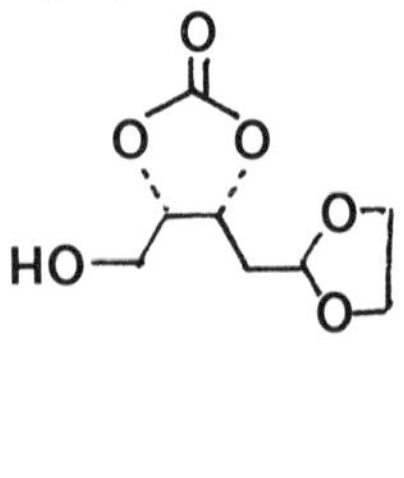

(85)

The α-acetoxyketones recovered from the preparation of (78)-(80) resulted optically active. The materials recovered at the end of several runs with yeast (from which (78), (79) and (80) were obtained with decreasing yields) were shown by ^{1}H NMR methods to contain 70-80% of a single enantiomer. From the latter products, on metal hydride reduction, the enantiomers of (78), (79) and (80) were obtained, thus increasing considerably the number of the chiral synthons available by this means.

The _re_ face hydrogen addition onto the carbonyl of (75), bearing two oxygen substituted carbon atoms in 1,2 and 1,4 relationship with the carbonyl is that expected by the Prelog rule. The prevalence of the _si_ face reduction of (77), possessing a pentacyclic acetale moiety in 1,3 relationship with the carbonyl, resembles, in some way, the mode of yeast reduction of 4-heterosubstituted 3-oxobutanoate esters.[27]

It has been shown[28] in professor's J.B.Jones laboratory that the α-acetoxyketones (75) and (76) are not accepted as substrates by purified yeast and horse liver alcohol dehydrogenases.

The latter observations thus show the advantage of the use, in the enzymic transformations of non-conventional substrates as source of chiral educts, of intact cells which allows the partecipation, within the bioconversion, of all the enzymic systems present. The work can be intrigued by the change in the stereochemistry even on subtle substrate structural modifications, but, when successful, offers the access to a variety of synthetically useful chiral synthons in a simple way, as in the present case.

Acknowledgements

This work has been supported by Farmitalia-C.Erba, spa, Milano, and by Piano Finalizzato Chimica Fine e Secondaria del CNR, Roma.

References and Footnotes

1 D.Seebach and E.Hungerbühler, in 'Modern Synthetic Methods 1980', R.Scheffold, Ed., Salle and Sauerländer, Frankfurt am Main, 1980, p 91

2 F.Arcamone, G.Franceschi, S.Penco, and A.Selva, 'Adriamycin, a novel antitumor antibiotic', Tetrahedron Letters,1969,1007

3 F.Arcamone, S.Penco, A.Vigevani, S.Redaelli, G.Franchi, A.Di Marco, A.M.Casazza, T.Dasdia, F.Tornelli, A.Necco, C.Soranzo,'Synthesis and antitumor properties of new glycosides of daunomycinone and adriamycinone', J.Med.Chem.,1975,18,703

4 D.Horton and W.Weckerle, 'Preparative synthesis of 3-amino-2,3,6-tri deoxy-L-lyxohexose (daunosamine) hydrochloride from D-mannose' Carbohydr.Res.,1976,46,227

5 J.P.Marsh, C.W.Mosher, E.M.Acton, L.Goodman,'The synthesis of daunosamine', Chem.Comm.,1967,973

6 K.Kieslich,'Microbial transformations of non-steroid cyclic compounds' G.Thieme Verlag, Stuttgard,1976

7 C.Neuberg,'Biochemical Reductions at the Expense of Sugars', Adv. Carbohydr.Chem.,1949,4,75

8 C.Neuberg and H.Ohle,'Biosynthetic carbon chain union in fermentation processes',Biochem.Z.,1922,128,610

9 H.C.Brown and G.Ganesh Pai,'Asymmetric Reduction of Prochiral Ketones with B-(3-Pinanyl)-9-borabicyclo[3.3.1]nonane',J.Org.Chem.,1985,50, 1384

10 C.Fuganti and P.Grasselli,'Transformations of non-conventional substrates by fermenting baker's yeast: production of optically active methyl-diols from aldehydes', Chem Ind(London),1977,983

11 F.S.Fisher and O.Wiedeman,'Ueber die Hydrierung ungesättigter α-Ketosäuren, Aldehyde und Alkohole durch gärende Hefe',Annalen,1934, 513,260

12 G.Bertolli, G.Fronza, C.Fuganti, P.Grasselli, L.Majori and F.

Spreafico,'On the α-ketols obtained from α,β-unsaturated aromatic aldehydes and baker's yeast', Tetrahedron Letters,1981,965

13 C.Fuganti, P.Grasselli and G.Marinoni,'Further studies on the transformations of unsaturated aldehydes by fermenting baker's yeast: a facile synthesis of L-olivomycose', Tetrahedron Letters,1979,1161

14 C.Fuganti, P.Grasselli, S.Servi, F.Spreafico, C.Zirotti and P.Casati, 'On the steric course of baker's yeast reduction of α-hydroxyketones' J.Org.Chem.,1984,49,4087

15 C.Fuganti and P.Grasselli,'Stereospecific synthesis from non-carbohydrate precursors of the deoxy- and methyl-branched deoxy-sugars L-amicetose, L-mycarose and L-olivomycose', J.C.S.Chem.Commun., 1978,299

16 G.Fronza, C.Fuganti and P.Grasselli,'Synthesis of N-trifluoroacetyl-L-acosamine and -L-daunosamine', J.C.S.Chem.Commun.,1980,442

17 C.Fuganti, P.Grasselli and G.Pedrocchi-Fantoni,'Stereospecific synthesis of N-benzoyl-L-daunosamine and L-ristosamine', J.Org.Chem., 1983,48,909

18 G.Fronza, C.Fuganti, P.Grasselli and G.Pedrocchi-Fantoni,'Synthesis of the four configurational isomers of N-benzoyl-2,3,6-trideoxy-3-C-methyl-3-amino-L-hexose from the (2S,3R)-diol obtained from α-methyl cinnamaldehyde by fermentation with baker's yeast', J.Carbohydr.Chem. 1983,2,225

19 C.Fuganti, S.Servi and C.Zirotti,'Non-carbohydrate based synthesis of natural LTB_4', Tetrahedron Letters,1983,5285

20 C.Fuganti, P.Grasselli, G.Pedrocchi-Fantoni, S.Servi and C.Zirotti, 'Carbohydrate-like chiral synthons. Preparation of (R)γ-hexanolide, (5R,6S,7S) 6,7-isopropylidendioxy-δ-octanolide and (+)-exobrevicomin from (2S,3S,4R) 2,3-isopropylidendioxy-4-benzyloxyhept-6-ene', Tetrahedron Letters,1983,3753

21 G.Fronza, C.Fuganti, P.Grasselli, G.Pedrocchi-Fantoni and C.Zirotti, 'On the steric course of the addition of diallylzinc onto α,β-dialkokoxy chiral carbonyl compounds: stereospecific synthesis of 2,6-dideoxysugars of the L-series', Tetrahedron Letters,1982,4143

22 G.Fronza, C.Fuganti, P.Grasselli, G.Pedrocchi-Fantoni,'Carbohydrate-like chiral synthons: synthesis of the N-trifluoroacetyl derivatives of 4-amino-2,4,6-trideoxy-L-lyxo, -L-arabino, -L-ribo-hexose from the (2S,3R)-2,3-diol formed from cinnamaldehyde in fermenting baker's yeast', Carbohydr.Res.,1985,136,115

23 C.Fuganti, P.Grasselli and S.Servi,'Synthesis of (-)-frontalin from the (2S,3R)-diol prepared from α-methylcinnamaldehyde and fermenting baker's yeast', J.Chem.Soc. Perkin I,1983,241

24 C.Fuganti and P.Grasselli,'Synthesis of the C_{14} chromanyl moiety of natural α-tocopherol', J.C.S.Chem.Commun.,1982,205

25 C.Fuganti, P.Grasselli, S.Servi and C.Zirotti,'Synthesis of the enantiomeric forms of cis and trans 1-benzyloxy-2,3-epoxybutane and of (3S,4S) 4-methyl-3-heptanol', Tetrahedron Letters,1982,4269

26 C.Fuganti, P.Grasselli, F.Spreafico, and C.Zirotti,'Synthesis of the enantiomeric forms of α- and β-alkoxy carbonyl compounds from the (2S,3R)-2,3-diol prepared in fermenting baker's yeast from α-methyl cinnamaldehyde', J.Org.Chem.,1984,49,543

27 Bing-nan Zhou, A.S.Gopalan, F. VanMiddlesworth, W-R Shieh, and C.Sih, 'Stereochemical control of yeast reductions. I. Asymmetric synthesis of L-carnitin', J.Amer.Chem.Soc.,1983,105,5925; C.Fuganti, P.Grasselli, P.Casati and M.Carmeno,'On the steric course of baker's yeast mediated reduction of alkyl 4-azido- and 4-bromo-3-oxobutyrate. Synthesis of (R)- and (S)-carnitin', Tetrahedron Letters,1985,101

28 The experiments have been performed by dr.Stefano Servi in professor's J.B.Jones laboratory in Toronto.

ALCOHOL DEHYDROGENASE CATALYSED OXIDOREDUCTION REACTIONS IN ORGANIC CHEMISTRY.

G.L. Lemière.
University of Antwerp (R.U.C.A.),
Laboratory of Organic Chemistry,
Groenenborgerlaan 171,
B-2020 Antwerp, Belgium.

ABSTRACT: A review is given of the alcohol dehydrogenases most used in organic chemistry. Especially the use of horse liver alcohol dehydrogenase (HLADH) is illustrated. The most important variables in HLADH-catalysed reductions with ethanol as a coupled NADH-regenerating substrate are discussed. The first promising results of HLADH-catalysed oxidations with acetaldehyde as a coupled NAD^+-recycling substrate are presented. In contrast to what is predicted in the literature the latter method shows to be a feasible one. It can be used for the oxidation of structurally different alcohols, including some cyclopentanols.

In organic chemistry the carbonyl and hydroxyl group, and their interconversion play a central role. A lot of useful and relatively cheap reductants and oxidants are available for the preparation of achiral and racemic alcohols and ketones or aldehydes. However, in recent years an increasing interest in chiral products and hence an increasing need for enantioselective reagents exists. In this domain alcohol dehydrogenases catalysing the reaction

can be very useful.

I will give a very short review of the alcohol dehydrogenases most used by organic chemists. For more complete information I refer to some reviews in the literature (1,2). Further I will go into more details about the use and applicability of horse liver alcohol dehydrogenase. I will speak about our own work in this field and present some of our recent results on HLADH-catalysed oxidations with acetaldehyde as an NAD^+ regenerating coupled substrate.

Many different alcohol dehydrogenases have been identified, but which one a synthetic chemist will choose for his goal? His choice will be determined first of all by the constitutional specificity and stereo-

M. P. Schneider (ed.), Enzymes as Catalysts in Organic Synthesis, 19–34.

Table 1.
Alcohol Dehydrogenases (ADH).

	YADH	HLADH	CFADH	ETH-ADHs MJADH	PLADH
Commercial availability	+	+	-	-	-
Constitutional specificity	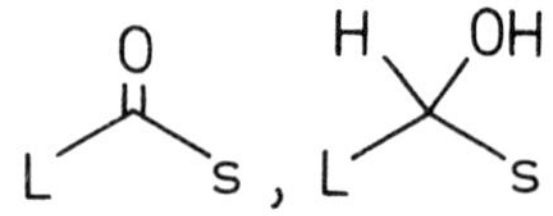 s = H, CH_3	Cyclic and acyclic primary and secondary alcohols aldehydes and ketones.	Only 6-membered ring ketones and alcohols are investigated.	Cyclic and acyclic primary and secondary alcohols aldehydes and ketones.	Cyclic ketones and alcohols.
Stereospecificity	S, Re	S, Re	S, Re	R, Si	S, Re
Coenzyme	NAD(H)	NAD(H)	NADP(H)	NADP(H)	NADP(H)

Figure 1.
Steric course of most ADH-catalysed oxidoreduction reactions.

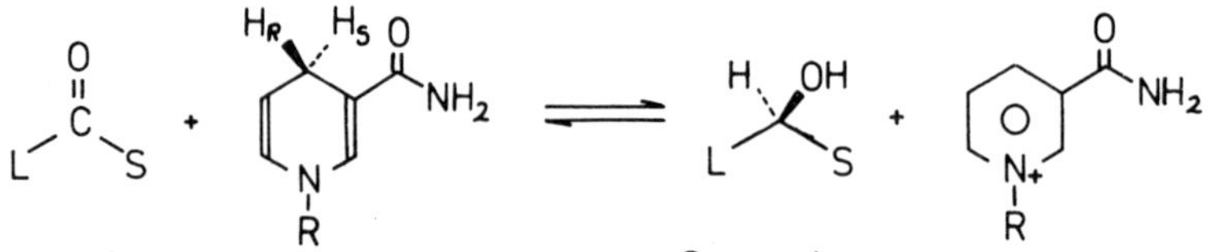

Attack at Re-face by H_R

S-alcohol

specificity of the enzymes but also by their commercial availability, since chemists generally do not want to prepare isolated enzymes. Table I gives a review of the alcohol dehydrogenases most used by organic chemists. Yeast alcohol dehydrogenase (YADH) and horse liver alcohol dehydrogenase are both commercially available and exhibit the same stereospecificity, as shown in figure 1 (3-5). The carbonyl group is attacked at the Re-face by the pro-R hydrogen of carbon atom number 4 of the dihydronicotinamide ring of the coenzyme to form an alcohol with S-configuration. The constitutional specificity of HLADH is very broad (6), this means it accepts a great variety of alcohols and aldehydes or ketones as substrates. Several examples will be given further on. The constitutional specificity of YADH is much narrower (6). This enzyme accepts only a hydrogen atom or a methyl group as the smaller group s of the substrate. This means that all alcohols and ketones in which both carbon moieties are bigger than a methyl group, especially all cyclic compounds, are excluded as substrates. As a result this enzyme is only of limited use for the preparation of chiral products. YADH has been used for the preparation of enantiomeric $(1-{}^2H)$alcohols as shown in figure 2 . In 1-propanol the 1-pro-R proton is stereospecifically exchanged for a deuterium atom to give the $(R)-(1-{}^2H)$-1-propanol. In $(1,1-{}^2H_2)$-1-propanol the 1-pro-R deuterium atom is exchanged for a proton to give the $(S)-(1-{}^2H)$-1-propanol (7).

Figure 2.
Preparation of both enantiomers of $(1-{}^2H)$propanol by YADH (7).

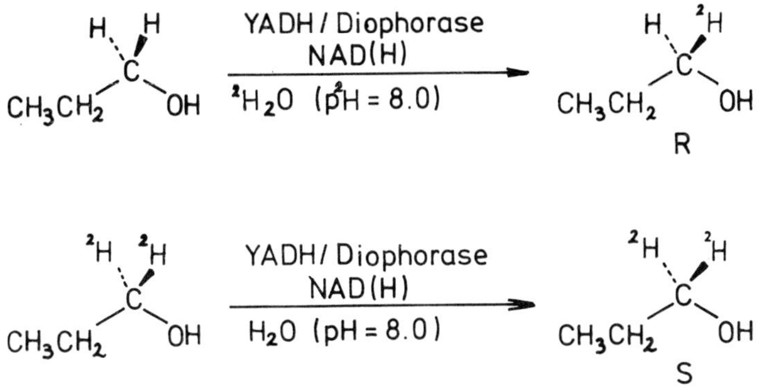

The ETH-alcohol dehydrogenases are dehydrogenases isolated from *Curvularia falcata* (Tehon) Boedijn (CFADH), from *Mucor javanicus* (Wehmer) (MJADH) and from pig liver (PLADH). These enzymes are studied in the groups of V. Prelog and H. Dutler (ETH Zürich). They are not commercially available and hence nearly unaccessible to synthetic chemists. Nevertheless these enzymes could be of great use since they have a fairly broad constitutional specificity and a well defined stereospecificity.

The CFADH could be a good alternative for HLADH since it has the same Re,S stereospecificity. Its constitutional specificity has been studied with six-membered ring ketones and alcohols (8).

The MJADH is an interesting enzyme because of its opposite stereospecificity: carbonyl groups are attacked from the Si side to form R-alcohols. Further it has a rather broad constitutional specificity (9,10).

The PLADH again has Re,S stereospecificity and it accepts cyclic ketones and alcohols (11,12).

In table II these three ETH-enzymes are compared in a qualitative way by considering the reduction of racemic trans-1,4-decalindione (9). In each of the three cases one enantiomer of the substrate is reduced faster than the other enantiomer, which results in a kinetic resolution of the racemic ketone, and only one of the two possible diastereoisomeric alcohols is formed. Such an overall stereochemical course is found in the reduction of many other ketones. Thus when the reaction is carried out carefully one enantiomerically pure alcohol and/or one enantiomerically pure ketone can be prepared. Which ketone and which alcohol are obtained depends on the choice of the enzyme.

Table II.
Qualitative comparison of the ETH-ADHs (9).

	ETH-Alcohol Dehydrogenases		
	Mucor javanicus	*Curvularia falcata*	Pig liver
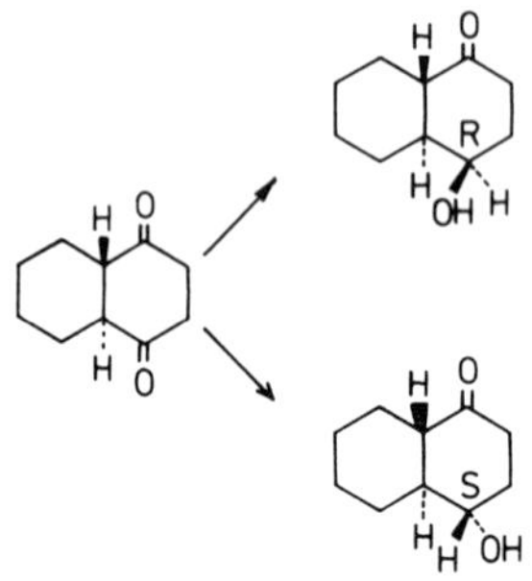	fast		
		fast	slow
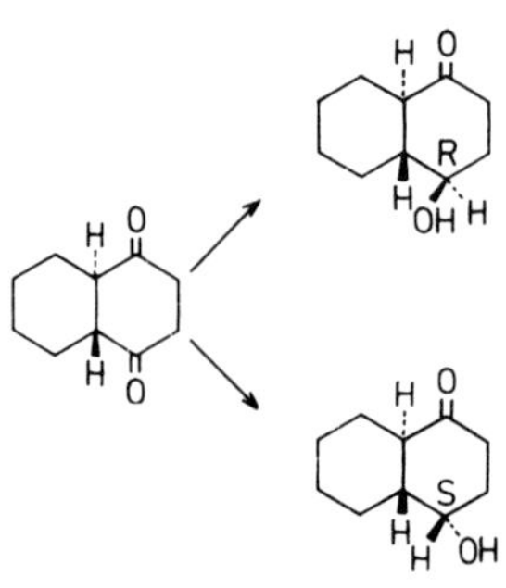	slow		
		slow	fast

In my opinion in the future mould enzymes such as CFADH and MJADH, and other alcohol dehydrogenases from microbiological sources will take the place of enzymes from animal sources such as HLADH. Indeed in general microorganisms can be cultured in bulk quantities in a relatively short time and the alcohol dehydrogenases can be isolated by standard methods.

Until now we have not considered the second partner in the oxido-reduction reactions, *i.e.* the coenzyme NAD(H) or NADP(H). These coenzymes are far too expensive to be used in stoichiometric or greater amounts in preparative-scale reactions. This economic problem can be overcome by using catalytic quantities of the coenzyme and by regenerating it *in situ* to the active form. In principle this can be done in three different ways: non-enzymatically, enzymatically by a coupled-enzyme system or enzymatically by a coupled-substrate system.

Let us first look to HLADH-catalysed reductions with coenzyme regeneration. Only a few non-enzymatic NADH regenerating systems have been developed. For the electrochemical reduction methods I refer to the lectures of H. Simon and S.Y.R. Pugh (13). Further, sodium dithionite is the only chemical reducing agent for which exclusive and quantitative 1,4-reduction of NAD^+ to NADH occurs. This NADH recycling reagent has been used by J.B. Jones. Figure 3 shows the HLADH-catalysed reduction of bicyclo[3.2.1]-2-octanone with sodium dithionite regeneration of NADH (14). In this experiment the coenzyme is recycled 13.5 times. Up to 105 recyclings of coenzyme have been achieved in this type of reductions (15).

Figure 3
Non-enzymatic NADH regeneration (14).

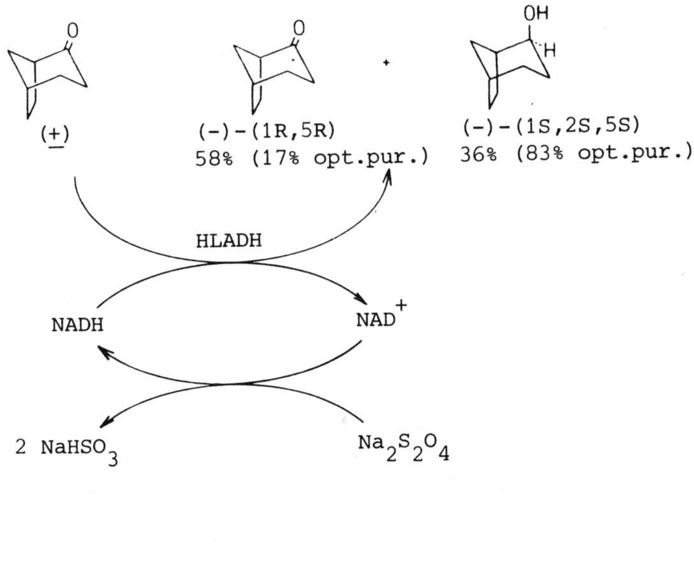

Figure 4.
Coupled-enzyme NADH regeneration (16).

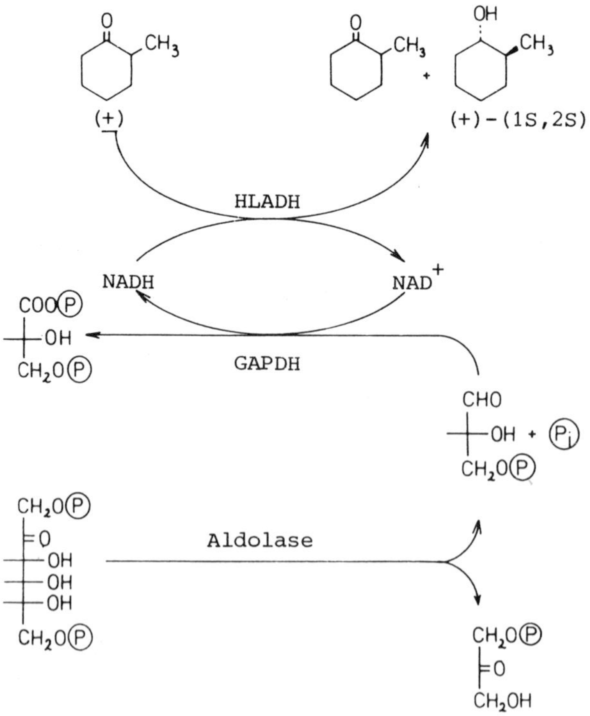

In coupled-enzyme recycling methods an auxiliary enzyme system is used to regenerate the coenzyme. This auxiliary enzyme system can be one single enzyme or a multi-enzyme system. In figure 4 the HLADH-catalysed reduction of 2-methylcyclohexanone is shown. The coenzyme is recycled by a glyceraldehyde-3-phosphate dehydrogenase (GAPDH)-catalysed oxidation of glyceraldehyde-3-phosphate. This auxiliary substrate is formed from fructose-1,6-diphosphate by an aldolase-catalysed cleavage. In this example 86 coenzyme recyclings are obtained (16).

In a coupled substrate recycling system only one enzyme is used catalysing as well the desired oxidoreduction reaction as the coenzyme recycling reaction. This is accomplished by adding a second compound which is a good substrate for this enzyme but in the opposite direction. In figure 5 the HLADH-catalysed reduction of 2-(4-imidazolyl)-[1-^{3}H]acetaldehyde to (S)-[1-^{3}H]histaminol is shown. The NADH is regenerated by the same enzyme and the coupled substrates are ethanol or cyclohexanol. The chiral histaminol has been used by A.R. Battersby for the preparation of chirally labeled histamine (17).

Let me turn now to our own research. We chose for the ethanol coupled NADH regeneration in HLADH-catalysed reduction reactions for the following reasons. We have tried out non-enzymatic NADH regeneration with sodium dithionite but this reagent turned out to be too

Figure 5.
Coupled-substrate NADH regeneration (17).

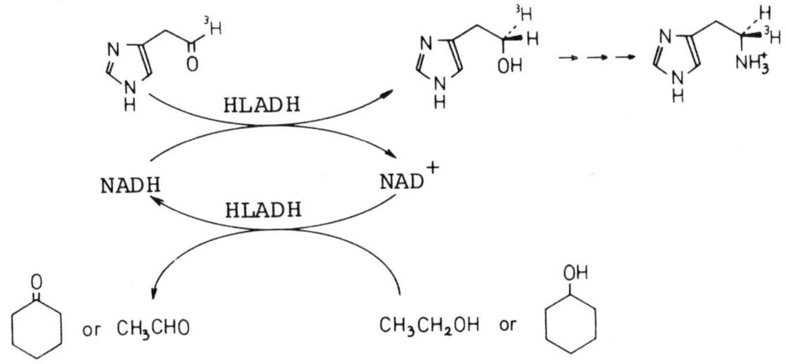

deleterious. Further, we ruled out coupled-enzyme coenzyme recycling systems because of the additional cost of the coupled enzyme and because of the apparent complexity of this system. The most important reason why we have chosen for the coupled-substrate system is that the rate limiting dissociation of the coenzyme from the binary HLADH-NAD(H) complexes can be avoided here (scheme 1). The HLADH-bound NAD^+ formed in the reduction step, is recycled by ethanol, without dissociation from the enzyme, to HLADH-bound NADH. This assures faster preparative-scale reactions, and simplifies the reaction kinetics. Indeed this enables a direct measurement of the reaction rate constant of the catalytic reaction step.

Scheme 1.

$$\text{HLADH-NADH} + \text{ketone} \underset{}{\overset{k_{3B}}{\rightleftharpoons}} \text{HLADH-NAD}^+ + \text{alcohol}$$

$$\text{HLADH-NAD}^+ + \text{ethanol} \underset{}{\overset{k'_{3A}}{\rightleftharpoons}} \text{HLADH-NADH} + \text{acetaldehyde}$$

$$\text{HLADH-NADH} + \text{ethanol} \underset{}{\overset{K_I}{\rightleftharpoons}} \text{HLADH-NADH-ethanol}$$

$$\frac{Et}{v_o} = \frac{1}{k'_{3A}(\text{ethanol})} + \frac{1}{k_{3B}(\text{ketone})}\left(1 + \frac{(\text{ethanol})}{K_I}\right)$$

A very simple initial rate equation (scheme 1) has been elaborated to describe the kinetics of this system (18). As can be seen the reaction rate is independent from the coenzyme concentration. This is true as long as the coenzyme concentration is higher than 10^{-6}M. Beneath this concentration the reaction rate slows down with decreasing coenzyme concentration. Although at such low coenzyme concentrations more than 50,000 NAD^+ recyclings are obtained, these conditions are not recommended because of the very long and unpractical reaction times required. On the other hand increasing the coenzyme concentration far above this value is of no use since the reaction rate will not increase, only the number of coenzyme turnovers will decrease. In our preparative-scale experiments we normally realise 80 to 800 recyclings (19).

Further the reaction rate is proportional with the ketone concentration. The dependence on the coupled substrate, ethanol, is a little more complicated. The first term of the rate equation is determined by the recycling reaction, the last one by the formation of a complex HLADH-NADH-ethanol. Figure 6 shows that this last term is the most important one. For ethanol concentrations from 0.10M up to 2.00M a linear correlation between the reciprocal reaction velocity $1/v_o$ and the ethanol concentration is found. This means the reaction rate slows down for increasing ethanol concentrations, since catalytically active enzyme is taken away from the reaction cycle by the formation of a catalytically inactive complex. Thus, chemists using HLADH with ethanol as a coupled substrate have to be aware of the fact that the reaction can not be accelerated by adding more and more ethanol since this only slows down the reaction.

Figure 6.
Dependence of the HLADH-catalysed reduction rate on the concentration of the coupled substrate ethanol (18).

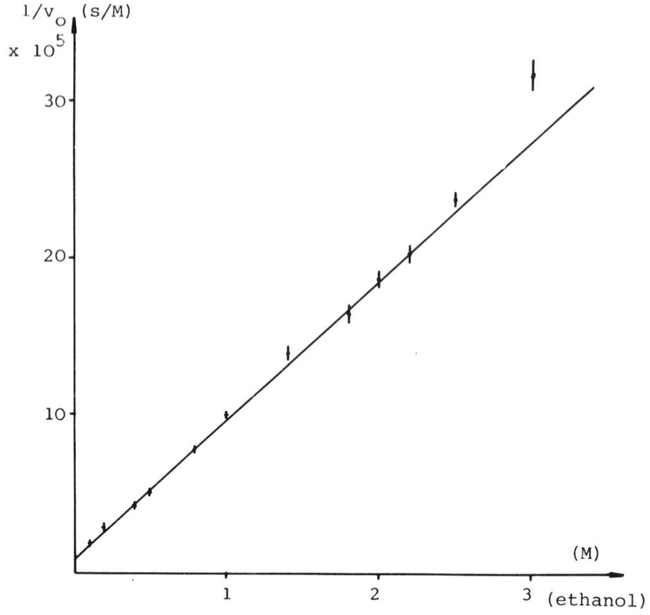

Nevertheless in many cases it may be necessary to drive the reaction to completion since the overall equilibrium of the reaction is unfavourable. This can be achieved by building in a trap for the formed acetaldehyde. For this purpose we ran our reductions in Tris/HCl buffer. The tris-hydroxymethylaminomethane catches the acetaldehyde by the formation of a Schiff-base. Another trap was set up by running the reaction in an evacuated desiccator containing a dimedone solution (20).

Figure 7.
Ring-size effects (24).

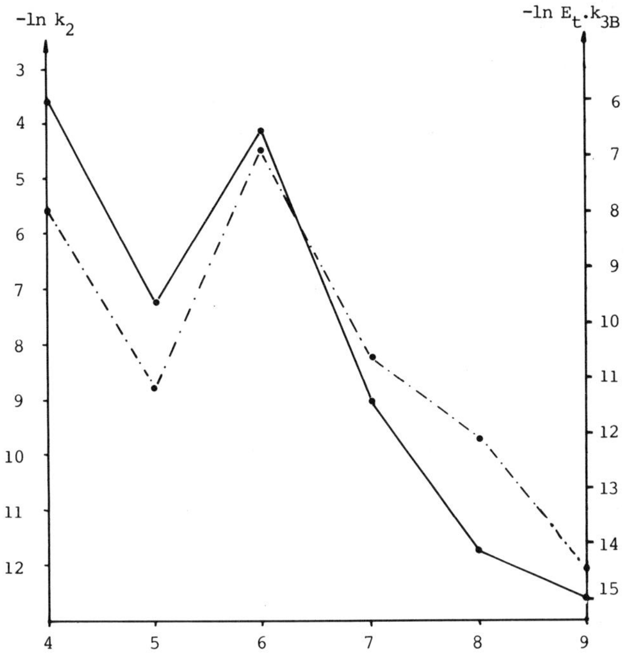

—— $NaBH_4$ reductions.
-.-. HLADH-catalysed reductions.

Table III.
Reduction rate constants k_{3B} of some substituted cyclohexanones.

Substrate		k_{3B}	Substrate	k_{3B}	Ref.
(2S)-2-methyl-CHN		6.6	(2R)-2-methyl-CHN	0	25
(3S)-3-methyl-CHN		276	(3R)-3-methyl-CHN	5.1	26
4-methyl-CHN		258			26
(3S)-3-n-pentyl-CHN		1391	(3R)-3-n-pentyl-CHN	17	26
(3S)-3-tert-butyl-CHN		23	(3R)-3-tert-butyl-CHN	0.4	26
(3S)-3-cyano-CHN	1S,3S	285	(3R)-3-cyano-CHN	1.5	27
	1R,3S	112			27

CHN = cyclohexanone; k_{3B} is expressed in 1 $katal^{-1}$ sec^{-1}.

Of course the reaction rate is also determined by the rate constant k_{3B} which varies considerably from substrate to substrate. Aldehydes in general are reduced very easily (6,21). On the contrary acyclic ketones are rather poor substrates (22,23). The reduction rate of cyclic ketones varies considerably with the ring size (24,25) (fig. 7). Here a remarkable parallellism is found between the reaction rates of the HLADH-catalysed reductions and the sodium borohydride reductions, which means that both reactions are controlled by the same ring strain effects. As can be seen from this picture cyclohexanone is a relatively good substrate but substituents influence its reactivity (table III): 2-substituted cyclohexanones are poor substrates in comparison to 3- and 4-substituted cyclohexanones, lengthening of alkyl substituents increases and branching decreases the reduction rate. The (S)-enantiomers of 2- and 3-substituted cyclohexanones are reduced 50 to 270 times faster than the (R)-enantiomers. This and other stereochemical features are influenced by the polarity of the substituents (20). To rationalise this constitutional specificity and stereospecificity several models have been constructed: the diamond lattice model of V. Prelog (4,5), the cubic space model of J.B. Jones (28) and our transition state model (29). These models will not be discussed here.

The HLADH-catalysed reactions are reversible and in principle this enzyme can also be used for oxidation reactions. Nevertheless only one NAD^+-regenerating method in HLADH-catalysed preparative scale oxidations has been described. Figure 8 shows such an oxidation with non-enzymatic NAD^+-regeneration by flavin mononucleotide (FMN). This method has been fully exploited by J.B. Jones. Up to fifty coenzyme recyclings can be attained by this method. In the example given here a type of stereospecificity is shown, which was not yet seen in preceeding examples. The hydroxymethyl group attached at the S chiral center of meso-*cis*-1,2-bis(hydroxymethyl)cyclobutane is oxidised in an enantiotopically specific way to the corresponding aldehyde. This aldehyde cyclises spontaneously to a lactol which is further oxidised to the corresponding lactone. This lactone has been used to synthesise the enantiomerically pure pheromone grandisol (30,31).

Figure 8.
Non-enzymatic NAD^+ regeneration (30).

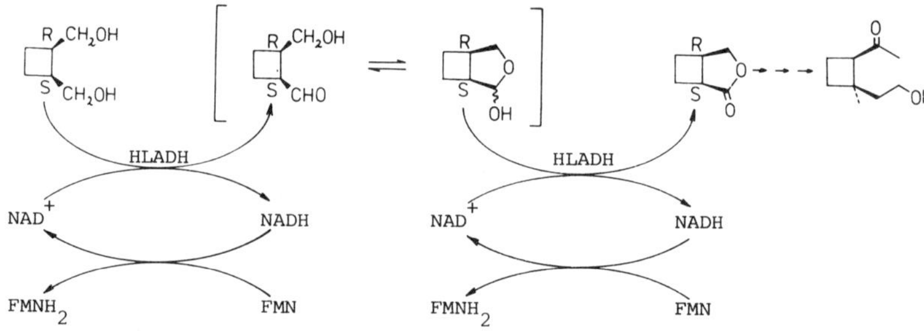

HLADH-catalysed oxidation with acetaldehyde as the coupled substrate is a logical inversion of the well known reduction reaction with ethanol as the coupled substrate. Nevertheless this reaction is allmost unknown. Its applicability is mentioned only once in the literature (32). Moreover, several arguments have been formulated why this reaction has little chance to proceed (1,2). These arguments are:

1. The CO ⇌ CHOH equilibrium is heavily in favour of the formation of the alcohol.
2. Very high concentrations of the carbonyl cosubstrate are needed to obtain significant amounts of the desired carbonyl product.
3. Problems of isolating the product will be aggravated by the high cosubstrate concentrations.
4. The more reactive carbonyl cosubstrates have the potential to inactivate or to modify the enzyme by reaction with nucleophilic acid residues.

The first argument of course is correct, but in a coupled-substrate recycling reaction one substrate-product pair is going from an alcohol to a carbonyl compound and the other pair from a carbonyl to an alcohol. The overall equilibrium will be determined by the redox-potentials of both oxidoreduction reactions. For example it is known that the equilibrium of the reaction cyclohexanol + acetaldehyde ⇌ cyclohexanone + ethanol is in favour of cyclohexanone and ethanol (18). When this overall equilibrium lies to the right no high concentrations of acetaldehyde are needed and the isolation problems will not be aggravated. Furthermore, when the physical properties such as volatility and water solubility of the cosubstrates are different enough from those of the desired products, the cosubstrates will not prohibit the isolation of these products. It is correct that acetaldehyde reacts with free amino groups, but only experiments will show wether acetaldehyde is too aggressive to the enzyme to be used as a coupled substrate.

To explore the possibilities of acetaldehyde-coupled oxidation reactions, we examined in identical reaction circumstances the oxidation of a set of structurally different alcohols (figure 9).

Cyclohexanol is oxidised nearly quantitatively within 1 hour, and 2-cyclohexenol within 6 hours. The smooth curve indicates that the oxidation rate of both enantiomers of this substrate are not very different.

Fifty percent of racemic 3-methyl-2-cyclohexenol is oxidised within one hour, after which the reaction stops. Clearly both enantiomers of this substrate are highly differentiated.

An analogous result is found for *trans*-3-methylcyclohexanol, whereas the *cis* isomer is oxidised very slowly. This is in full accordance with the selective reduction of 3-methylcyclohexanone to *trans*-(1S,3S)-3-methylcyclohexanol (see table III) (16,26).

For the oxidation of 3-methylcyclopentanol (*cis*-*trans* mixture) a smooth curve is found, indicating that no pronounced differentiation between the stereoisomers occurs. Nevertheless the overall oxidation rate is of the same magnitude as the oxidation rate of *trans*-3-methylcyclohexanol. This is not the case for the reduction of cyclopentanones, which are poor substrates in comparison with cyclohexanones (fig. 7).

Figure 9.
HLADH-catalysed oxidations with acetaldehyde as an NAD^+-regenerating coupled substrate.

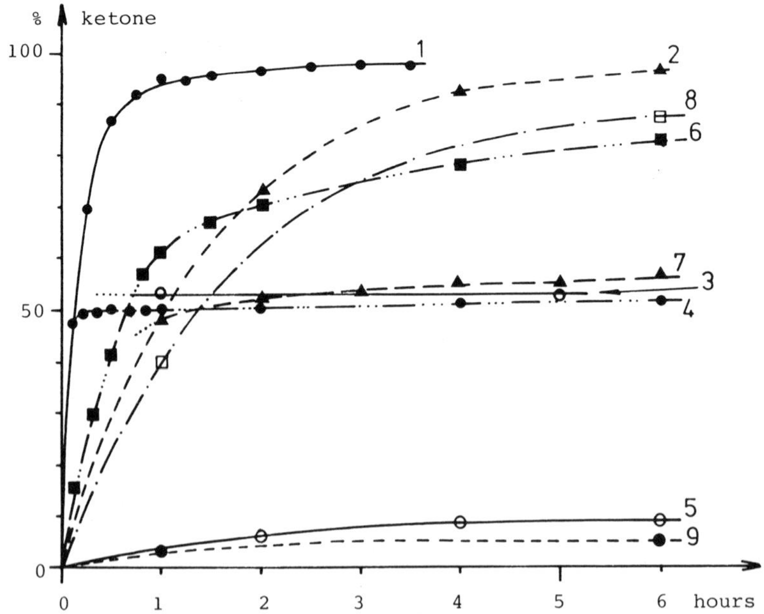

Reaction conditions: alcohol 0.010M, acetaldehyde 0.10M, NAD^+ 2.0×10^{-4}M, HLADH 1mg/10ml, in 0.050M MOPS buffer pH 7.0, temperature 25°C.
1 = cyclohexanol, 2 = 2-cyclohexenol,
3 = 3-methyl-2-cyclohexenol, 4 = *trans*-3-methylcyclohexanol,
5 = *cis*-3-methyl-cyclohexanol, 6 = 3-methylcyclopentanol (*cis* + *trans*),
7 = *cis*-2-methylcyclopentanol, 8 = (2S)-2-octanol,
9 = (2R)-2-octanol.

Furthermore *trans*-2-methylcyclopentanol is oxidised very slowly, but 50% of the *cis* isomer is oxidised within 1 hour, after which a sharp decrease of the oxidation rate is observed. Clearly here too both enantiomers are highly differentiated. This result is very unexpected since 2-methylcyclopentanone is not reduced at all. Indeed this ketone accumulates two unfavourable features: it is a cyclopentanone and it has a substituent on carbon 2.

Both enantiomers of 2-octanol were oxidised seperately. The (S)-enantiomer is oxidised ten times faster than the (R)-enantiomer, which confirms earlier observations (33).

I want also to mention that we tried to oxidise 1,2-hexanediol and 5,6-bis-(hydroxymethyl)bicyclo[2.2.1]hept-2-ene (34), but did not succeed. Very recently J.B. Jones published the HLADH-catalysed oxidation with NAD^+-regeneration by FMN of the latter diol (35).

Figure 10.
Dependence of the initial rate on the NAD^+ concentration.

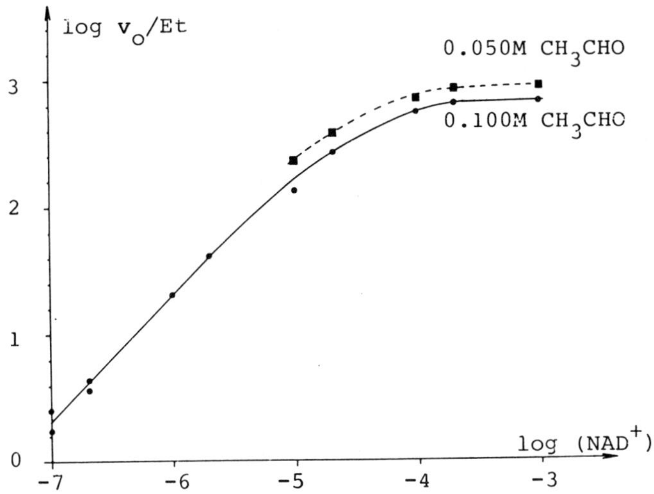

Reaction conditions: HLADH $7.00x10^{-7}$katal/1, cyclohexanol 0.010M in 0.050M MOPS buffer pH 7.0; temperature 25°C.

Figure 11.
Dependence of the initial rate on the acetaldehyde concentration.

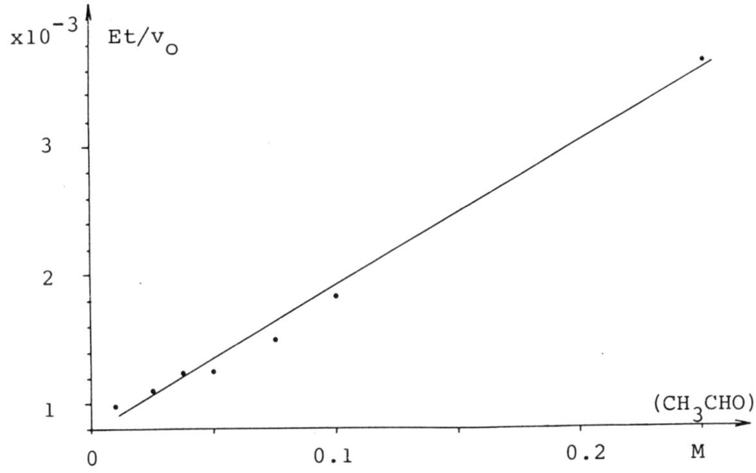

Reaction conditions: HLADH $7.00x10^{-7}$katal/1, NAD^+ $2.00x10^{-4}$M, cyclohexanol 0.010M in 0.050M MOPS buffer pH 7.0; temperature 25°C.

In order to optimize the recycling process we have started to study the oxidation of cyclohexanol with acetaldehyde as the coupled substrate.

As can be seen from figure 10 the oxidation rate is independent from the NAD^+ concentration when it is equal or higher than 10^{-4}M, below this level it decreases. For this reason further experiments were carried out at NAD^+ concentrations of $2x10^{-4}$M.

Further we studied the influence of the acetaldehyde concentration on the initial rate (figure 11). A nearly linear correlation is found between the reciprocal velocity $1/v_o$ and the acetaldehyde concentration. Thus the initial rate decreases with increasing acetaldehyde concentration. In the reduction reaction with ethanol as the coupled substrate an analogous correlation is found and it was ascribed to the well known formation of a complex HLADH-NADH-ethanol, but according to the literature a complex HLADH-NAD^+-acetaldehyde is not formed (36). Does this effect reflect the agressivity of acetaldehyde to the enzyme?

We further studied the influence of acetaldehyde by leaving the enzyme in contact with a 0.050M solution of acetaldehyde before starting the oxidation reaction, and by measuring the residual enzyme activity. Figure 12 clearly shows that the activity of the enzyme decreases when it is in contact with acetaldehyde. The half life time of the enzyme under these experimental conditions was approximately two days.

Figure 12.
Influence of acetaldehyde on the HLADH activity.

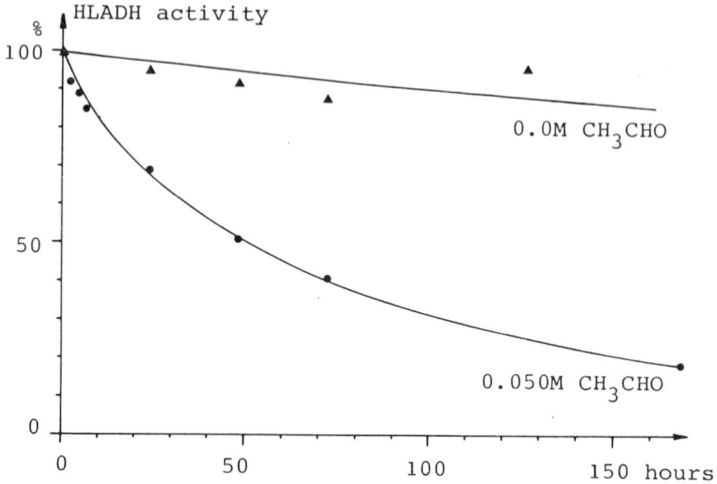

Reaction conditions: HLADH $7.00x10^{-7}$katal/l, NAD^+ $2.00x10^{-4}$M, cyclohexanol 0.010M in 0.050M MOPS buffer pH 7.0; temperature 25°C.

Finaly a linear correlation between the initial rate and the substrate concentration was found. The reaction could be run at temperatures up to 50°C without heat denaturation of the enzyme.

The results presented here show that NAD^+ recycling in HLADH-catalysed oxidations with acetaldehyde as the coupled substrate is a feasible method if the acetaldehyde concentration is kept low during the reaction. Especially for good substrates which can be oxidised in a relatively short space of time the aggressivity of acetaldehyde poses no great problems.

REFERENCES:

1. J.B. Jones, J.F. Beck in *Applications of Biochemical Systems in Organic Chemistry* Part I, Eds. J.B. Jones, Ch. Sih, D. Perlman, Techniques of Chemistry Vol X, Series Ed. A. Weissberger; J. Wiley, 1976, p. 107.
2. S.S. Wang, Ch.-K. King, Adv. Biochem. Eng. 12, 119 (1979).
3. R. MacLeod, H. Prosser, L. Fikentscher, J. Lanyi, H.S. Mosher, Biochemistry 3, 838 (1964).
4. V. Prelog, Pure Appl. Chem. 9, 119 (1964).
5. V. Prelog, Colloq. Ges. Physiol. Chem. 14, 288 (1963).
6. H. Sund, H. Theorell in *The Enzymes*, Vol. 7, 2nd ed., P.D. Boyer, M. Lardy, K. Myrback, Eds., Academic Press, N.Y. 1963, p. 25.
7. H. Günther, F. Biller, M. Kellner, H. Simon, Angew. Chem. 85, 141 (1973).
8. J. Rétey, Ph.D. thesis 3409, ETH, Zürich 1963.
9. Z. Kis, Ph.D. thesis 3684, ETH, Zürich 1965.
10. E. Hochuli, Ph.D. thesis 5284, ETH Zürich 1974.
11. G. Waldvogel, Ph. D. thesis 3681, ETH Zürich 1965.
12. A. Kull, Ph.D. thesis 4447, ETH, Zürich 1970.
13. a. H. Simon, this book p. 35
b. S.Y.R. Pugh, this book p. 217
14. A.J. Irwin, J.B. Jones, J. Amer. Chem. Soc. 98, 8476 (1976).
15. J.B. Jones, D.W. Sneddon, W. Higgins, A.J. Lewis, J. Chem. Soc. Chem. Commun. 856 (1972).
16. R. Helmchen-Zeier, Ph.D. thesis 4991, ETH, Zürich 1973.
17. A.R. Battersby, M. Nicoletti, J. Staunton, R. Vleggaar, J. Chem. Soc. Perkin I, 43 (1980).
18. T.A. Van Osselaer, G.L. Lemière, J.A. Lepoivre, F.C. Alderweireldt, J. Chem. Soc. Perkin II, 1181 (1978).
19. T.A. Van Osselaer, G.L. Lemière, E.M. Merckx, J.A. Lepoivre, F.C. Alderweireldt, Bull. Soc. Chim. Belg. 87, 799 (1978).
20. J.J. Willaert, G.L. Lemière, R.A. Dommisse, J.A. Lepoivre, F.C. Alderweireldt, accepted for publication in J. Chem. Res.
21. R. Pietruszko, K. Crawford, D. Lester, Arch. Biochem. Biophys. 159, 50 (1973).
22. K. Dalziel, F.M. Dickinson, Biochem. J. 100, 34, 491 (1966).
23. L. Kemps, University of Antwerp, graduate thesis, 1981.
24. T.A. Van Osselaer, G.L. Lemière, J.A. Lepoivre, F.C. Alderweireldt, Bull. Soc. Chim. Belg. 87, 153 (1978).
25. T.A. Van Osselaer, G.L. Lemière, J.A. Lepoivre, F.C. Alderweireldt, Bull. Soc. Chim. Belg. 89, 389 (1980).
26. T.A. Van Osselaer, G.L. Lemière, J.A. Lepoivre, F.C. Alderweireldt, Bull. Soc. Chim. Belg. 89, 133, (1980).

27. J.J. Willaert, G.L. Lemière, R.A. Dommisse, J.A. Lepoivre, F.C. Alderweireldt, Bull. Soc. Chim. Belg. 93, 139 (1984).
28. J.B. Jones, I.J. Jakovac, Can. J. Chem. 60, 19 (1982).
29. G.L. Lemière, T.A. Van Osselaer, J.A. Lepoivre, F.C. Alderweireldt, J. Chem. Soc. Perkin II, 1123 (1982).
30. I.J. Jakovac, H.B. Goodbrand, K.P. Lok, J.B. Jones, J. Amer. Chem. Soc. 104, 4659 (1982).
31. J.B. Jones, M.A.W. Finch, I.J. Jakovac, Can. J. Chem. 60, 2007 (1982).
32. J. Rétey, private communication (1966) to R.K. Hill, J.W. Morgan, J. Org. Chem. 33, 927 (1968).
33. F.M. Dickinson, K. Dalziel, Biochem. J. 104, 165 (1967).
34. Supplied by M. Vandewalle, R.U.Ghent.
35. K.P. Lok, I.J. Jakovac, J.B. Jones, J. Amer. Chem. Soc. 107, 2521 (1985).
36. Ch.S. Hanes, P.M. Bronskill, P.A. Gurr, J.T.-F. Wong, Can. J. Biochem. 50, 1385 (1972).

ON THE USE OF VIOLOGEN DYES FOR STEREOSPECIFIC BIOREDUCTION

H. Simon, H. Günther and J. Thanos
Institute for Organic Chemistry, Technical University
Munich, D-8046 Garching,
Federal Republic of Germany

ABSTRACT. Besides pyridine nucleotide-dependent reductases there exist reductases for C-C- and C-O-double bonds which accept electrons from reduced viologens such as 1,1'-dimethyl-4,4'-bipyridinium cation radical ($MV^{+\cdot}$). Reduced viologens can be regenerated electrochemically or enzymatically with hydrogenases or formate dehydrogenases. Many cells possess enzymes which reduce NAD or NADP at the expense of $MV^{+\cdot}$. Such an enzyme, which reduces NAD as well as NADP, has been purified from a thermophilic bacillus. The reduction rate of NADP is about 50% of that of NAD. Its stability is sufficient. Every microorganism having a methylviologen-dependent NAD or NADP reductase can be applied for electromicrobial reductions. Some practical aspects i.e. concerning the stability of enzymes and pyridine nucleotides under different conditions are reported.

INTRODUCTION

Most oxidoreductases which are able to reduce carbon-carbon or carbon-oxygen double bonds, i. e. which are actually reductases are pyridine nucleotide-dependent. Enzymatic reductions cannot be performed on a preparative scale using stoichiometric amounts of reduced pyridine nucleotides due to their high cost. So far NAD as well as NADP only can be reduced enzymatically if high recycling numbers are aimed at.

During the last few years we found reductases for carbon-carbon as well as for carbon-oxygen double bonds which are able to accept electrons from reduced viologen dyes such as methyl- or benzylviologen ($MV^{+\cdot}$ or $BV^{+\cdot}$) (1-5). The 2-enoate reductase as well as the 2-oxocarboxylate reductase can be used for a wide variety of substrates. Since we assume that more such enzymes exist, and due to the fact that many microorganisms possess mostly not yet characterized methylviologen-dependent NAD and/or NADP reductases (3,6,7), we studied a few aspects of the use of viologen dyes and looked for more viologen-dependent reductases.

M. P. Schneider (ed.), Enzymes as Catalysts in Organic Synthesis, 35–44.

ENZYMATIC REDUCTIONS PERFORMABLE WITH REDUCED VIOLOGENS

Table I shows some enzyme reactions which can be conducted with viologens. The enoate reductases from Clostridium tyrobutyricum DSM 1460 (C. spec. La 1) and from Clostridium kluyveri DSM 555 are not identical, however both reveal an extreme wide substrate specificity for nonactivated α,β-unsaturated carboxylates, aldehydes non cyclic and cyclic ketones (1,3). For kinetics and mechanism of the enzyme with NADH as electron donor see l.c. (8) and for the structure of the enzyme from C. tyrobutyricum see l.c. (9). In an immobilized form the enzyme is stable with a half life of about 150 h (10). The enzyme can also be used in systems containing only a few percent of water together with some detergent in hexane (11). Thus far we and other laboratories found in all cases an enantiomeric excess higher than 96% of the reduction products of enoates (3). This means that the ee-values may be even above 99%. Recently we reduced α-halogenated α,β-unsaturated crotonates with cells of C. La 1 in the presence of 1 mM methylviologen. For (2R)fluoro-, chloro- and bromobutyrate the ee-values as determined by derivatisation with (R)phenylglycinol and HPLC analysis were >98%. The same result was obtained for (2R)chlorobutyrate with purified enoate reductase in a system in which the $MV^{+\cdot}$ was regenerated with a crude extract of Proteus vulgaris and formate (12).

The 2-oxocarboxylate reductase from Proteus mirabilis and Proteus vulgaris which reacts neither with NADH nor NADPH shows also an extremly broad substrate specificity together with an excellent stereospecificity (3,5). It can be enriched up to 500 U x mg protein^{-1} x minute^{-1}. This reductase has a rather surprising kinetic behavior with methylviologen. The reduction rate of 2-oxocarboxylates increases with decreasing concentrations of reduced methylviologen at least down to values of 0.05 mM and increases with increasing concentrations of oxidized methylviologen. Therefore, using Proteus vulgaris cells in the electrochemical cell where the ratio $MV^{+\cdot}/MV^{++}$ can be set by the applied potential productivity numbers >100 000 can be observed. (Microbial reductions can be characterized by productivity numbers: PN = mmol product x kg catalyst (dry)$^{-1}$ x time (h)$^{-1}$). Usually values of 50-200 are observed for yeasts (3).

The enzymes 3-5 of Table I have not been characterized yet. However, in crude extracts reductions are observable only with $MV^{+\cdot}$ and not with NADH or NADPH. The observed productivity numbers are suitable for preparative reactions. The enal reductase from Methanobacterium thermoautotrophicum shows an interesting substrate stereospecificity as indicated in Table I.

The composition and physiological role of the methylviologen-dependent NAD or NADP reductases are often not clear. We purified such a reductase from the thermophilic Bacillus spec. DSM 406 to a purity >95% as judged by disc electrophoresis. The procedure shown in Table II is not yet optimized. This enzyme catalyzes in a reversible way not only the reactions 6 and 7 of Table I but also acts as an NADH dehydrogenase with dichlorophenol indophenol (DCPIP) or ferricyanide as acceptors. The relative rates of reactions are indicated in Table III. Bacillus spec. DSM 406 possesses also a membrane bound NADH dehydro-

Table I Some enzyme catalysed reactions applying reduced viologens[a)]

Number	Reaction	Enzyme	Remarks and Literature
1	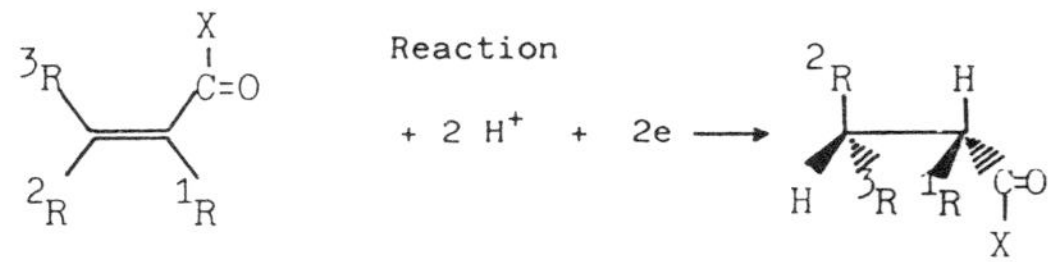	Enoate reductase from Clostridia species	Wide substrate specificity with aldehydes, ketones and carboxylates, ^1R = halogens, NHCHO, Me, OMe etc. ^2R many different groups, ^3R should not be too bulky, X = H, O^- or R
2	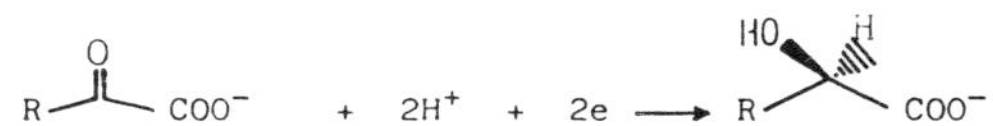	2-Oxocarboxylate reductase from Proteus species	Hardly any restrictions for R. Mono- as well as dicarboxylates are substrates. Productivity numbers > 100 000 can be obtained (3)
3	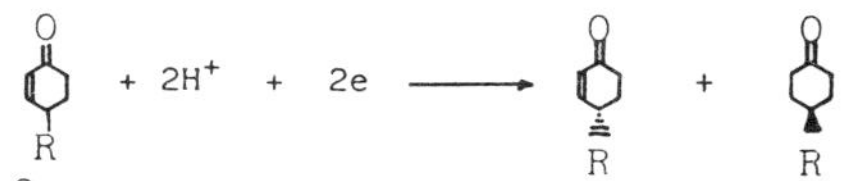	Unknown enal reductase in Methanogens	Productivity numbers of ~ 650 have been observed for $R=NHCOCH_3$[b)]
4	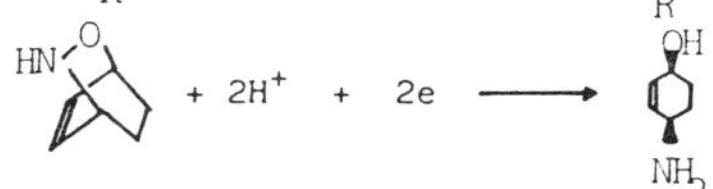	Unknown enzyme in Methanogens	Productivity numbers of ~200 have been ovserved[b)]
5	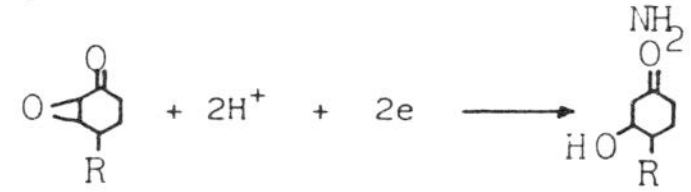 (?)	Unknown enzyme in Methanogens	Product partially further converted $R = NHCOCH_3$
6	$NAD + H^+ + 2e \longrightarrow NADH$	Activities observed in many microorganisms and mitochondria (3,6)	For enzymes known to catalyze this reaction and the stereochemical course see l.c. (3,6). For an enzyme from a thermophilic Bacillus see text.
7	$NADP + H + 2e \longrightarrow NADPH$	Activities observed in many microorganisms	See above. Especially high is the activity in Desulfurococcus mobilis
8	$2H^+ + 2e \longrightarrow H_2$	Well known reaction for all hydrogenases	
9	$HCOO^- \longrightarrow 2e + CO_2 + H^+$	Formate dehydrogenases	
			Enzyme from Proteus vulgaris shows at 37°C a half-life > 20 days and a specific activity of 0.7 U/mg protein

a) 2e stands for $2MV^{+\cdot}$, $2BV^{+\cdot}$ or other viologens.
b) Studies in cooperation with G. Kresze and K. Klier, to be published

Table II
Purification of a methylviologen-dependent NAD(P) reductase from the thermophilic Bacillus spec. DSM 406.

Purification step	Volume ml	Protein concentration mg/ml	Total activity U	Specific activity[a)] U/mg	Yield %
Supernatant of crude extract	124	21.5	682	0.25	100
Redesolved precipitate of 50% ammonium sulfate saturation	124	8.2	521	0.51	77[b)]
DEAE-Sepharose and ultrafiltration	99	2.0	379	1.9	55
Phenylsepharose	200	0.24	245	11.0	15
FPLC[c)]-Anion exchange chromatography	24	0.2	202	40.0	14

a) Activity at 28°C. b) 11% of the activity not precipitated.
c) Fast protein liquid chromatography.

Table III
Relative activities of a methylviologen-dependent NAD(P) reductase and a NAD(P)H dehydrogenase from the thermophylic Bacillus spec. DSM 406. (Part of the results obtained in cooperation with S. Nagata and J. Stadlmüller)

Enzyme	Reductase activity		Dehydrogenase activity			
			NADH		NADPH	
	NAD	NADP	Ferricyanide	DCPIP[b)]	Ferricyanide	DCPIP
Soluble	100[a)]	50	31	21	14	11
Membrane	2.5	--	1.5	100[b)]	--	--

a) 40 µmol NADH formation per min and mg protein at 28°C, pH 7 ≙ 100%.
b) 61 µmol NAD formation ≙ 100%.

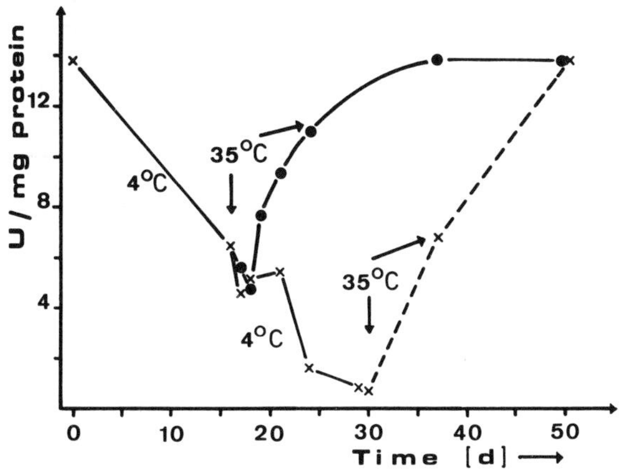

Fig. 1 Time course of the specific activity of partially purified methylviologen dependent NAD reductase (8.5 U/mg protein) from the thermophilic Bacillus spec. DSM 406.

A sample of the enzyme (0.016 mg/ml in 100 mM Trisacetate pH 7.0) is kept at 4°C. After 15 days 0.1 mM chlorhexidine diacetate was added, an aliquot of the sample was brought to 35°C and kept at this temperature. Another part remained at 4°C. Both samples were checked for their activity at the indicated points. x———x Sample at 4°C until day 30; x- - - - - -x Sample at 35°C. •———•———• Sample at 35°C after 15 days at 4°C. (Experiments conducted by S. Nagata.)

genase. This NADH dehydrogenase seems to be similar to that described by Kawada et al. (13.) As shown in Table III, not every NADH dehydrogenase is also an effective methylviologen-dependent NAD reductase. The soluble enzyme of Bacillus spec. DSM 406 seems to be useful from a practical point of view. As can be concluded from Fig. 1, no decrease of activity is observed when kept at 35°C in Tris-acetate, pH 7 in a rather dilute solution $<$ (0.02 mg/ml) for 20 days. However, at 4°C the activity decreases with a half-life of about 15 days. This inactivation is reversible at 35°C. The reductase is not sensitive to oxygen. The high stability at medium temperatures is not surprising for an enzyme from a thermophilic organism growing at 65°C. The enzyme reduces NAD and NADP. The rate for the latter is about 50% of the rate observed with NAD. In an electrochemical cell using D_2O-buffer, preparative amounts of (S) [4-^{2}H] -NAD(P)H can be prepared (3,6).

The extremely thermophilic archaebacterium Desulfurococcus mobilis also possesses a very stable and high activity of a methylviologen dependent NADP reductase. Studies with crude extracts revealed in the presence of reduced methylviologen at 35°C a half-life for the activity of at least 20 days.

All hydrogenases (reaction 8 of Table I) seem to be able to reduce methylviologen at the expense of molecular hydrogen. Their presence is a prerequisite for microbial hydrogenations which we applied to the reduction of more than 80 different 2-enoates and carbonyl compounds (3). Often catalytic concentrations of methylviologen increase the rates of hydrogenation. However, there are (rare) cases in which reduced methylviologen reacts spontaneously with the unsaturated substrate (vide infra).

Viologen-dependent formate dehydrogenases (reaction 9 of Table I) are rather useful for the regeneration of reduced methyl- or benzylviologen. Proteus vulgaris grown in the presence of formate shows about 0.7 U formate dehydrogenase per milligram of protein in the crude extract. Its stability at 37°C is astonishingly high (Table I).

Fig. 2 gives an example for the combination of crude extracts from two microorganisms. C. kluyveri catalyzes the formation of NADH at the expense of hydrogen gas and Enterobacter agglomerans possesses an NADH dependent reductase for 2-oxo-3-methyl-3-hydroxybutanone. Under non optimized conditions a productivity number of about 1450 has been observed.

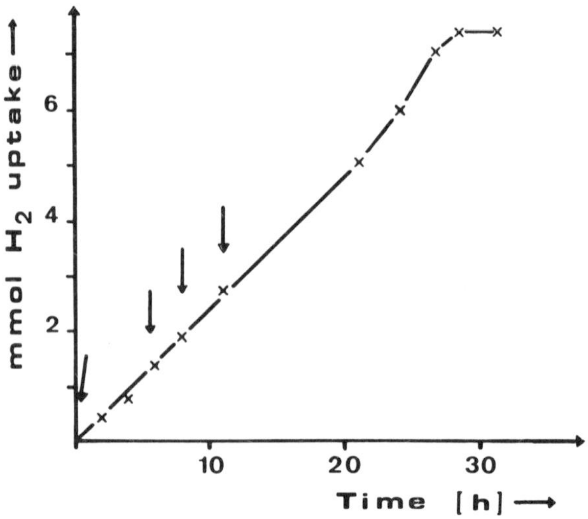

Fig. 2 Hydrogenation of 2-hydroxy-2-methyl-3-butanone. In a 160 ml bulb connected with a mercury filled manometer 11 ml 50 mM potassium phosphate buffer containing 15.3 mg protein of a crude extract of Enterobacter agglomerans (27.5 U 2-hydroxy-2-methyl-3-butanone reductase), 81 mg protein of a crude extract of Clostridium kluyveri, 100 µmol NAD and 100 µmol MV^{++} were stirred under a hydrogen atmosphere at 35°C. The 4 arrows indicate the successive addition of a total of 7.8 mmol 2-hydroxy-2-methyl-3-butanone in equal portions. The hydrogen uptake was followed manometrically. At the beginning of the 24th hour the stirring rate was increased. This caused a faster hydrogen uptake, showing that the transfer of hydrogen to the liquid phase was rate limiting. After hydrogen uptake ceased all the substrate was converted to the product.

PRACTICAL ASPECTS AND EXPERIENCES USING VIOLOGENS

When working with reduced viologens, oxygen has to be strictly excluded. Molecular oxygen reacts extremely fast with the cation radical $MV^{+\cdot}$ ($k = 8 \times 10^{8}\ M^{-1}\ sec^{-1}$) (14) to the oxygen superoxide anion radical. This and its products may impair the viologen, enzymes, substrates and products.

There are several possibilities for the regeneration of reduced viologens. Chemical methods using dithionite (15) or metals such as zinc (16) are hardly useful under preparative aspects. The increasing concentrations of sulfite or zinc ions would seriously impair most enzyme activities. The enzymatic reduction of methyl- or benzylviologen with hydrogen gas or formate, catalyzed by hydrogenase or a formate dehydrogenase, respectively is suitable. Both enzymes seem to be rather stable as long as oxygen is carefully excluded. Immobilized *C. tyrobutyricum* showed a half-life of about 12 days in a hydrogenation experiment in the presence of 2 mM methylviologen. In this experiment it was not determined whether the hydrogenase or the enoate reductase was responsible for the slow decline of activity. So far we have not determined the half-life of formate dehydrogenase under operational conditions. However, as mentioned above the storage stability at 37°C is very high. *Proteus vulgaris* catalyzes the reduction of 2-oxo-4-methylpentanoate with formate under an atmosphere of nitrogen and with hydrogen gas, respectively. The productivity numbers are in the order of 7000. In the meantime we observed that the simultaneous application of hydrogen gas and formate leads to an increase of the productivity number to 27000.

The electrochemical reduction of viologens has some advantages and we used it in many electromicrobial or electroenzymatic reductions (1-3). For the design of an electrochemical cell (10) a cathode of high hydrogen overpotential is desirable. The substrate must be only enzymatically reduced. Suitable cathode materials for operation near physiological pH values and potentials set lower or equal than -460 mV versus the saturated calomel electrode[a] include Ag, C and Hg. The potential is set by a standard three electrode arrangement. A low power potentiostat is adequate.

For the diaphragm 0.5 mm cationic foils of Nafion[R] (Du Pont) or 0.15 mm Selemion[R] (Asahi Glass) are suitable materials. These diaphragms fulfil the following requirements: Easy proton transfer from the anolyte to the catholyte where protons are consumed (reactions 1-8 of Table I), sufficient conductivity, and practically no oxygen transfer from the anolyte to the catholyte.

The symmetry of the cathode should be chosen in such a way that potential variations higher than 30 mV compared with the potential value in the vicinity of the reference electrode are avoided.

The cells should operate under predominantly diffusion controlled conditions. The diffusion layer is decreased by stirring the solution by means of a magnetic stirring bar or rotating electrodes.

The electrochemistry of viologens is well established (17).

---

a) All potentials are measured versus this reference.

Electrochemically reduced methylviologen is sufficiently stable. A 5 mM solution of $MV^{+\cdot}$ in 0.1 M Tris-acetate buffer pH 7.0 stored in a Schlenk-tube under an atmosphere of nitrogen at 25°C showed a half-life of about 3 months.

Two types of application can be differentiated: (i) The reduced viologen donates electrons directly to a reductase, or (ii) it is used as an electron donor for the enzymatic regeneration of NAD(P)H. Therefore, the stability of pyridine nucleotides in the presence of methylviologen is an important aspect. Table IV shows that the stability of NAD and NADH depends on the buffer, which is a well known fact (18), and on the presence of reduced methylviologen. According to Table V the stability of NAD is also a function of the cathode potential of the electrochemical cell. A potential of -740 mV corresponds to a ratio $MV^{+\cdot}/MV^{++}$ of 9 under equilibrium conditions. Under operational conditions this ratio may be much lower, depending on the activity of the enzymes catalyzing the methylviologen consuming reactions.

Table IV
Stability of NAD and NADH in different buffers (potassium phosphate or Tris-sulfate 0.1 M, pH 7.0) and in the presence of oxidized or reduced methylviologen at room temperature.

[Pyridine nucleotide] = 0.8 mM, [methylviologen] = 2 mM

	Buffer	Viologen	half-life h
NAD	K-phosphate	MV^{++}	>5000[a)]
NAD	K-phosphate	$MV^{+\cdot}$	123
NAD	Tris-sulfate	MV^{++}	920
NAD	Tris-sulfate	$MV^{+\cdot}$	124
NADH	K-phosphate	MV^{++}	31
NADH	K-phosphate	$MV^{+\cdot}$	37
NADH	Tris-sulfate	MV^{++}	>5000
NADH	Tris-sulfate	$MV^{+\cdot}$	234

a) Values partially extrapolated.

After taking up a second electron methylviologen becomes extremely unstable making this reduction practically irreversible. Therefore any overreduction which starts at potentials lower than -800 mV should be avoided. The redox potential of the second reduction step is -1040 mV.

Regarding the stability of unsaturated substrates against reduced viologens we observed neither a spontaneous reaction of 2-enoates with reduced methylviologen nor a non enzymatic reduction in the electrochemical cell, so far at potentials not more negative than -790 mV. This

Table V
Dependence of NAD stability on cathode potential of an electrochemical cell under practically currentless conditions in the presence of 2 mM viologen in 0.1 M Tris-sulfate buffer pH 7.0 at 24°C.

Potential[a)]	Time of measurement(h)	half-life (h)
792[b)]	22	12
690[b)]	117	170
620[c)]	116	570
590[c)]	117	1000

a) Measured against saturated calomel electrode, b) Methylviologen, c) Benzylviologen

is in agreement with the results of cyclic voltammetric measurements. The following carbonyl compounds show spontaneous reduction with reduced methyl- and benzylviologen: 4-acetylpyridine, ethyl 4-chloro-3-oxo-butyrate and 2,6,6-trimethyl-2-cyclohexene-2,4-dion. Phenylglyoxylate reacts spontaneously with reduced methylviologen. However, with benzylviologen optically pure mandelic acid has been obtained. This can be explained by the fact that the redox potential of benzylviologen is about 120 mV less negative than that of methylviologen.

CONCLUSION

The methodology of biocatalytic, enantioselctive reductions of unsaturated compounds has made considerable progress. Efficient procedures for NADH regeneration are available, even for technical applications (see C. Wandrey, this volume). It can be assumed that further microorganisms and reductases suitable for preparative purposes will be found. Electromicrobial and electroenzymatic reductions as well as microbial hydrogenations often have the advantage of high productivity numbers. Also their time course can be continously followed by the current or hydrogen consumption, respectively.

Acknowledgements. We are grateful to Mrs. C. Frank and Mr. L. Riesinger for skillful technical assistance, Dr. S. Nagata and Dr. J. Souppe for providing unpublished results. The work was supported by Deutsche Forschungsgemeinschaft (Sonderforschungsbereich 145) and by Fonds der Chemischen Industrie.

REFERENCES

1 H. Simon, J. Bader, H. Günther, S. Neumann, J. Thanos (1984) Enzyme Engineering 7, A. I. Laskin et al. (Eds.) Ann. N. Y. Acad. Sci., 434 171-185.

2 H. Simon, H. Günther, J. Bader, S. Neumann (1985) Enzymes in organic synthesis, Ciba Foundation Symposium 111, 97-111, Pitman, London.

3 H. Simon, J. Bader, H. Günther, S. Neumann, J. Thanos (1985) Angew. Chem. 97, 541-555. Angew. Chem. Int. Ed. Engl. 24, 000-000.

4 S. Neumann, H. Simon (1984) FEBS LETTERS 167, 29-32.

5 S. Neumann, H. Günther, H. Simon (1984) Third European Congress on Biotechnology I, 377-382, Verlag Chemie, Weinheim.

6 J. Bader, H. Günther, S. Nagata, H.J. Schuetz, M.L. Link, H. Simon (1984) J. Biotech. 1, 95-109.

7 J. Bader, H. Günther, S. Nagata, H. Simon (1984) Third European Congress of Biotechnology I, 371-376, Verlag Chemie, Weinheim.

8 M. Bühler, H. Simon (1982) Hoppe Seyler's Z. Physiol Chem. 363, 609-625.

9 S. Kuno, A. Bacher, H. Simon (1985) Biol. Chem. Hoppe Seyler 366, 463-472.

10 J. Thanos, J. Bader, H. Günther, S. Neumann, H. Simon. Methods in Enzymology, in press.

11 J. Thanos, H. Simon to be published

12 J. Souppe, H. Simon to be published

13 N. Kawada, K. Takeda, Y. Nosoh (1981) J. Biochem. 89, 1017-1027.

14 J.A. Farrington, M. Ebert, E.J. Land (1978) J. Chem. Soc. Faraday Trans 1, 74, 665-675.

15 K. Tsukahara, R.G. Wilkins (1985) J. Am. Chem. Soc. 107, 2632-2635 and literature cited therein.

16 P.A. Trudinger (1970) Anal. Biochem. 36, 222-225.

17 C.L. Bird (1981) Chem. Soc. reviews 10/1, 49-82 and literature cited therein.

18 C.H. Wong, G.M. Whitesides (1981) J. Am. Chem. Soc. 103, 4890-4899.

APPROACHES TO CHIRAL BUILDING BLOCKS FOR NATURAL PRODUCT SYNTHESIS

E.Winterfeldt
Institut für Organische Chemie
Universität Hannover
Schneiderberg 1 B
D-3000 Hannover

Substituted cyclopentenones of type 1 [1] are to be considered extremely useful chiral building blocks. Pure enantiomers of this material open up two general roads to a number of highly interesting compounds depending on the fate of the enolate ion 2 generated by the substituent directed Michael addition.

Scheme I

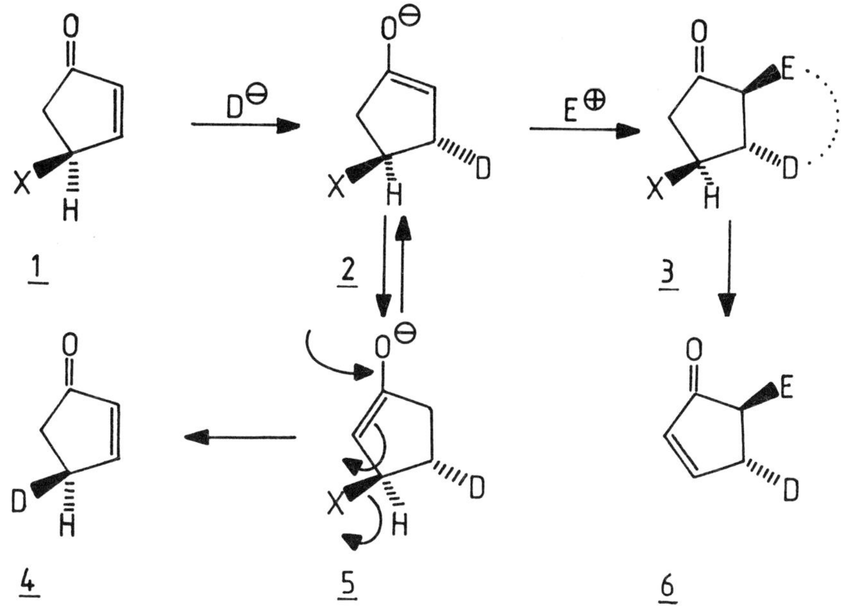

M. P. Schneider (ed.), Enzymes as Catalysts in Organic Synthesis, 45–54.

Electrophilic capture or protonation may either take place directly thus generating trisubstituted cyclopentanones of type 3 which may lead to prostaglandines[2] or to disubstituted cyclopentenones if X is an appropriate leaving group. No electrophile being immediately available enolate equilibration may generate 5 which then on elimination will lead to 4, a process that amounts to substitution with retention. Depending on the structure of D 4 can be channeled into various reaction products either by intermolecular or intramolecular processes and in this lecture after discussing a few important methods to generate pure enantiomers of the starting material 1 I shall concentrate completely on this left part of Scheme I and discuss some synthetic possibilities emerging from further chemical manipulation of cyclopentenone (4). The most important enzymatic roads leading to 1 are either the enantioselective hydrolysis of diacetate 7 or ketoacetate 1 (X = OAc).

Scheme II

7

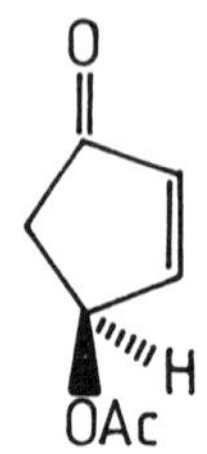

1

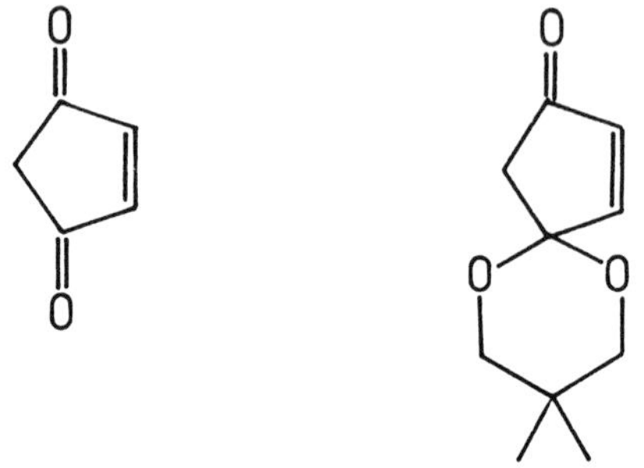

In both cases the separation of the two enantiomers is, owing to a substantial difference in polarity, comparatively easy. Having these compounds available one can investigate chirality transfer in the course of Michael additions and Scheme II is indicating the range of possibilities. Even with a symmetric Michael donor excellent chirality transfer may be achieved. Depending on the choice of reaction conditions completely different products can be obtained.

Scheme III

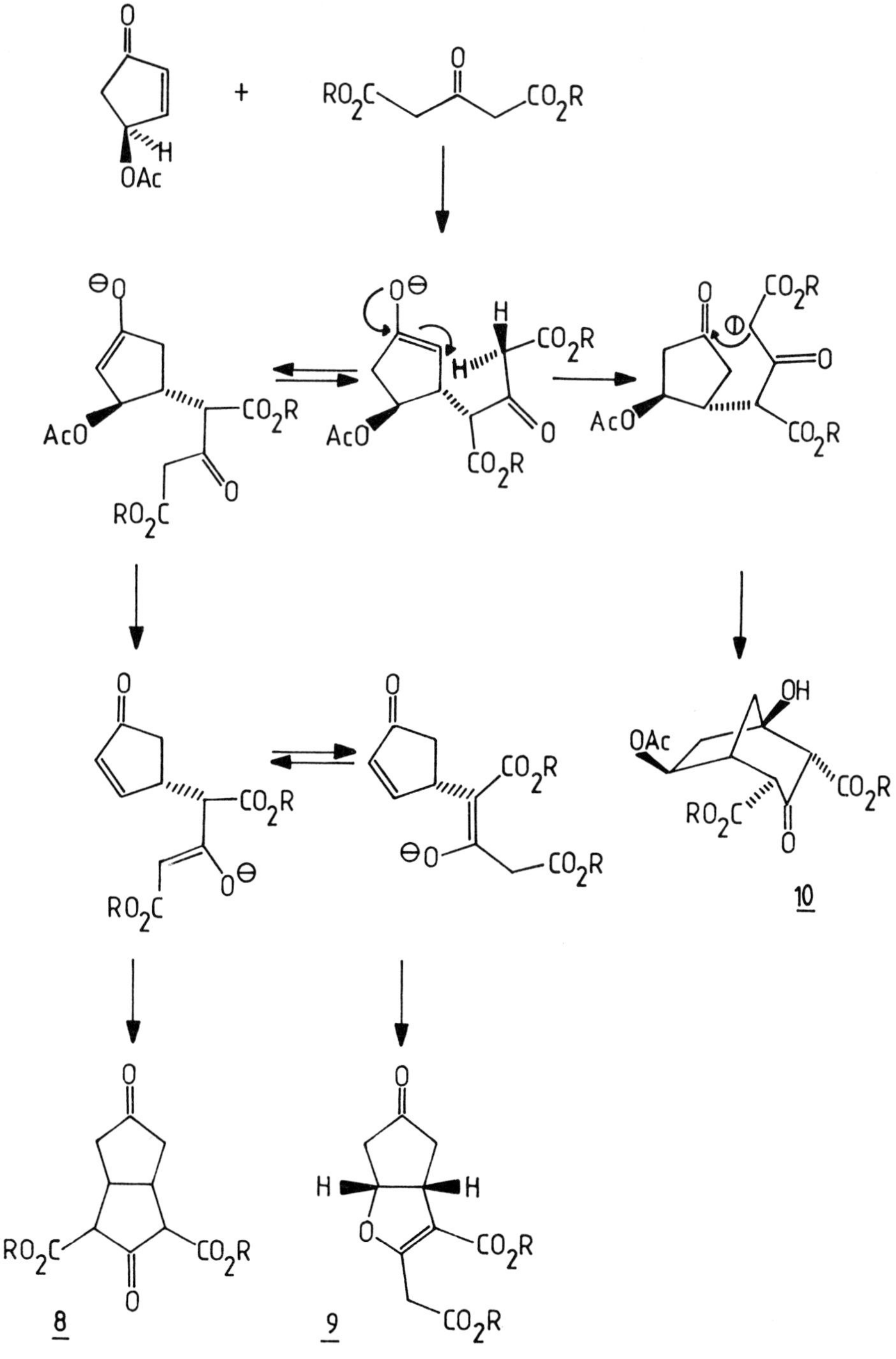

The enolether 9 resulting from lithiumhydride deprotonation is of some general importance as it can be obtained from various β-keto esters as for instance 11 and 9 can in all cases by deprotonation and reprotonation be converted stereoselectively into the enolethers of type 12 in high chemical yield.

Scheme IV

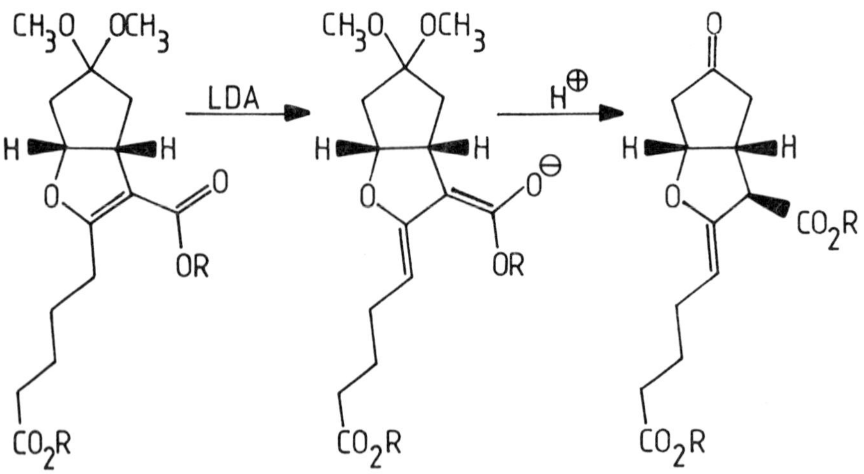

11 12

With compounds like 8, 9 or 11 cis annellation is guaranteed and it is also expected for six-membered rings like 13. With seven- or eight-membered rings however, prediction is not that easy as the steric bulk of the ester groups could well direct the second nucleophile into the trans position (14). With 14 the trans configuration was proven by NMR techniques. With the eight-membered ring the direct formation could not be achieved with this simple double addition process owing to severe transanulare interactions in the transition state. This situation can be considerably improved however, by inserting a cis double bond into the double donating species. This way 15 was prepared easily and in high yield and as the results from the seven-membered series could not be simply transferred to this molecule and as additionally this particular bicycle was considered as a starting material for terpene-synthesis the configuration was secured by X-ray structure determination.

Scheme V

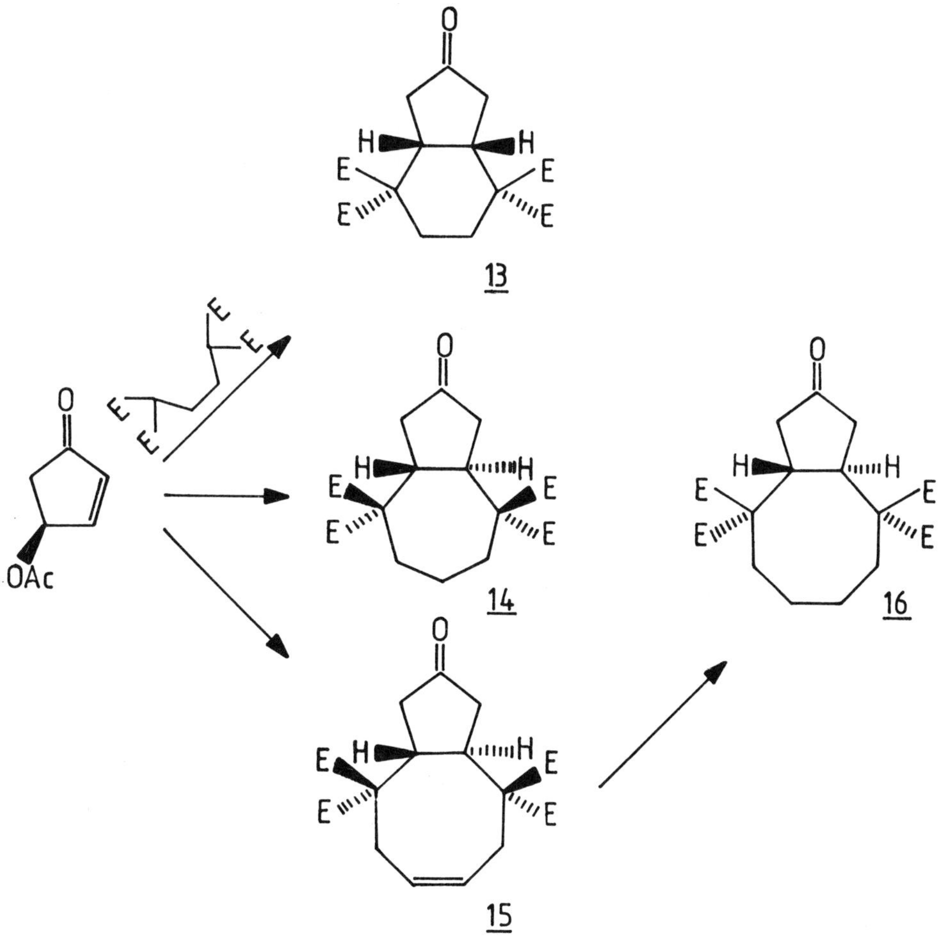

Of course this intramolecular double Michael addition can also be used for the annellation of heterocyclic ring systems, a few of which are given in Scheme VI.

Scheme VI

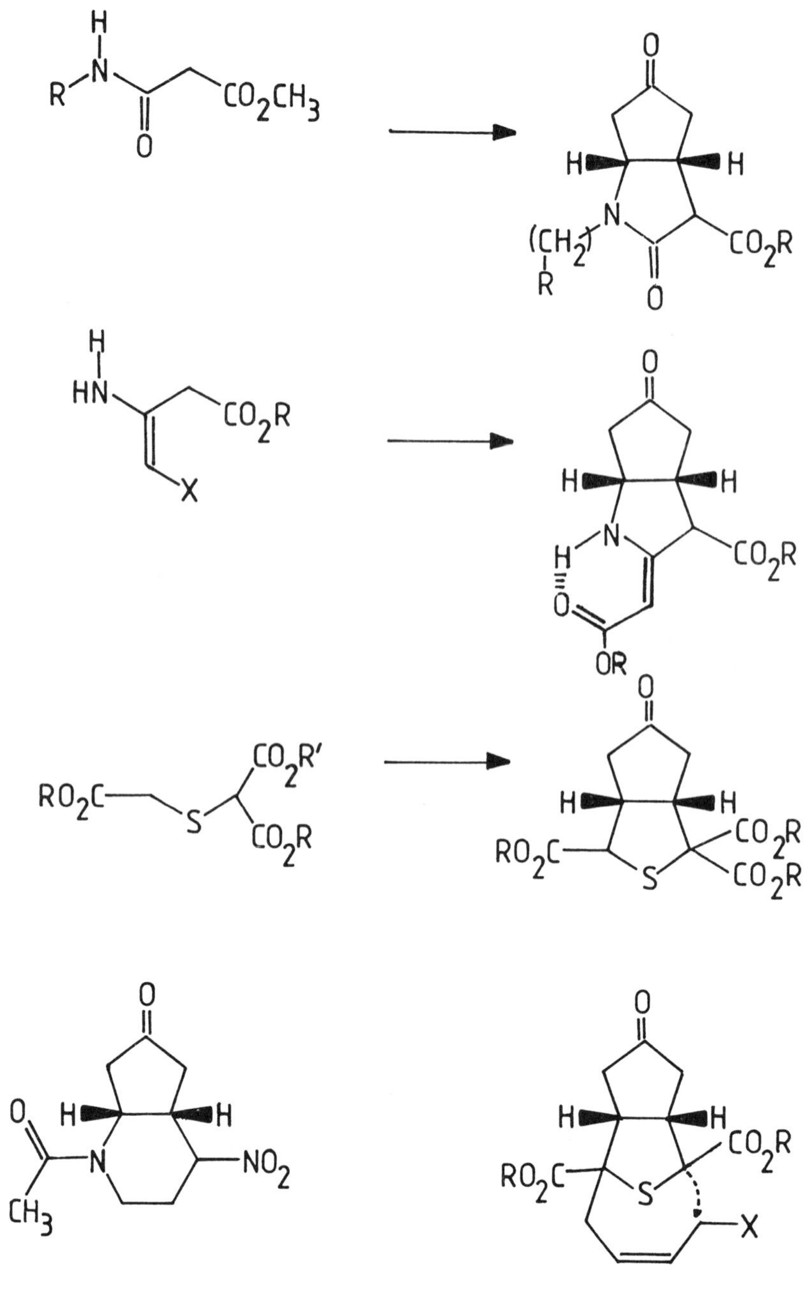

Scheme VII

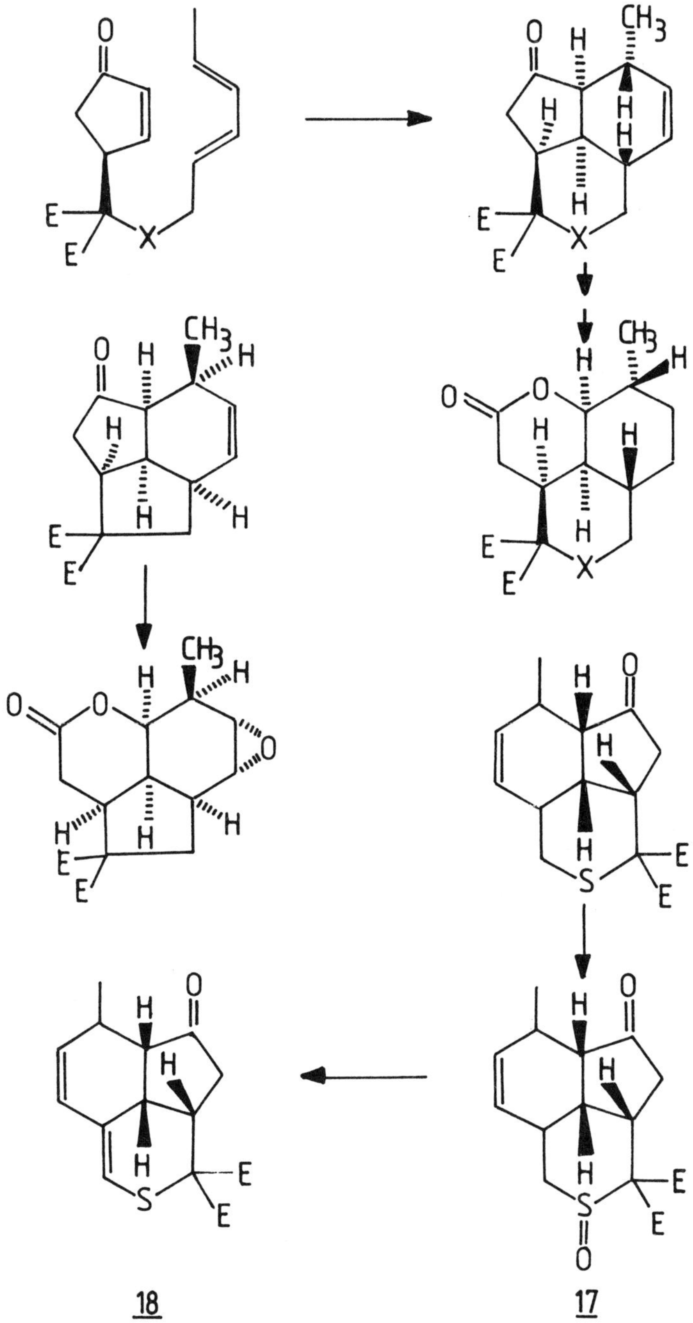

18 17

Scheme VIII

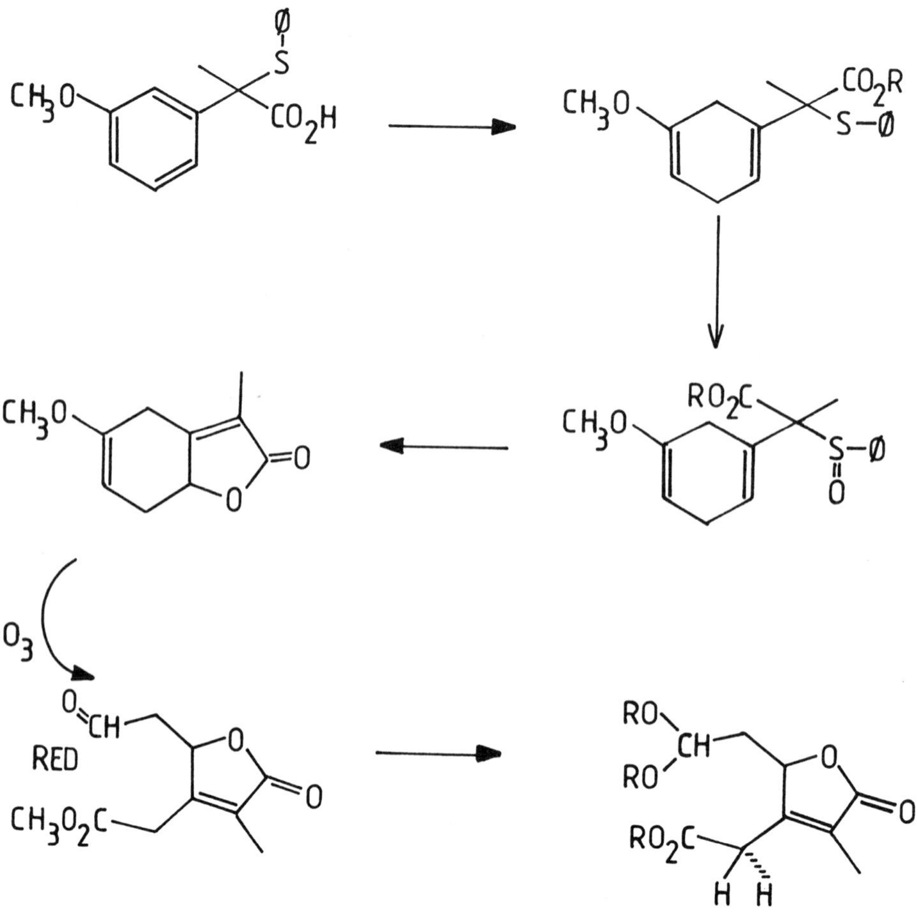

19

As all double donatig species are non-symmetric and will give rise to step-wise annellation processes the product configuration is nicely linked to the starting material. The cyclization reactions discussed so far exclusively consume the electron poor double bond of 4 by Michael addition processes. A π-system of this type should however, also give rise to Diels-Alder cycloaddition products and a few general possibilities are listed in Scheme VII.

Of special interest ist the corresponding sulfur compound (X = S). Formation of the sulfoxide 17 is of particular ease at the dihydro-product but can also be achieved with the cycloaddition product directly and this substance then is an interesting starting material for a regioselective Pummerer reaction (see 18).

This way the centres linking the donor and the 4π-systems can be removed without completely loosing functionality at the termini.

In the last part of the lecture I want to at least describe one example of a cyclization sequence that leaves the cyclopentenone double bond completely untouched. For this experiment we choose the building block 19 which was prepared according to Scheme VIII.

In this compound the CH_2-group indicated has the donating capacity of a vinylogue malonate and Michael addition proceeded as expected (see 20).

Scheme IX

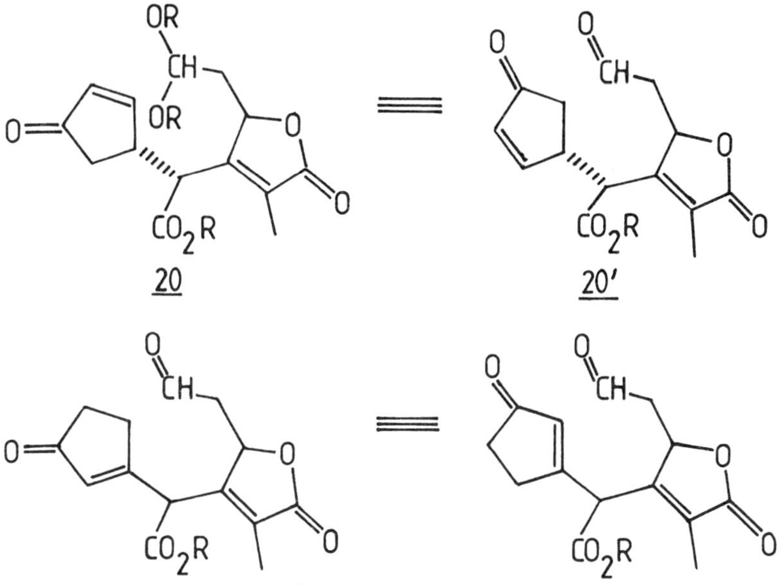

20 20' 20'' 21

As can be judged from conformation 20' the cyclization step in this case will of course have to be of the aldol type. But as again the isomerization of the double bond can be easily achieved one also has compound 21 available which as 21' clearly shows is well prepared for a vinylogue aldo cyclization. Thus opening the road to various substitution patterns in the five-membered ring with high constitutional flexibility.

References

1 M.Harre, P.Raddatz, R.Walenta und E.Winterfeldt, Angew.Chem. 94, 496 (1982); Angew.Chem., Int.Ed.Engl. 21, 480 (1982).

2 R.Noyori und M.Suzuki, Angew.Chem. 96, 854 (1984).

USE OF MICROORGANISMS FOR THE RESOLUTION OF SYNTHETICALLY USEFUL BICYCLO[3.2.0]HEPT-2-EN-6-ONES

Stanley M. Roberts,
Department of Microbiological Chemistry,
Glaxo Group Research Ltd.,
Greenford,
Middlesex. UB6 OHE
England.

ABSTRACT: The resolution of synthetically useful, racemic cyclobutanone derivatives using whole-cell systems has been investigated. Bicyclo[3.2.0]hept-2-en-6-one was reduced using actively fermenting bakers' yeast to give the corresponding 6endo-ol and 6exo-ol. In both cases the S-configuration was prevalent at the newly created chiral centre. The same bicyclic ketone was converted into optically active 6S-6endo-ol and enantiomerically enriched starting material using *Mortierella ramanniana*. The alcohols and optically active ketone produced in this way were efficiently converted into naturally occurring prostaglandins-$F_{2}\alpha$ and -A_2. 7,7-Dimethylbicyclo[3.2.0]hept-2-en-6-one was transformed by *M. ramanniana* into 6(S)-7,7-dimethylbicyclohepten- 6endo-ol and 6(S)-7,7-dimethylbicyclohepten-6exo-ol with excellent enantioselectivity. The former optically active alcohol was converted into the pheromone (+)-eldanolide and both alcohols were converted into the arachidonic acid metabolite (+)-leukotriene-B_4. Thus it has been demonstrated that microorganisms can be used to resolve bicyclo[3.2.0]heptenones, enhancing the usefulness of these compounds for the synthesis of a variety of chiral natural products.

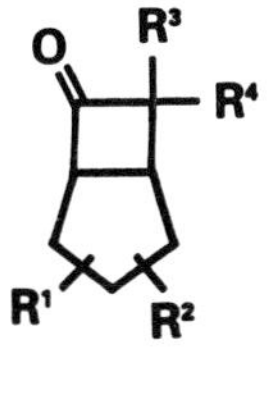

(1)

1. INTRODUCTION

The synthetic utility of bicyclo[3.2.0]heptanones (1) has been recognised for some time[1]. The reasons for the usefulness of this bicyclic ring-system in the synthesis of molecules of interest from a biological and/or structural viewpoint can be summarised as follows:-

M. P. Schneider (ed.), Enzymes as Catalysts in Organic Synthesis, 55–75.

a) Bicycloheptan-6-ones are simply prepared by [2+2]cycloaddition of a keten or keteniminium salt and a cyclopentene or a cyclopentadiene[2].

b) The angular, relatively inflexible shape of the system ensures that attacking reagents will usually modify the simple cycloadducts in a highly regioselective fashion.

c) The modified bicyclo[3.2.0]heptanones can be transformed into a wide range of mono-, bi-, and tri-cyclic ring systems thanks to the reactivity of the four-membered ring. The most common modifications of the cyclobutanone moiety involve ring-expansion, ring-contraction and ring-cleavage.

Recent examples of the use of bicycloheptanones in the synthesis of natural products and analogues include the preparation of hybridalactone[3], hirsutic acid C[4], pentalenene[5], pentalenolactones E and F[6], prostaglandins[7], boonein[8], multifidene[9] and the ophiobolin nucleus[10].

In order to enhance the usefulness of the bicyclo[3.2.0]-ring system for the preparation of optically active natural products, a general method of resolution of these bicyclic ketones would be valuable. Over the years, a number of procedures have been employed to prepare bicycloheptanones in optically active form. These methods include:-

a) Reaction of the racemic ketone with a chiral amino alcohol (e.g. ephedrine) and separation of the diastereoisomeric oxazolidines[11].

b) Formation of an azomethine using excess of racemic ketone and an optically active amine, followed by distillation of the volatile ketone and hydrolysis of the residual Schiff's base[12].

c) Use of a chiral ketenimine[13] (or chiral ketenacetal[14]) to induce asymmetry into the [2+2]cycloaddition reaction leading to the bicyclic ketone.

d) Formation of the bisulphite addition complex of the bicyclic ketone and resolution of the adduct using an optically active base[15].

e) Reaction of the ketone with resolved N,S-dimethyl-S-phenylsulfoximine followed by thermolysis of the separated diastereomers[16].

Since none of the above methods can be used for the preparation of a <u>wide range</u> of optically active bicycloheptanones, we decided to investigate the use of microorganisms for the stereo-selective reduction of these ketones.

2. RESOLUTION OF BICYCLO[3.2.0]HEPT-2-EN-6-ONE

Initial attempts to reduce (±) bicycloheptenone (2) using bakers' yeast on a small scale showed that two major products were formed. These products were identified as bicyclo[3.2.0]hept-2-en-6*endo*-ol (3) and bicyclo[3.2.0]hept-2-en-6*exo*-ol (4). Both products were optically active. The optical purity of the *endo*-ol (3) was assessed by conversion into the bromohydrin (5) using N-bromo-acetamide in aqueous acetone, formation of the menthylcarbonate derivative, and analysis by g.l.c. The ratio of enantiomers was judged to be 94:6. Similarly the *exo*-ol (4) was converted into the same bromohydrin and the ratio of enantiomers, assessed in the above manner, was shown to be 8:92. The absolute configuration of the major enantiomer in the *exo*-alcohol was shown to be (4b) by oxidation to the known γ-lactone (6b). The major enantiomer in the *endo*-alcohol was shown to be (3a) by the same Jones/Baeyer-Villiger double oxidation procedure[17].

(a)............ 2 .... (b)

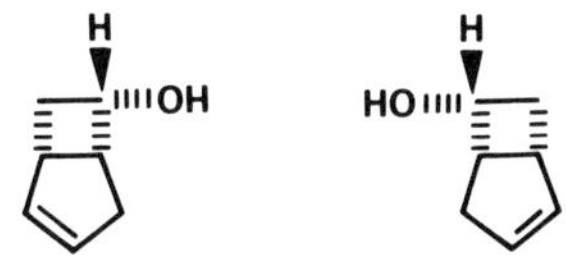

(a) .... 3 .... (b)

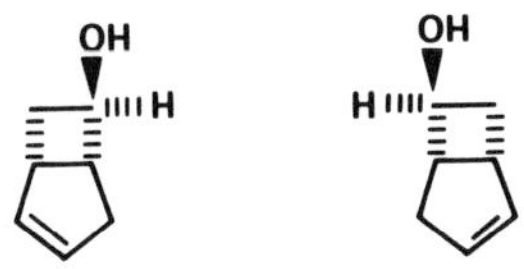

(a) .... 4 .... (b)

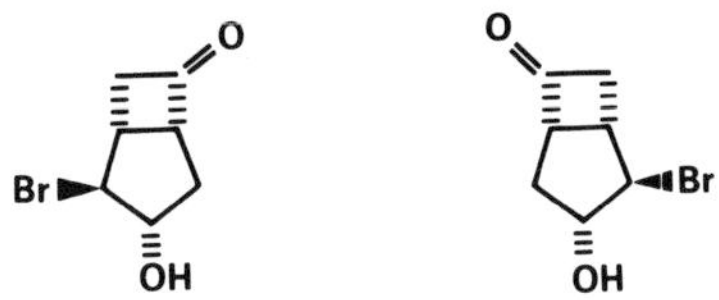

(a)............. 5 .... (b)

(a) .... 6 .... (b)

Thus both the *endo*-alcohol (3a) and the *exo*-alcohol (4b) were produced with a satisfactory optical purity. The derived bromohydrins [(5a) and (5b) respectively] were obtained in an optically pure state after one recrystallization from carbon tetrachloride.

The reduction of (±)-bicyclohept-2-en-6-one (2) with bakers' yeast was compared with the corresponding process using dehydrogenase enzymes from other yeasts and fungi (Table)[18].

TABLE: REDUCTION OF (±)-BICYCLO[3.2.0]HEPT-2-EN-6-ONE WITH MICROORGANISMS

Yeast or Fungus	Percentage conversion after one day	
	endo-ol	exo-ol
Fermipan bakers' yeast	7	3
Sacch. acidfaciens C856	20	11
Sacch. cerevisiae C1989	14	4
Curvularia lunata C2100	22	trace
Mortierella ramanniana C2506	32	-

The fungus M. ramanniana gave an interesting result in that the 6endo-ol was found to be the only alcohol formed. In addition the reduction was considerably faster than with the yeast preparation. On a small scale, it was shown that the 6endo-alcohol produced using M. ramanniana had a high optical purity.

The reductions using bakers' yeast and M. ramanniana were scaled up to provide material sufficient for our synthetic studies. The ketone (2) (30 g) was treated with actively fermenting yeast: after filtration of the mycelium, 6endo-ol (5.6 g), 6exo-ol (4.2 g) and starting material (3.8 g) were obtained by extraction and chromatography. Treatment of the racemic ketone (25 g) with M. ramanniana and work-up in the prescribed manner gave 6endo-ol [7.0 g; enantiomer ratio (3a):(3b) = 95:5] and ketone [3.4 g; enantiomer ratio (2a):(2b) = 10:90].

With multigram quantities of useful optically active materials in hand the synthesis of prostaglandins possessing the natural configuration was explored.

3. SYNTHESIS OF (+)-PROSTAGLANDIN-$F_2\alpha$ AND (+)-PROSTAGLANDIN-A_2

The bromohydrin (5a), derived from the alcohol (3a), was protected to give the tert-butyldimethylsilyl ether (7) (Scheme 1)[19]. Treatment of the ketone (7) with a non-nucleophilic base caused abstraction of a proton from the active methylene group, intramolecular S_N2 displacement of the bromine atom, and formation of the strained

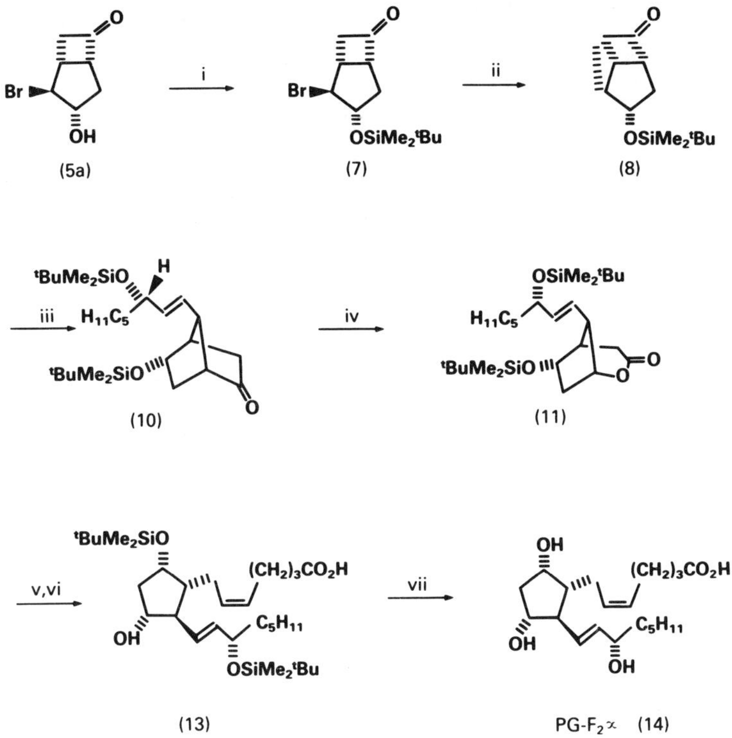

Reagents:

i. $ClSiMe_2{}^tBu$, imidazole, dimethylformamide (DMF);
ii. KO^tBu, HO^tBu;
iii. reagent (9), ether, -78°C;
iv. $MeCO_3H$, $MeCO_2H$ then NaOH, H_2O;
v. HAl^iBu_2;
vi. $Ph_3PCH(CH_2)_3CO_2^-$;
vii. H_2O, $MeCO_2H$, tetrahydrofuran (THF)

Scheme 1

(9) (12)

tricyclic ketone (8). The ketone (8) reacts with a wide variety of nucleophiles to give 7-substituted bicyclo[2.2.1]heptan-2-ones through a Michael-type cleavage of the three-membered ring[20]. In particular,

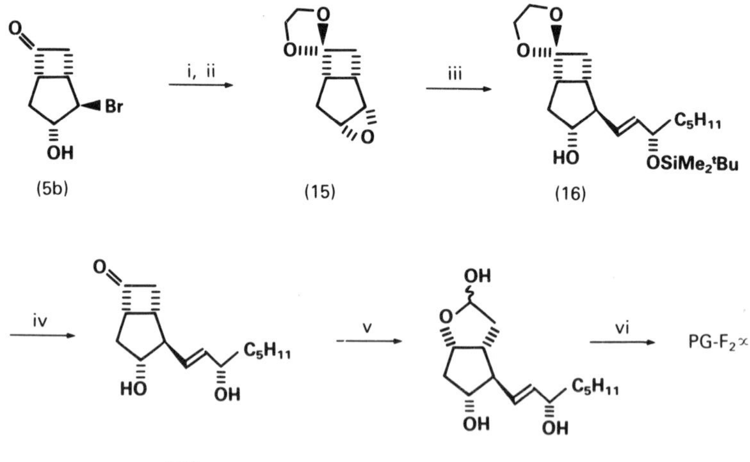

(5b) (15) (16) (18) (19) (14)

Reagents:

i. $(CH_2OH)_2$, H^+;
ii. K_2CO_3, MeOH;
iii. reagent (9), ether <u>ca</u> -40°C then chromatography;
iv. H^+, H_2O;
v. hν, H_2O, MeCN;
vi. $Ph_3PCH(CH_2)_3CO_2^-$ then chromatography

Scheme 2

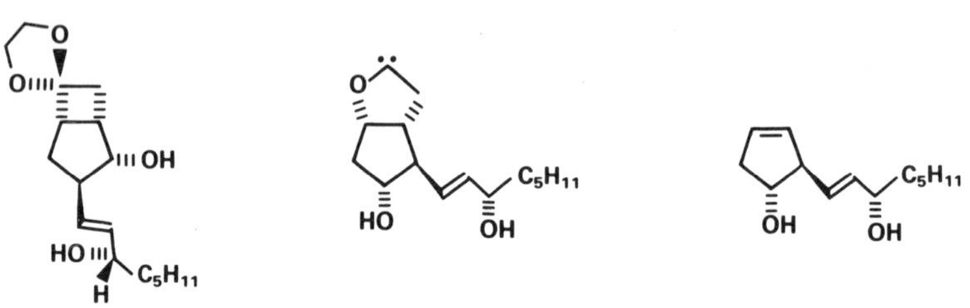

(17) (20) (21)

reaction of the ketone (8) with the cuprate reagent (9) gave the prostaglandin synthon (10). Baeyer-Villiger oxidation of the ketone (10) gave a good yield of the lactone (11): the work up of this oxidation reaction involved washing the crude product with a dilute aqueous solution of sodium hydroxide to remove the small quantity of the isomeric δ-lactone (12) that was present[21]. Reduction of the lactone (11) to the corresponding lactol, followed by a Wittig reaction using the appropriate C_5-ylid, gave the acid (13). This compound was deprotected to give optically pure prostaglandin-$F_{2}\alpha$ (PG-$F_{2}\alpha$) (14).

The bromohydrin (5b), derived from the alcohol (4b) or the ketone (2b), was converted into the epoxyacetal (15) by protection of the carbonyl group and treatment with base (Scheme 2)[19]. Reaction of the epoxide with the cuprate reagent (9) occurs in a highly regio-selective fashion to give the alcohol (16) (58%), after chromatography to remove the small quantity of the isomeric alcohol (17) that is formed concurrently[22]. Deprotection furnished the dihydroxyketone (18). Photolysis of the ketone (18) in aqueous acetonitrile gave the lactol (19) (ca. 70%) [via the oxacarbene (20)] and the alkene (21) (12%). The lactol (19) was converted into PG-$F_{2}\alpha$ (14) by the prescribed Wittig reaction.

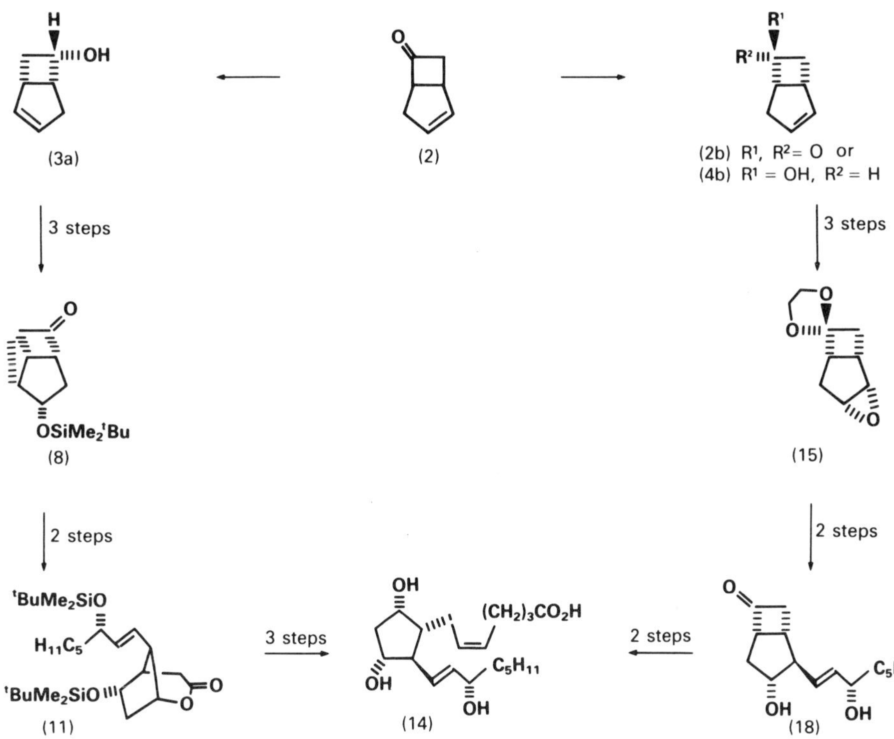

Scheme 3

Schemes 1 and 2 describe short efficient routes to PG-$F_{2}\alpha$. The yield of the natural product from the bromohydrin (5a) was 15% and from the bromohydrin (5b) the overall yield of prostaglandin $F_{2}\alpha$ was 18%. The complementary nature of the two routes is shown in Scheme 3.

Slight modifications to the processes give access to optically active prostaglandins-E_2 (22) and -D_2 (23)[19].

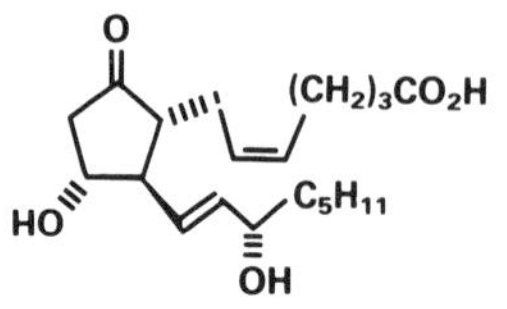

(22)

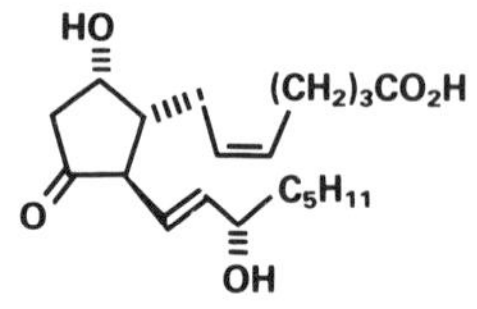

(23)

The synthesis of prostaglandin-A_2 (PGA_2) (26) from the alcohol (3a) (Scheme 4) bears a strong resemblance to the route to PG-$F_{2}\alpha$ described in Scheme 1. An intramolecular 1,3-acyl shift gives access to the desired γ-lactone (25) from the δ-lactone (24)[23].

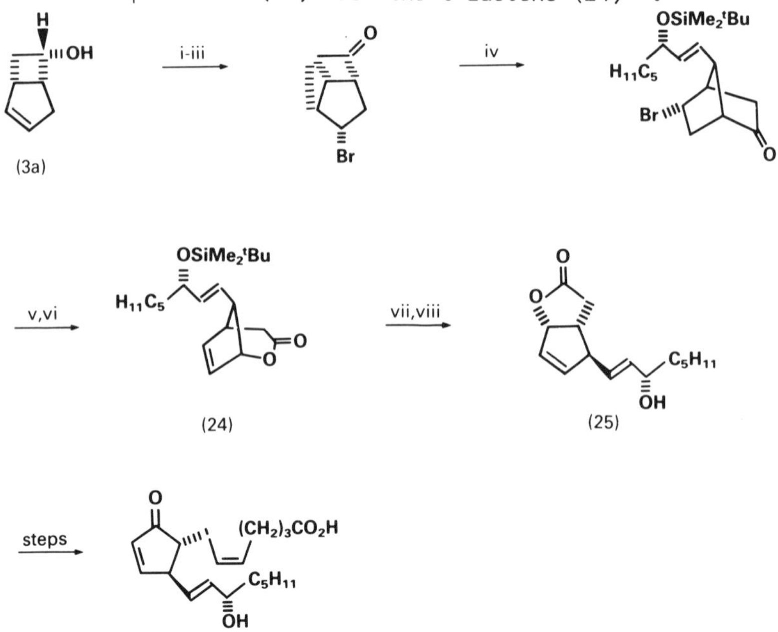

(3a) (24) (25) PG-A_2 (26)

Reagents:

i. Br_2, $NaHCO_3$, CCl_4;
ii. N-bromosuccinimide (NBS), H_2O, acetone, $MeCO_2H$;
iii. $NaN(SiMe_3)_2$;
iv. reagent (9), -78°C, ether;
v. m-$ClC_6H_4CO_3H$;
vi. diazabicycloundecene;
vii. Me_2NCHO, Δ, then HF, H_2O,MeCN

Scheme 4

The preparation of PG-A_2 from the ketone (2b) required a new strategy (Scheme 5). Baeyer-Villiger oxidation of the ketone gave the lactone (6b). Light-induced radical bromination gave the required bromolactone (27) (44%) after chromatography to remove the isomeric bromolactone (28). Treatment with methoxide ion gave the epoxyester (29). Previous studies had indicated that cuprate reagents show a pronounced tendancy to react with cyclopentene epoxides, halides and acylates _via_ an S_N'_anti_ pathway even when the S_N2 reaction pathway is sterically unhampered[24]. Therefore it was not surprising to find that the epoxide (29) reacted with the cuprate reagent (9) to give the γ-lactone (30) as the major product (34%). Chromatography served to separate the γ-lactone (30) from the isomer (31) (21%). Deprotection of (30) gave the lactone (25) which was converted into PG-A_2 in the manner described above.

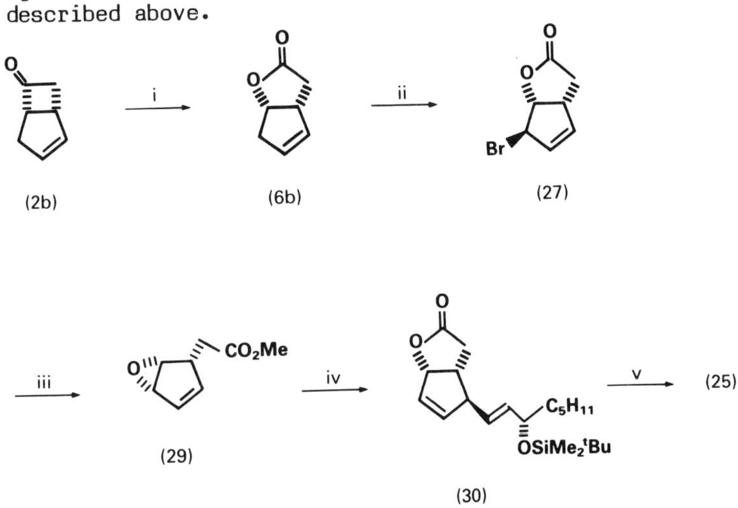

Reagents:

i. $MeCO_3H$, $MeCO_2H$;
ii. NBS, CCl_4, hν;
iii. K_2CO_3, MeOH, ether;
iv. reagent (9), -78°C, CH_2Cl_2;
v. HF, H_2O, MeCN

Scheme 5

(28) (31)

The two routes to PG-A_2 are, once again, "enantio-complementary" (Scheme 6). The overall yield of the lactone (25) from the alcohol (3a) was 10.3% while from the ketone (2b) the overall yield of compound (25) was 7.0%.

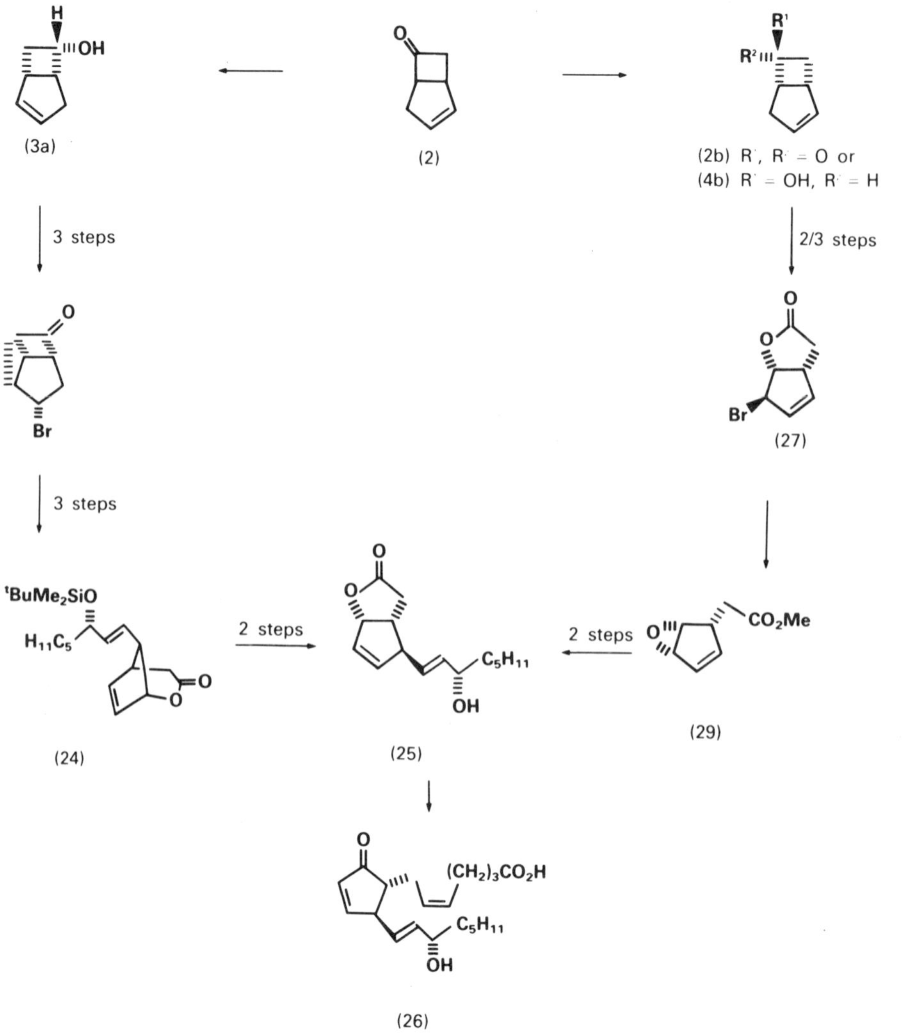

Scheme 6

The high level of effort put in to secure the various routes to the prostaglandins[25] reflects the intense amount of interest in the preparation of prostaglandin analogues (prostanoids) in industrial laboratories. Some analogues of the primary prostglandins act as thromboxane antagonists and hence may be of future importance in the treatment of certain cardiovascular disorders[26]. It seems likely that some prostglandin-E_2 analogues will become therapeutically important cytoprotective agents[27]. PG-A_2 and analogues have been shown to be active against certain types of cancer[28]. The above studies have led to two different routes to prostaglandins A, D, E, and F. Different analogues are provided (in optically active form) by the two routes, thus broadening the range of prostanoids available for biological testing.

4. RESOLUTION OF 7,7-DIMETHYLBICYCLO[3.2.0]HEPT-2-EN-6-ONE

In more recent months, attention has been focussed on two different, biologically active natural products. Leukotriene-B_4 (LT-B_4) (32) is a compound produced from arachidonic acid *in vivo* by the action of lipoxygenase enzymes. It is a potent chemotactic agent for polymorphonuclear leukocytes and may mediate inflammatory reactions *in vivo*. Evidence for LT-B_4 involvement in a number of pathological conditions including diseases such as psoriasis, ulcerative cholitis, rheumatoid arthritis, and gout is accumulating[29]. The second compound of current interest is eldanolide (33). This compound is an attractant pheromone produced by the male African sugar-cane borer *Eldana saccharina* (Wlk), a major pest on sugar-cane and maize in many African

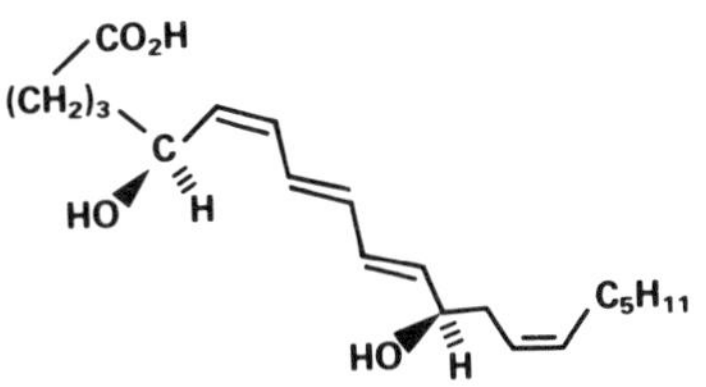

(32)

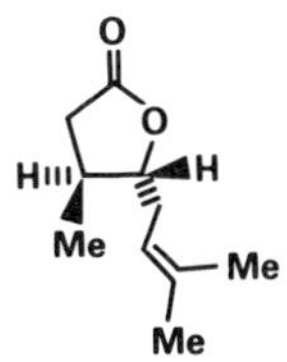

(33)

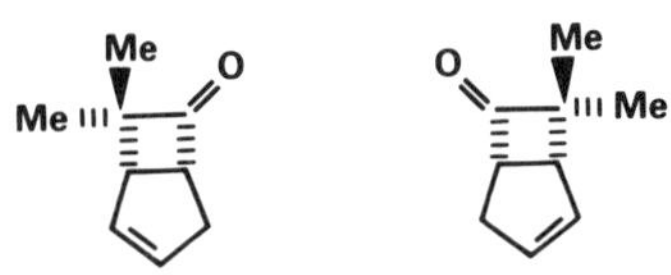

(a)............(34)............(b)

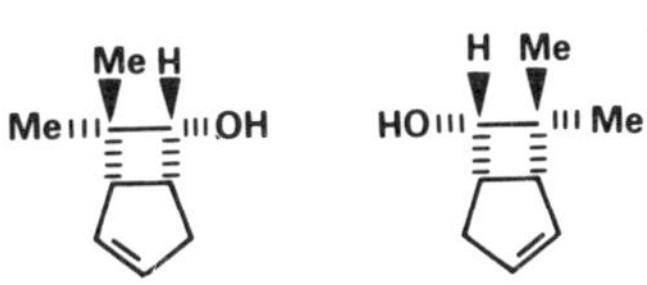

(a)............(35)...........(b)

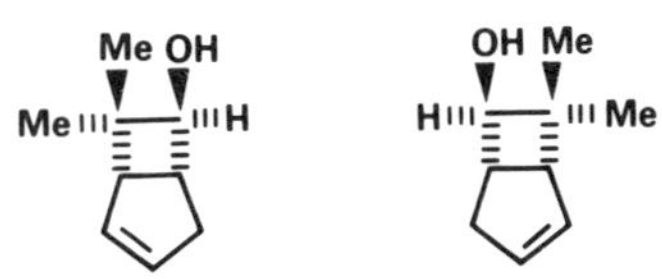

(a)............(36)...........(b)

countries. We have recently prepared both these naturally occurring compounds from 7,7-dimethylbicyclo[3.2.0]-hept-2-en-6-one (34). Since both natural products are chiral and since only one enantiomer displays potent biological activity, a method of resolution of 7,7-dimethylbicyclo[3.2.0]hept-2-en-6-one was required initially.

Small scale experiments comparing the reduction of bicycloheptenone (2) and 7,7-dimethylbicycloheptenone (34) by *M. ramanniana* under identical conditions gave interesting results (Figure). At high dilution (1 g substrate/litre), both the 6*endo*-alcohol *and* the 6*exo*-alcohol were produced from 7,7-dimethylbicyclohepten-6-one. The rate of the enzymic reduction of ketone (34) [as measured by the production of alcohol(s)] was approximately the same as that observed for the ketone (2). At relatively high substrate concentrations (5 g/litre), the rate of reduction of the 7,7-dimethylbicycloheptenone was significantly lower and only the 6*endo*-ol was detected. A larger scale experiment utilising the 7,7-disubstituted bicycloheptenone (34) (15 g) *M. ramanniana* (*ca* 3 kg) in 15 litre of medium gave, after two days and a normal work-up procedure, roughly equal quantities of the two alcohols (35) and (36) in about 50% yield. Formation of the isopropylurethane derivative of (35) and g.l.c. over a Chirasil-Valine column showed that the enantiomer ratio was 90:10. The enantiomer (35b) was shown to be the major component (*vide infra*). Similar treatment of the *exo*-alcohol (36) showed that it was produced in a high state of optical purity, and the ratio of enantiomers (36a) and (36b) was ⩾95:⩽5.

It is interesting to compare and contrast the results from the reduction of ketones (2) and (34) with *M. ramanniana*:

a) At low substrate concentrations (*ca*. 1 g/litre) the rate of production of bicycloheptenol(s) *is* approximately the same for both ketones. At higher substrate concentrations the rate of production of dimethylbicycloheptenol is markedly lower than the rate of production of bicycloheptenol. (The high rate of disappearance of the dimethylbicycloheptenone from the broths may be due to greater absorption of this ketone into the mycelial mass as a consequence of enhanced lipophilicity).

b) At relatively low concentrations of substrate, both 6*exo*-ol and 6*endo*-ol were produced from the dimethylbicycloheptenone. Only the corresponding 6*endo*-ol was obtained from bicycloheptenone at all substrate concentrations employed.

c) All three alcohols were formed with the *S*-configuration at C-6 predominant. This means that the absolute configuration of the C-5 position is *R* for the major bicyclo[3.2.0]hept-2-en-6*endo*-ol enantiomer formed and *S* for the major 7,7-dimethylbicyclo[3.2.0]-hept-2-en-6*endo*-ol enantiomer formed.

FIGURE Conversion of Bicycloheptenone (2) (-·-·-) and Dimethylbicycloheptenone (34) (----) from the ketone (▲) to the 6endo-alcohol (o) and the 6exo-alcohol (◻) using M.ramanniana (125 g/l)

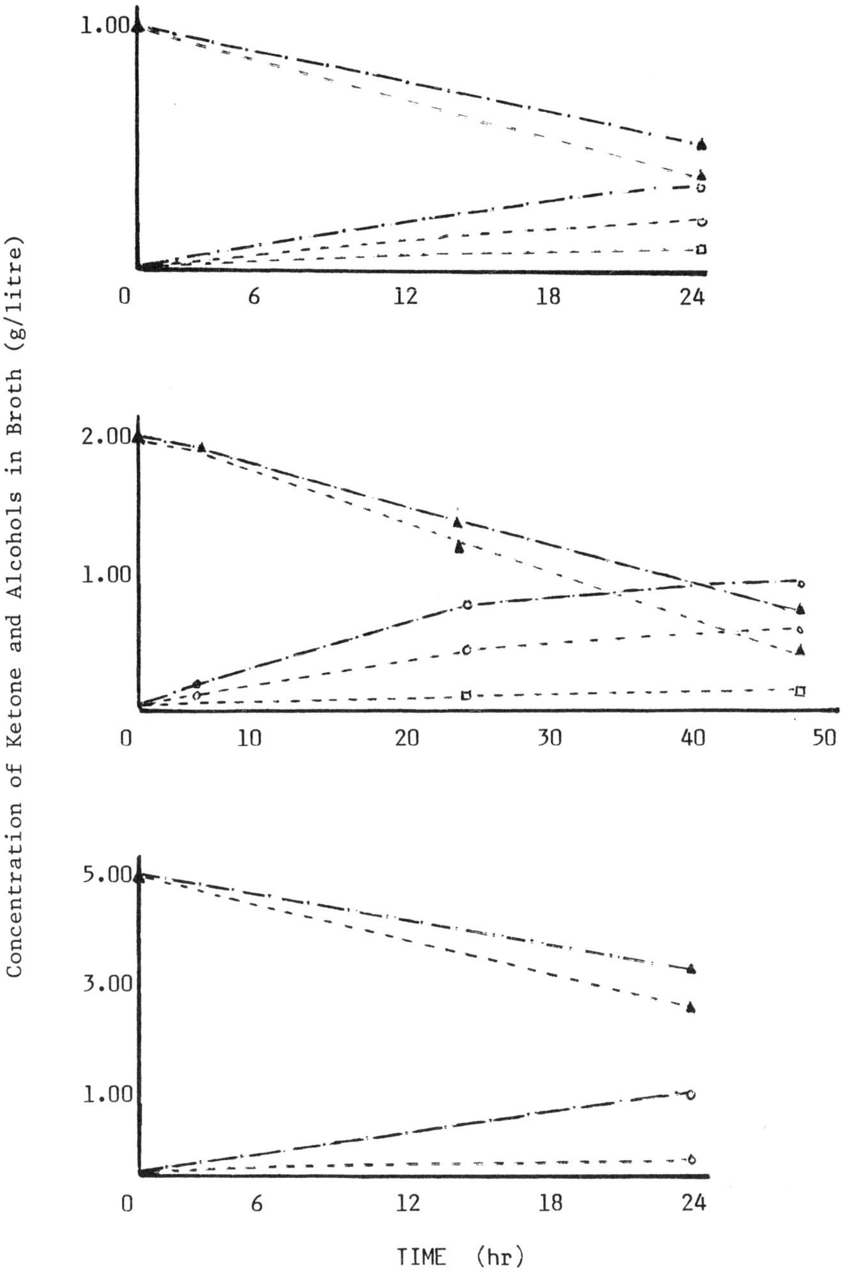

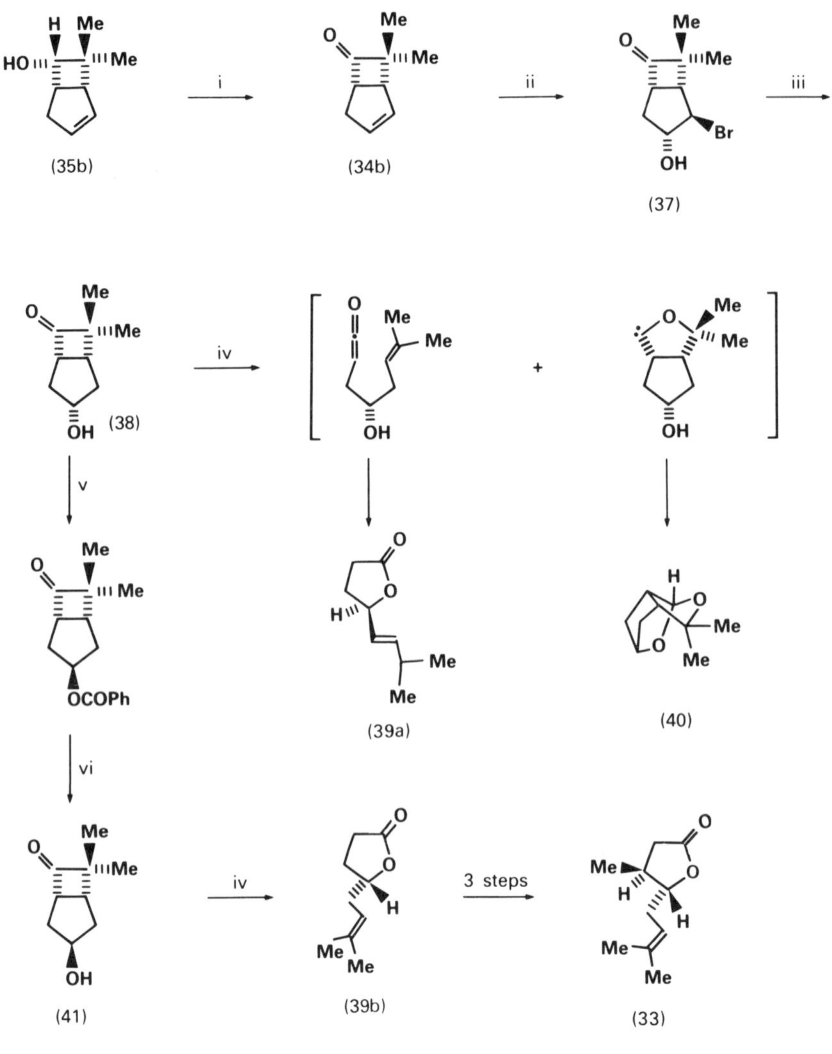

Reagents:

i. pyridinium dichromate;
ii. MeCONHBr, H_2O, acetone;
iii. $(n\text{-}Bu)_3SnH$, AIBN, toluene;
iv. hν pentane;
v. Ph_3P, diethyl azodicarboxylate, $C_6H_5CO_2H$, THF;
vi. K_2CO_3, MeOH

Scheme 7

d) The variation in the amounts of 6exo-ol and 6endo-ol produced from the dimethylbicycloheptenone after 24h, depending on substrate/fungus concentrations, suggests two or more dehydrogenase enzymes are active in the reduction of this ketone.

5. SYNTHESIS OF (+)-LEUKOTRIENE-B_4 (32) AND (+)-ELDANOLIDE (33)

Treatment of 7,7-dimethylbicyclo[3.2.0]hept-2-en-6-one (34b) with N-bromoacetamide in aqueous acetone gave the bromohydrin (37) in

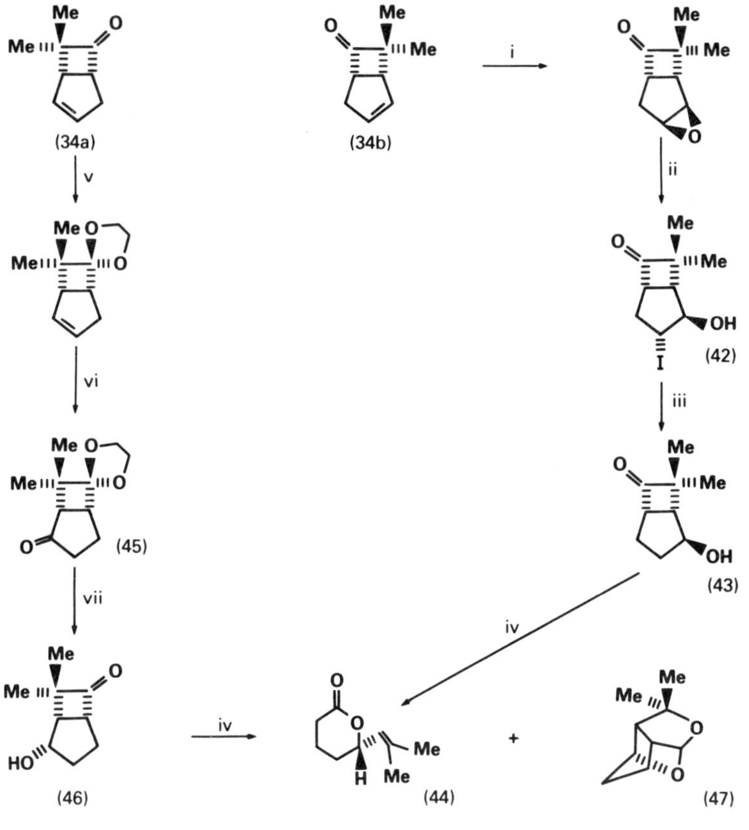

Reagents:

i. m-$ClC_6H_4CO_3H$;
ii. HI, H_2O;
iii. $^nBu_3SnH$, C_6H_6;
iv. hν, pentane;
v. $(CH_2OH)_2$, H^+;
vi. B_2H_6, THF, then NaOH, H_2O_2, then pyridinium dichromate;
vii. $NaBH_4$ then H^+, H_2O

Scheme 8

excellent yield (Scheme 7). Hydrodebromination of the bromohydrin (37) with tri-n-butyltin hydride furnished the hydroxyketone (38). Photolysis of a solution of recrystallized ketone (38) in pentane using a medium- pressure mercury lamp and quartz apparatus gave the γ-lactone (39a) (34%) (through intermediate formation of the alkenylketen) and the acetal (40) (30%) (though intramolecular attack of the pendant hydroxyl group on the oxacarbene moiety). The lactone (39a) can be converted into 3R,4S-eldanolide using a standard three-step procedure[30]. Furthermore the hydroxyl group in the bicycloheptenone (38) was inverted using the Mitsunobu procedure to give the 3exo-hydroxybicycloheptanone (41). Photolysis of this compound in pentane gave the lactone (39b) which was converted into naturally occurring 3S,4R- eldanolide (33)[31]. The overall yield of lactone (39b) from ketone (34b) was approximately 20%.

Epoxidation of the alkenone (34b) followed by reaction with hydroiodic acid gave the iodohydrin (42). Hydrodeiodination gave the 2exo-ol (43). Photolysis of hydroxyketone (43) afforded the δ-lactone (44) as the sole product albeit in a modest 40% yield (Scheme 8).

The δ-lactone (44) is also available from the alkenone (34a). Thus the bicycloheptenone is protected as the acetal derivative prior to hydroboration using diborane in tetrahydrofuran. The alcohol(s)

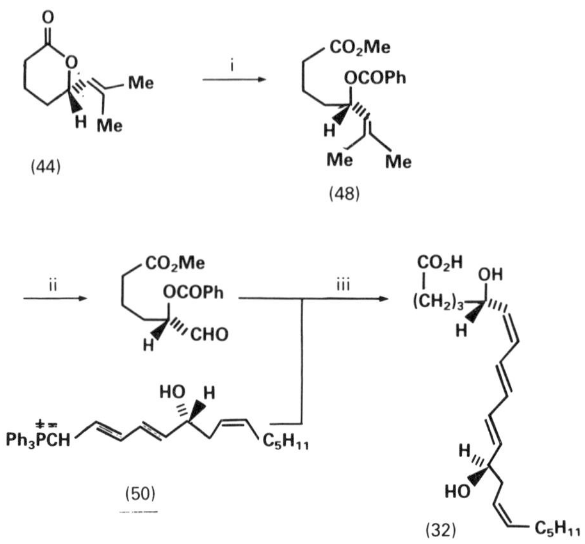

Reagents:

i. Et_3N, MeOH then PhCOCl, base;
ii. O_3 then DMS;
iii. Wittig reaction then NaOH then chromatography

Scheme 9

produced were oxidised with pyridinium dichromate to give the bicycloheptan-2-one (45) as the only isolable product. Reduction of the ketone moiety using sodium borohydride proved to be highly regioselective to furnish, after removal of the acetal group, the 2endo-alcohol (46). Photolysis of this compound gave the lactone (44) and the acetal (47) (Scheme 8). The overall yield of lactone (44) from the bicycloheptenone (34b) was 17.5%, while from the enantiomer (34a) the yield of lactone (44) was 16.0%. Methanolysis of the lactone (44) followed by benzoylation of the allylic hydroxyl group gave the diester (48) (85%) (Scheme 9). Ozonolysis coupled with a reductive work-up procedure gave the aldehyde (49) (100%) which was reacted with the readily prepared ylid (50)[32] to furnish, after chromatography and distillation, leukotriene-B_4 (32) possessing the natural configuration.

6. CONCLUSION

We have found that various microorganisms will reduce bicyclo[3.2.0]heptenone and 7,7-dimethylbicyclo[3.2.0]heptenone. The fungus *M. ramanniana* was found to be particularly useful in providing multigram quantities of products having satisfactory optical purities. In addition, the bicyclo[3.2.0]heptenone (51) has been resolved using bakers' yeast by Lowe *et al.*, in the elegant studies towards the synthesis of cyclobutanone analogues of the β-lactam antibiotics[33].

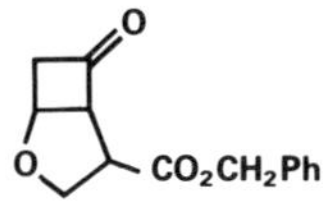

(51)

Obviously, the cyclobutanone ring can be quite heavily substituted in the vicinity of the carbonyl group without destroying the ability of the dehydrogenase enzymes to perform reduction. It is possible that the intrinsic reactivity of a strained four-membered ring ketone may mean that such a substrate does not have to approach as closely to the activating ion (e.g. zinc ion) at the active site of the enzyme. The constraints that usually dictate preferred substitution patterns adjacent to a carbonyl moiety[34] may be somewhat relaxed for the cyclobutanone system.

The major obstacle in our work has been the inability of the microorganisms of choice to tolerate high concentrations of the substrates. For example, at a concentration of 15 g/l the bicycloheptenone (2) tends to show a toxic effect to bakers' yeast and M. ramanniana: the more lipophilic dimethylbicycloheptenone (33) underwent slow reduction at 5 g/litre using M. ramanniana (Figure). It is likely that the hydrophobic bicycloheptenones cause changes in the cell membrane leading to a loss of viability of the organism. Despite this drawback we are continuing to investigate the use of microorganisms and isolated dehydrogenase enzymes for the provision of useful, optically active starting materials for organic synthesis.

7. ACKNOWLEDGEMENTS

This work is reported on behalf of members of research teams at Glaxo Group Research (Greenford and Ware) and the University of Salford. Much of the work is based on a series of novel, important, and wide-ranging ideas generated by Professor R.F. Newton (Director of Chemistry, Ware). Professor Newton and Drs. Wakefield and Wallace (Salford) personally supervised most of the chemical "bench work". The majority of the bench chemistry was performed by Dr. J. Davies (syntheses of PG-$F_{2}\alpha$ and PG-A_2), Dr. P.B. Kay (synthesis of PG-A_2) and Mr. J.A. Winders (syntheses of eldanolide and leukotriene-B_4). Most of the microbiological studies were performed by Dr. M.J. Dawson and Mr. W.F. Wall (Greenford). Other important contributors include Miss S. Butt, Mr. H.G. Davies (Greenford) and the co-authors of articles listed in references 18 and 24.

REFERENCES

1. S.M. Ali, T.V. Lee, and S.M. Roberts, Synthesis, 1977, 155.

2. L. Ghosez and M.J. O'Donnell in 'Pericyclic Reactions', eds. A.P. Marchand and R.E. Lehr, Academic Press, New York, 1977, Volume II, p. 79; J.B. Falmagne, J. Escudero, S. Taleb-Sahraoui and L. Ghosez, Angew Chem. Int. Ed. Engl., 1981, **20**, 879.

3. E.J. Corey and B. De, J. Am. Chem. Soc., 1984, **106**, 2735.

4. A.E. Greene, M.-J. Luche and J.-P. Deprés, J. Am. Chem. Soc., 1983, **105**, 2435.

5. G.A. Annis and L.A. Paquette, J. Am. Chem. Soc., 1982, **104**, 4504.

6. D.E. Cane and P.J. Thomas, J. Am. Chem. Soc., 1984, **106**, 5295.

7. R.F. Newton and S.M. Roberts, Tetrahedron, 1980, 36, 2163; see also T.W. Hart and M.-T. Comte, Tetrahedron Letters, 1985, **26**, 2713.

8. J. Toczek, T.V. Lee, and S.M. Roberts, J. Chem. Soc., Chem. Commun., 1985, 371.

9. J.E. Burks and J.K. Crandall, J. Org. Chem., 1984, **49**, 4663.

10 L.A. Paquette, D.R. Andrews, and J.P. Springer, J. Org. Chem., 1983, **48**, 1147.

11. B. Riefling, Tetrahedron Letters, 1985, 2063.

12. U.A. Huber and A.S. Dreiding, Helv. Chim. Acta., 1970, **53**, 495.

13. C. Houge, A.M. Frisque-Hesbain, A. Mockel, L. Ghosez, J.P. Declercq, G. Germain, and M. Van Meerssche, J. Am. Chem. Soc., 1982, **104**, 2920.

14. K. Bruneel, D. De Keukeleire, and M. Vandewalle, J. Chem. Soc. Perkin Trans. 1, 1984, 1697.

15. E.W. Collington, C.J. Wallis and I. Waterhouse, Tetrahedron Letters, 1983, **24**, 3125; H. Greuter, J. Dingwall, P. Martin and D. Bellus, Helv. Chim. Acta, 1981, **64**, 2812.

16. C.R. Johnson and J.R. Zeller, Tetrahedron, 1984, **40**, 1225.

17. R.F. Newton, J. Paton, D.P. Reynolds, S.N. Young, and S.M. Roberts, J. Chem. Soc. Chem. Commun., 1979, 908.

18. M.J. Dawson, G.C. Lawrence, G. Lilley, M. Todd, D. Noble, S.M. Green, S.M. Roberts, T.W. Wallace, R.F. Newton, M.C. Carter, P. Hallett, J. Paton, D.P. Reynolds and S. Young, J. Chem. Soc., Perkin Trans. I, 1983, 2119.

19. J. Davies, S.M. Roberts, D.P. Reynolds, and R.F. Newton, J. Chem. Soc.,Perkin Trans. I, 1981, 1317.

20. Other interesting reactions of this tricyclic ring system have been reported: A.B. Smith III, B.H. Toder, R.E. Richmond, and S.J. Branca, J. Am. Chem. Soc., 1984, **106**, 4001.

21. D.P. Reynolds, R.F. Newton and S.M. Roberts, J. Chem. Soc., Chem. Commun., 1979, 1150.

22. For analogous regioselective epoxy-ring opening reactions see D.J. Kertesz, A.F. Kluge, P. Kuenzler, T. Li, M.M. Marx, J.J. Bruno, and L. Chang, Prostaglandins, 1984, **27**, 851; R.J. Ferrier, P.C. Tyler, and G.J. Gainsford, J. Chem. Soc., Perkin Trans. 1, 1985, 295.

23. R.F. Newton, D.P. Reynolds, J. Davies, P.B. Kay, S.M. Roberts, and T.W. Wallace, J. Chem. Soc., Perkin Trans 1, 1983, 683.

24. S.M. Roberts, G.T. Woolley, and R.F. Newton, J. Chem. Soc., Perkin Trans 1, 1981, 1729.

25. Reviews on prostaglandin synthesis are available:- New Synthetic Routes to Prostaglandins and Thromboxanes, eds. S.M. Roberts and F. Scheinmann, Academic Press, London, 1982; Prostaglandin Synthesis by J.S. Bindra and R. Bindra, Academic Press, New York, 1977.

26. E.W. Collington in Proceedings of the Second SCI/RSC Medicinal Chemistry Symposium, ed. J.C. Emmett, Royal Society of Chemistry, London. 1984, p. 299.

27. R.J. Ancill, Med. Actual, 1984, **20**, 213.

28. M. Fukushima and T. Kato in Icosanoids and Cancer, eds. H. Thaler-Dao et al., Raven Press, New York, 1984, p. 275.

29. A.W. Ford-Hutchinson, Fed. Proc., 1985, **44**, 25; C.L. Malmsten, Critical Revs. in Immunology, 1984, **4**, 307; T. Ruzicka, T. Simmet, B.A. Peskar, and O. Braun-Falco, Lancet, 1984, 222.

30. J.P. Vigneron, R. Meric, M. Larcheveque, A. Debal, J.Y. Lallemand, G. Kunesch, P. Zagatti, and M. Gallois, Tetrahedron, 1984, **40**, 3521.

31. H.G. Davies, S.M. Roberts, B.J. Wakefield, and J.A. Winders, J. Chem. Soc., Chem. Commun., in press.

32. L.S. Mills and P.C. North, Tetrahedron Letters, 1983, **24**, 409.

33. G. Lowe and S. Swain, J. Chem. Soc., Perkin Trans. 1, 1985, 391.

34. J.B. Jones in Enzymes in Organic Synthesis, (Ciba Foundation Symposium III), eds. R. Porter and S. Clark, Pitman, London, 1985, p.3, and references therein; G.L. Lemiere, T.A. Van Osselaer, J.A. Lepoivre, and F.C. Alderweireldt, J. Chem. Soc., Perkin Trans II, 1982, 1123.

THE SCOPE OF BIOCATALYSTS IN ORGANIC CHEMICAL PROCESSING

S.E.Godtfredsen, K.Ingvorsen, B.Yde, and O. Andresen
Novo Industri A/S
Novo Alle
DK-2880 Bagsvaerd
Denmark

The diversity and complexity of organic molecules in nature is a remarkable reflection of the power of enzymes as catalysts for organic chemical reactions. Almost any conceivable structural type of organic compounds has been constructed in nature thanks to the enzymes; yet these often highly intricate and complex structures have been generated under the mildest conditions of temperature and pH. Even, the careful regulation of complex biosynthetic networks in living matter rely, to a large extent, on inherent control mechanisms of the enzymes which, clearly, are the most powerful and efficient catalysts known, besides being most subtle and delicate (7).

TABLE I
Advantages of enzymes as catalysts in organic chemical processing.

- **Enzymes catalyze a wide variety of organic reactions**
- **Enzymes are very efficient catalysts**
- **Enzymes are enantioselective**
- **Enzymes are regioselective**
- **Enzymes are chirale catalysts**
- **Enzymes are effective under mild conditions**
- **Enzymes are selective in regard to their substrates**
- **Enzymes can be prepared by fermentation**

M. P. Schneider (ed.), Enzymes as Catalysts in Organic Synthesis, 77–95.

In view of the lavish use of the enzymes made by nature for synthesis of organic compounds and the extraordinary efficiency of these catalysts it would seem obvious to make use of these catalysts also in preparative organic chemistry. In fact, the properties of the enzymes would, indeed, fulfil many requirements of the preparative organic chemist, cf. Table I. Most important, the regio- and enantioselectivity of the enzymes, besides their substrate specificity, can be invaluable to the organic chemist like the ability of the enzymes to function under the mildest conditions compatible with even the most fragile organic molecules. Moreoverone would expect the price of these extraordinary catalysts to be rather modest due to the possibility of preparing the enzymes by microbial fermentation.

Fig. 1.
Distribution of industrial enzymes by volume.

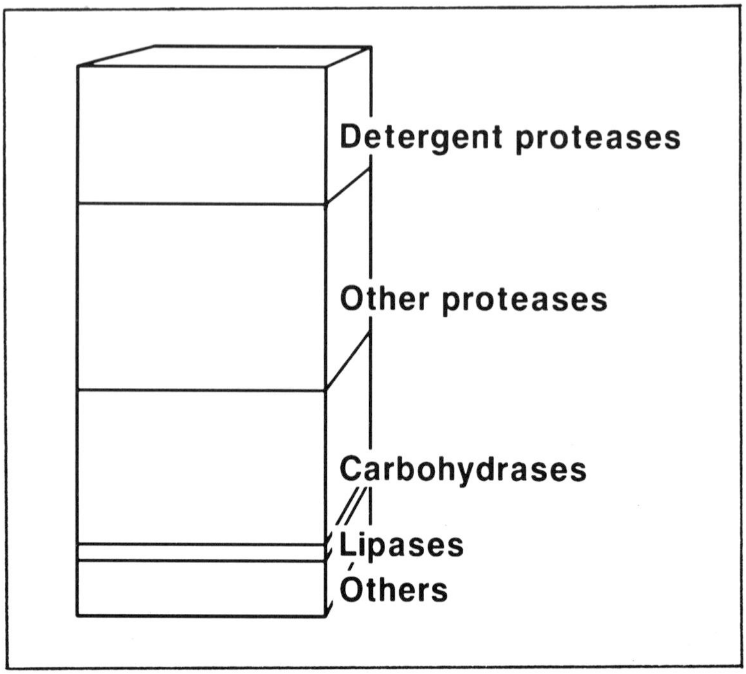

The, potentially, exceedingly wide scope of the enzymes in preparative organic chemistry thus easily envisioned is in striking contrast to the present situation concerning the use actually made of the enzymes in industrial chemical processing. From Figures 1 and 2, which summarize the distribution of the current enzyme production on enzyme types and application areas, it is thus apparent that the proteases, carbohydrases and lipases account for more than 95% of the total production of technical enzymes which still find essentially all their applications within the traditional areas listed in Figure 2 (2). It is noteworthy that only about a dozen of the several thousand enzymes known is actually produced in sizeable quantities and that the production of enzymes for the organic chemical industry is still too small to be visible in the overall picture.

One may well wonder whether this deep discrepancy between promise and practice is a reflection of some kind of inherent disadvantage of the enzymes in regard to organic chemical processing or whether other circumstances have limited their use within this area, so far. Some possible disadvantages of the enzymes as organic catalysts may, in fact, readily be mentioned, cf. Table II. Many enzymes are thus readily inactivated under extreme conditions as, for example, at high temperatures or under strong acidic or basic conditions. Likewise, many enzymes are readily inactivated by exposure to reactive organic compounds of well-established use in preparative organic chemistry, and they are usually thought to function optimally in aqueous solutions only, rather than in organic solvents often needed to solubilize the majority of organic chemicals of interest. Moreover, enzymes are often inhibited by their substrates and their products at concentrations far below those usually considered necessary for an economical process.

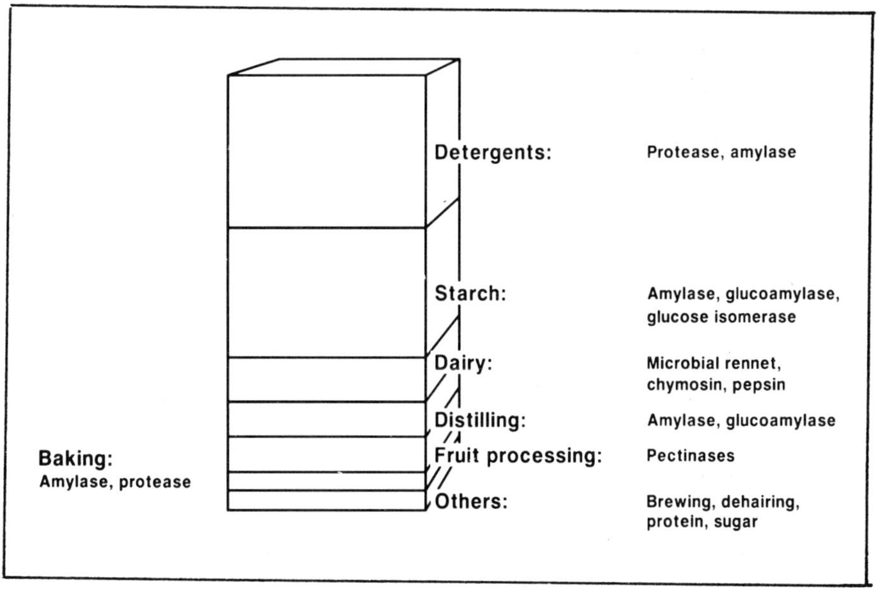

Fig. 2. Distribution of enzyme sales on industries.

Although clearly valid for the large majority of known enzymes these possible disadvantages of the enzymes can by no means be considered as owing to some inevitable, inherent property of the enzymes. Rather, experience shows that enzymes active under the most unusual conditions may in fact be found, if desired. As a respond to a requirement of the starch industry enzyme producers have, for example, provided vast quantities of carbohydrases active in water at temperatures close to 100 °C. Likewise, as a respond to a requirement of the detergent industry the enzyme producers have

TABLE II
Disadvantages of enzymes as catalysts in organic chemical processing.

- **Enzymes are inactivated by many organic reagents, under strong acidic or basic conditions, and at high temperatures**
- **Enzymes exhibit a limited activity in most organic solvents**
- **Enzymes are subject to product- and substrate inhibition**
- **Enzymes may provoke allergic reactions**

furnished proteases active in water at temperatures as high as 90 °C under strongly alkaline conditions never before thought to be compatible with enzyme activity. Work on lipases, which catalyze hydrolytic cleavage of triglycerides and transacylations as visualized in Figure 3, has, similarly, revealed that these remarkable enzymes function under highly unusual conditions, for example as suspensions in their pure substrates or even in organic solvents like hexane at temperatures as high as 100 °C (9). Since, in addition, dramatic improvements of the stability and reactivity of enzymes can often be achieved by chemical modification, immobilization, and other techniques, and since, for example, the potentials of enzyme engineering and the occurrence of microbial life under the most extreme conditions hold great promise in regard to the development of highly resistant enzymes, there is clearly little justification for the view that enzymes are fragile catalysts with properties which will necessarily seriously limit their use in preparative organic chemistry. Rather, it is quite likely that, given sufficient efforts, enzymes can, in most instances, be found and made to function in a satisfactory manner in accordance with the imagination of even the most daring organic chemist.

However attractive and versatile the enzymes are as organic catalysts it should, of course, not be forgotten that, for many years, synthetic organic chemistry has undergone an intense development. For decades, new synthetic methods, reagents and catalysts have been developed at an amazing rate. The very limited use made of enzymes in organic chemical processing may, therefore, be a mere reflection of the existence of suitable, classical chemical approaches or other techniques for preparation of organic chemicals of industrial interest.

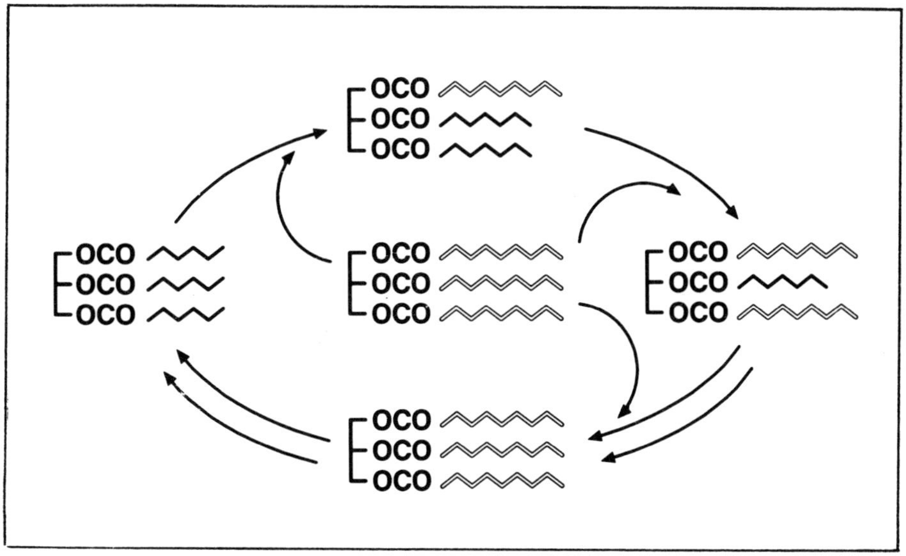

Fig. 3. Lipase catalyzed transacylation of triglycerides.

Production of L-amino acids represents an example of this situation in which the producer has available at least three optional techniques: he may synthesize the amino acids by pure classical organic chemistry, he may generate the amino acids by fermentation or, as a third option, he may apply one or more enzymes in a synthetic sequence, in this particular instance, for example as an aid to resolve a racemic mixture of some intermediate in the reaction sequence or of the final product.

An example of an enzyme which could be applied in this manner is the amino peptidase, an enzyme which catalyzes conversion of L-amino acid amides into the corresponding L-amino acids leaving the D-amino acid amides untouched (3,4). As illustrated in Figure 4, this enzyme may be applied for production of optically pure D-phenylglycine from a racemic mixture of the amino acid amide. The method applies, however, equally well for most amino acids, and is particularly attractive in view of the easy availability of the required amino acid amides via the classical Strecker synthesis. Moreover, yields achieved in the enzyme catalyzed step are excellent as is the optical purity of the resulting amino acids. In spite of these salient features of the process, the enzymatic method is by no means necessarily the method of choice for production of amino acids. As is apparent from Figure 5, the actual flow-sheet for the enzymatic synthesis, the enzyme dependent step in this sequence represents, however attractive, only a minor part of the

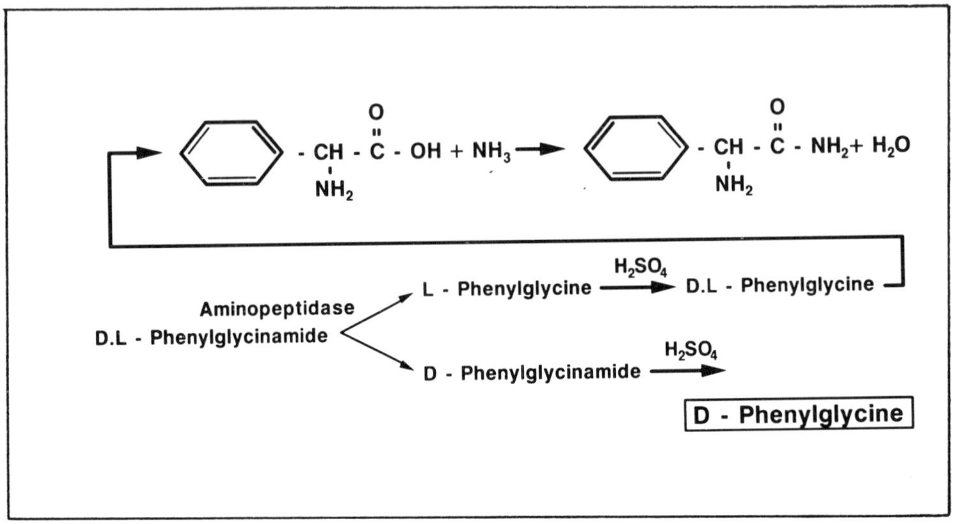

Fig. 4. Enzymatic process for preparation of D-phenylglycine.

overall process scheme, the contribution of the enzyme catalyzed step to the overall economy of the process being correspondingly small. Moreover, methods for production of natural amino acids by fermentation have proved very successful and highly competitive as compared to classical chemical synthesis. This is readily apparent from Figure 6 which indicates the cost structure of a natural amino acid prepared by these two techniques. Although, as indicated in Table III, enzymatic synthesis offers several advantages as compared to a fermentative process, it does not seem likely that these advantages are of sufficient value to justify a switch from fermentation to enzymatic synthesis. In this particular case where the advantage of enzymatic synthesis relies almost exclusively on the application of an enzyme for separation of enantiomers, the limited use of enzymes is due not to any particular drawback of the enzymes but rather to a lack of sufficient impact of the enzyme on the economy of the overall process. In general, use of enzymes within the amino acid field may be justified, for example, for production of unnatural isomers of amino acids as D-phenylglycine and related compounds but, with some exceptions, not

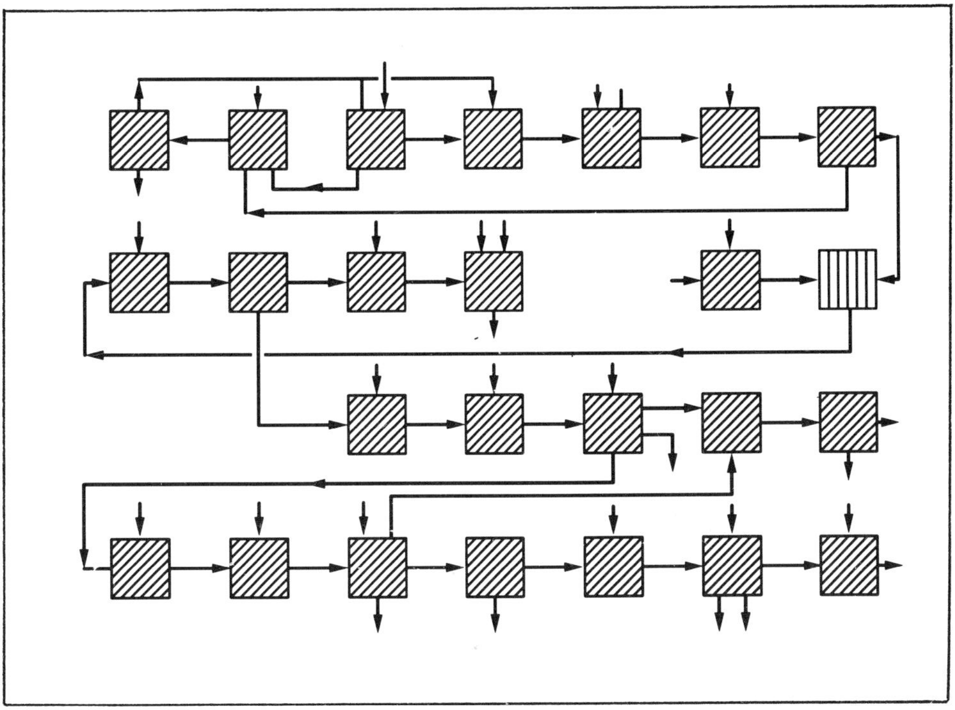

Fig. 5. Flow sheet for enzymatic synthesis of an amino acid. ▥ illustrates the enzyme dependent step.

for large scale synthesis of the naturally occurring amino acids.

The situation is somewhat different in regard to synthesis of 6-APA by side-chain cleavage of either penicillin-V or penicillin-G, both of which are produced by fermentation, cf. Figure 7. As in the case of the amino acids this conversion can be achieved by chemical as well as by enzymatic synthesis both of which proceed in quite satisfactory yields and provide the required product in comparable qualities. Owing to the sensitivity of the penicillin nucleus, the chemical method must, however, be carried out at a low temperature and by use of organic reagents and solvents which are rather costly and give rise to environmental problems. In contrast, the enzyme catalyzed cleavage proceeds without any use of such reagents and solvents and proceeds smoothly at room temperature. In this

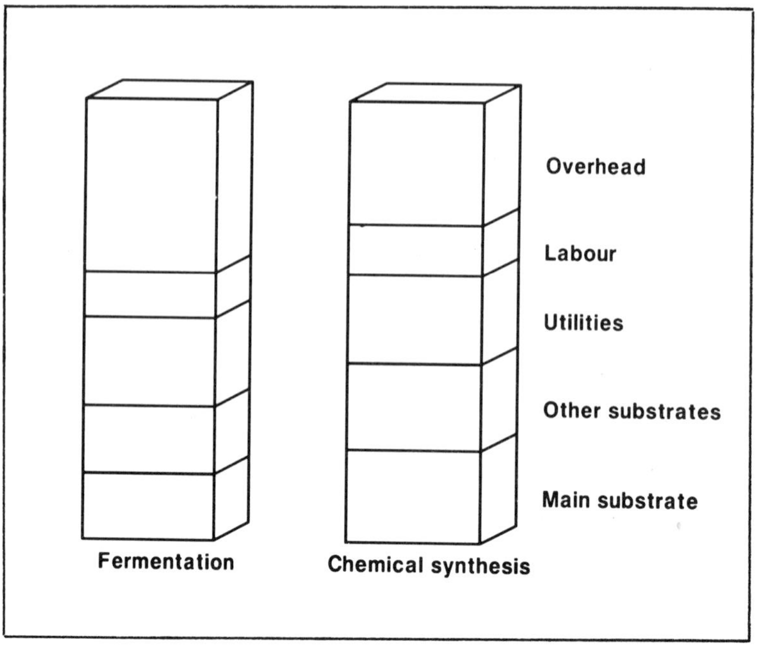

Fig. 6. Cost structure of a typical amino acid.

particular case, application of an enzyme has, therefore, some clear advantages. Although the reaction in question may be carried out by pure chemical methods, enzymatic synthesis is, accordingly, the method of choice for preparation of

6-APA which is currently produced in a quantity of 4,000 t/y. The enzyme applied for this process may, as indicated in Figure 7, also be applied for enzymatic synthesis of 7-ADCA by cleavage of the cephalosporine obtained by chemical conversion of penicillin-V. Oddly enough, it has proved exceedingly difficult to isolate an analogous enzyme for cleavage of the product of fermentation of Cephalosporium species, cephalosporine-C, into 7-ACA (5,8), a compound which could serve as an intermediate for synthesis of semisynthetic cephalosporins, like 6-APA is utilized in penicillin chemistry. It has daringly been proposed that this is an example of an enzyme which cannot be found in nature.

TABLE III
Comparison of amino acid production by fermentation and enzymatic synthesis.

	Enzymatic synthesis	Fermentation
Feasible conversion	one step	complex
Raw material	relatively expensive	cheap
Price of catalyst	relatively expensive	cheap
Downstream processing	simple	complex
Reaction medium	water and organic solvents	water
By-product formation	none	very high
Required equipment	simple	complex

TABLE IV
Comparison of chemical process and enzyme catalyzed process in regard to formation of by-products and co-products.

By-products and co-products in percent	0–1	1–10	10–50	50–100
Enzyme catalyzed process	Hardly competitive	Possible environmental advantage	Possibly competitive	Often advantageous
Chemical process	Impurities accepted	Separation and discharge	Separation and possibly recirculation	Usually not used

The possibility of avoiding formation of by-products in the broadest sense of the word, - i.e. solvents, salts, and other materials which must be discharged, products arising from parallel reactions, etc., etc. - by application of enzymes is most probably an aspect of enzyme catalysis which, in many instances, will render enzymatic synthesis decisively advantageous as compared to classical chemical methods. As indicated in Table IV enzymes may, in principle, be considered particularly attractive as catalysts for reactions which, when carried out by conventional means, provide more than 50% of by-products, while, a priori, enzymatic synthesis must be expected to be of less interest and of limited potential only in situations where a classical, chemical synthesis proceeds with only low or negligible formation of by-products.

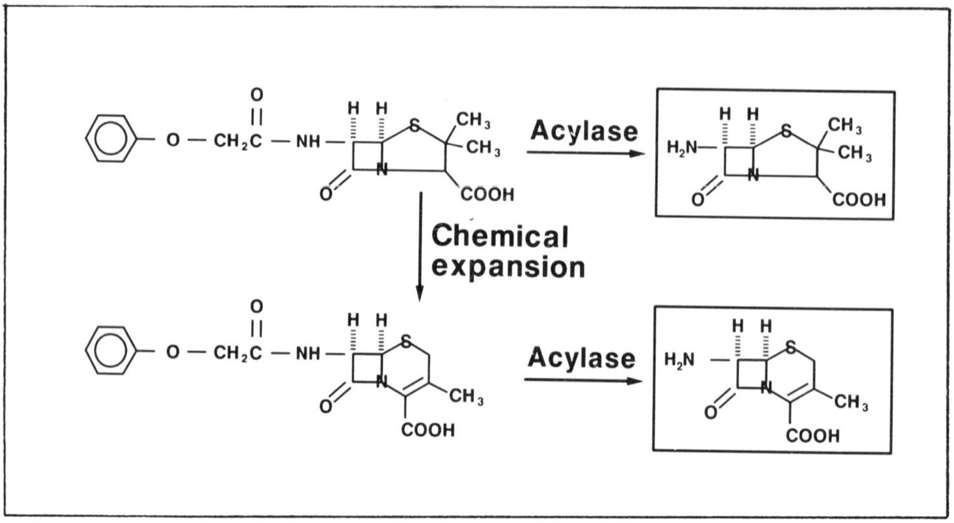

Fig. 7. Enzymatic process for preparation of 6-APA and 7-ADA.

Nitrile converting enzymes constitute a group of enzymes which, on this basis, must be expected to have an interesting potential as catalysts in organic chemical processing. The reactions catalyzed by this group of enzymes are summarized in Figure 8. As indicated, some of these enzymes, the nitrilases, catalyze conversion of nitriles directly into the corresponding carboxylic acids while other nitrile converting enzymes, the hydratases, catalyze conversion of nitriles into amides which, by the action of amidases usually present in preparations of hydratases, are readily converted further into carboxylic acids (1).

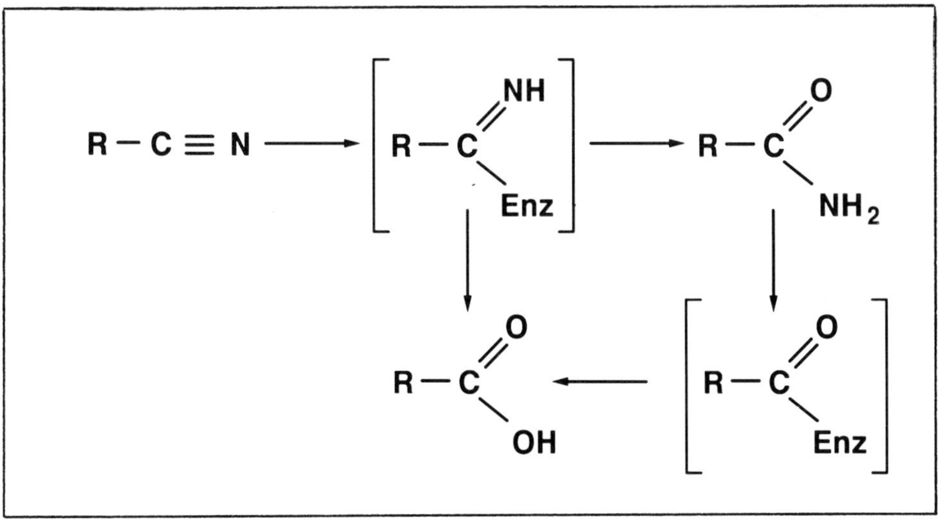

Fig. 8. Proposed reaction mechanism for nitrile converting enzymes

The nitrile converting enzymes are produced by microorganisms which, most commonly, produce hydratases and amidases rather than nitrilases. In some instances, such hydratases and amidases produced by a single microorganism have proved to exhibit a remarkable, broad substrate specificity (6). Accordingly, these enzymes may be utilized, for example, for synthesis of optically pure amino acids and hydroxy acids from the corresponding amino nitriles or cyanohydrins, respectively, or, by using amidase-free preparations of hydratases, for conversion of nitriles into the corresponding amides (6). In contrast to the conventional methods for hydrolysis of nitriles which are accompanied by formation of vast quantities of salts thereby making purification and isolation of the desired products difficult and creating environmental problems, the enzymatic conversion of nitriles proceeds smoothly under mild conditions and without formation of unwanted salts. Since, in addition, isolation of the reaction products of the enzymatic conversion is extremely easy and since optically active products can be generated by the enzymatic hydrolysis, the nitrile converting enzymes clearly deserve close attention as catalysts in organic chemical processing.

One further utility of these enzymes is apparent from Figure 9 which, in a schematical fashion, indicates the types of products in principle obtainable by hydration of dinitriles. As indicated, hydrolysis of dinitriles may give rise to five different products among which, however, only two - the diamide and the diacid - are accessible in high yields by acid or base catalyzed hydrolysis.

During a screening for nitrile converting enzymes we have checked a wide variety of such enzymes in regard to the distribution of products obtained by hydrolysis of a series of dinitriles by these enzymes. In this context we have found that some enzymes tested give rise to almost exclusive formation of mono-cyano-mono-carboxylic acids from the corresponding dinitriles while other enzymes provide high yields of diacids from dinitriles. Moreover, by applying amidase-free mutants of some strains tested, we have been able to convert, in almost quantitative yields, some dinitriles into the corresponding mono-amido-mono-cyano compounds. Since, in addition, a dinitrile may be converted into the corresponding mono-amido-mono-carboxylic acid by applying, in sequence, a mono-acid generating strain and an amidase-free preparation from a di-acid generating strain, these experiments demonstrate how application of enzymes makes accessible certain compounds not amenable for preparation in high yields by conventional means: all of the five structural types listed in Figure 9 can thus be prepared in high yields using enzymes as catalysts while only two are accessible in high yields by chemical synthesis.

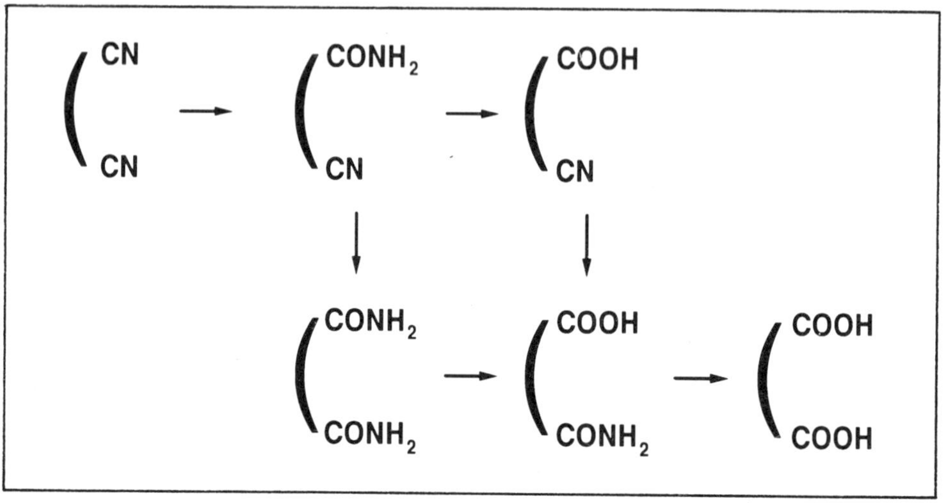

Fig. 9. Possible products of conversion of dinitriles with nitrile converting enzymes.

One particularly interesting use of these findings relates to the preparation of precursors for nylon-6 and nylon-6,6. As indicated in Figure 10, adiponitrile may be converted by one group of nitrile converting enzymes into adipic acid, a precursor for nylon-6,6. By applying another group of nitrile converting enzymes, adiponitrile can be converted into either the corresponding mono-amido-mono-cyano compound or mono-cyano-mono-carboxylic acid. Both of these compounds may, in turn, be converted by catalytic hydrogenation into the amines which can be further converted into caprolactam, the precursor of nylon-6.

This new route to caprolactam is of interest for several reasons. For example, the classical route to caprolactam, which involves conversion of cyclohexanone into its oxime and subsequent Beckman rearrangement, is accompanied by generation of vast quantities of by-products: typically, 4 kg of ammonium sulfate are generated for every kg of caprolactam synthesized. Moreover, adiponitrile is an attractive starting material for several reasons. First of all, it can be generated by electrochemical dimerization

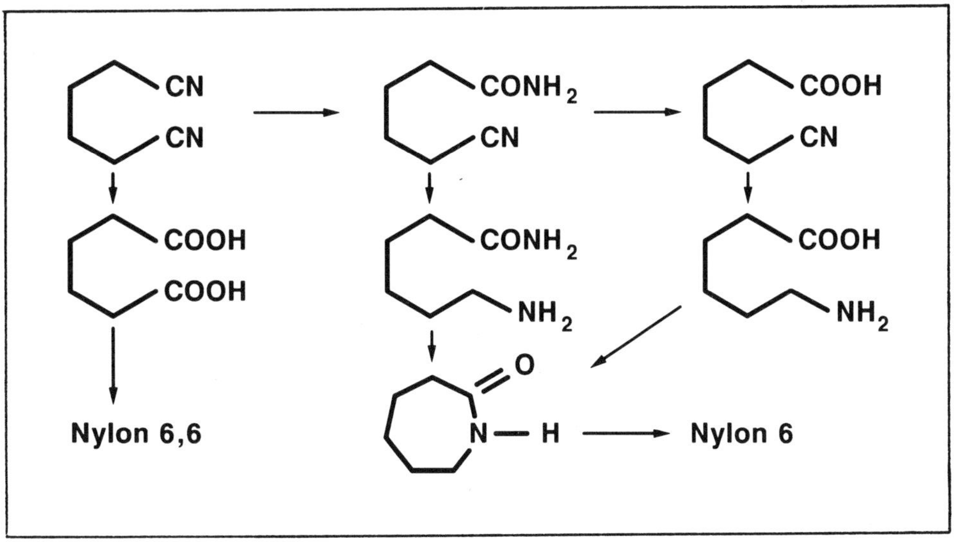

Fig. 10. Preparation of nylon precursors by enzymatic synthesis.

of acryloitrile thereby making posible utilization of nuclear power for production of organic chemicals rather than crude oil. Secondly, this process may readily be combined with the subsequent enzymatic step thereby reducing purification costs and, thirdly, the price, not only of the starting material for the enzymatic synthesis, but also of other substrates required in this process, compares quite favourably with that of the classical chemical approach. Finally, this route to caprolactam utilizes the C2 - C3 fraction of crude oil in contrast to the classical route which makes use of aromatic fractions in great demand as booster for lead-free gasoline.

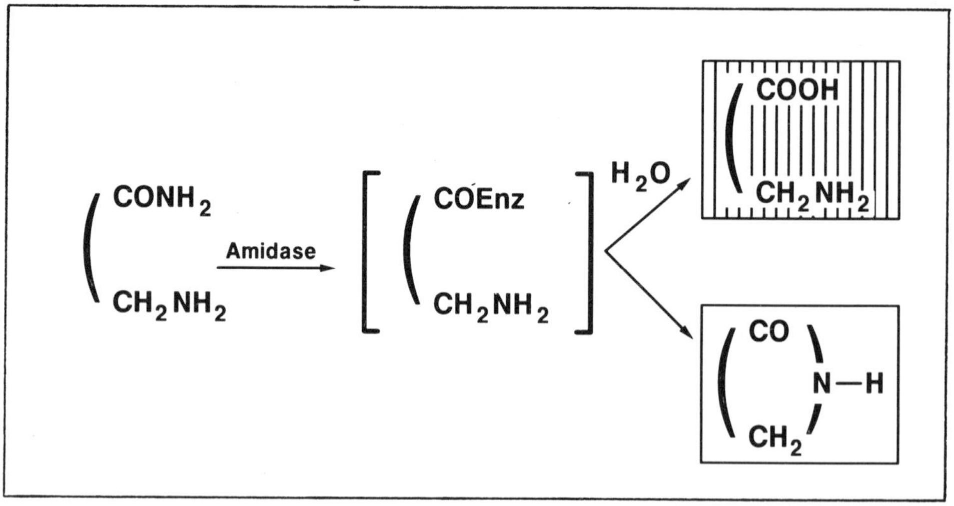

Fig. 11. Possible courses for amidase catalyzed conversion of aminoamides.

One further, potentially interesting, application of nitrile converting enzymes in caprolactam synthesis is outlined in Figure 11. As indicated, conversion, with the aid of an amidase, of the amino amido compound generated by successive enzymatic hydration and catalytic hydrogenation of adiponitrile may, in principle, give rise either to the corresponding amino acid or, as an interesting alternative, the corresponding lactam. In fact, we have identified amidases which, in potentially interesting yields, catalyze the cyclization of amino amides rather than their clevage to the free acid. Accordingly, the synthesis of caprolactam summarized in Figure 12 can be considered as a quite interesting alternative to the classical procedure. As is apparent from the figure, the enzymatic synthesis is exceedingly simple, it requires few raw materials only, and it does not give rise to any appreciable formation of by-products. This, and many other applications of nitrile converting enzymes, cf. Figure 13, may, therefore, very well turn out to be quite useful in preparative organic chemical processing.

In all of the preceeding examples of possible applications of enzymes as organic catalysts the final products can be made available by conventional organic chemical methods as well as by enzymatic synthesis. A choice between the two methods depends, therefore, in the final analysis on economic considerations, environmental questions, security in regard to supply of raw materials, and a large number

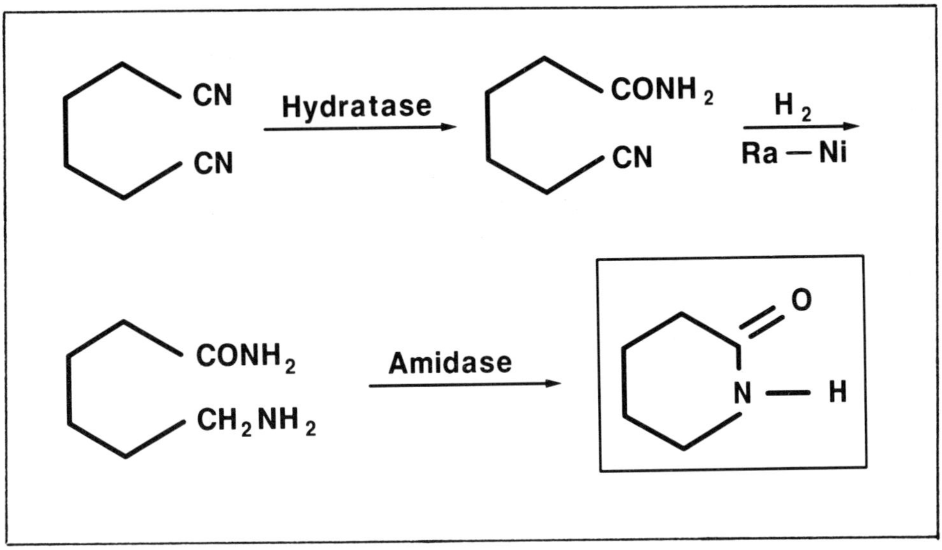

Fig. 12. Combined enzymatic and chemical synthesis of caprolactam.

of other related questions. In some instances, however, a technically realistic organic chemical method is not available for preparation of quite interesting compounds nevertheless readily available by enzymatic synthesis. This situation arises particularly often in connection with preparation of complex, biologically active compounds as, for example, peptide hormones of large molecular weights. A well-known and commercially important example is conversion of pork insulin into its human counterpart by enzyme catalyzed exchange of the terminal amino acid residue in the B-chain of the insulin molecule. This process proceeds exceedingly well with the aid of enzymes, but could hardly be effected by classical chemical means.

Besides providing access to the wide variety of structural types indicated in Figure 13 the mono-nitrilases offer the possibility of preparing chirale building blocks of these structural types. This possibility is particularly attractive for conversion of pro-chirale dinitriles which may be quantitatively converted into optically pure compounds with the aid of mono-nitrilases. Often, the substituted dinitriles needed in this connection will be readily available due to the activation of the unsubstituted dinitriles provided for by the nitrile groupings. In comparison to other enzymatic methods for preparation of optically active building blocks as e.g. enzymatic resolution of racemates this method is highly attractive, among other things, because no recycling of undesired antipodes need to be considered.

One further aspect of application of enzymes in organic chemical processing is apparent from Figure 14. In this figure, for the purpose of a rough illustration only, five of the most important products or product types mentioned in the preceeding discussion as amenable to enzymatic synthesis are listed and ordered according to their approximate volume. Accordingly, caprolactam which is currently produced in a quantity of approximately one mill. t/y is listed to the extreme left while a typical peptide hormone is listed

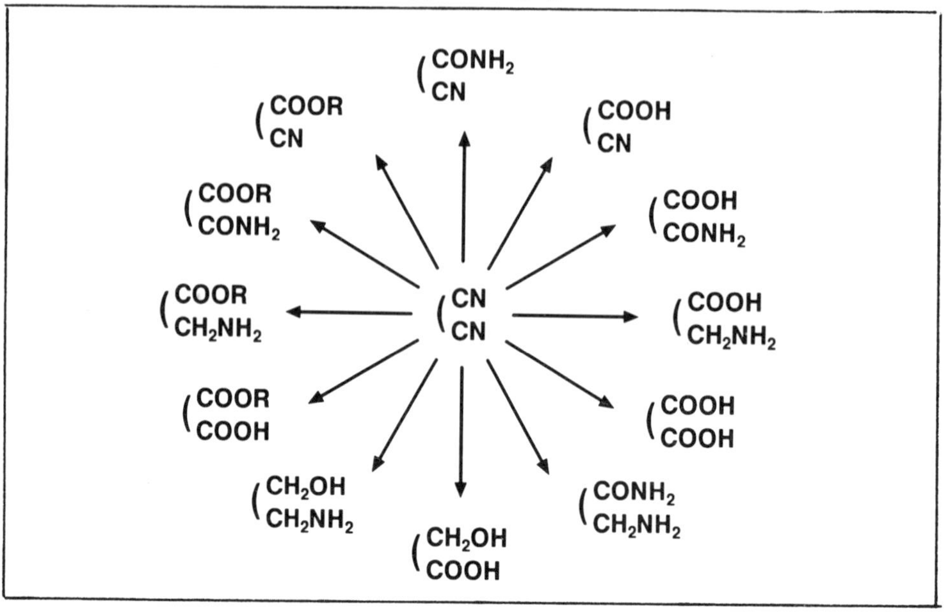

Fig. 13. Structural types accessible through enzymatic conversion of dinitriles.

to the extreme right. A very rough estimate of the potential value of enzymes needed for preparation of these various compounds by enzymatic synthesis can be made on the basis of the figures in Tables V and VI which indicate the prices of some enzymes currently produced on a large scale and the degree of utilization of these enzymes in some very well

TABLE V
Cost of some commercial enzymes.

ENZYME	Appr.price DM/Kg pure Enzyme protein	% Enzyme in commercial preparations	Physical form
Glucoamylase	70	5-10	Liquid
α Amylase	550	1-2	Liquid
Det.protease	350	3-6	Granules
Glucose isomerase	825	5-10	Immobilized granules
Calf rennet (Chymosin)	14000	0.1-0.2	Liquid
Microbial rennet	1800	0.2-0.4	Liquid

TABLE VI
Efficiency of enzymes used in high fructose corn syrup manufacture.

	α-Amalyse	Glucoamylase	Glucose isomerase
Turnover number under conditions of application mol product/mol enzyme/ minute	$3x10^4$	10^4	$2x10^2$
Relative price of enzyme	67	8	100
Amount of enzyme used ppm of product	12	150	10-20
Relative cost of enzyme	80	133	100
Ton product produced per Kg enzyme	80	7	2,5-5

established and optimized processes (2). It is probably safe to assume that owing to, for instance, the application of organic solvents, reactive organic chemicals, and forcing reaction conditions, the degree of utilization of enzymes in organic chemical processing will be inferior to the very efficient use made of enzymes within the traditional application areas. In enzymatic synthesis the requirement for a typical enzyme preparation containing about 5% of pure enzyme protein will, therefore, probably be around 1% per weight of the product produced with the aid of the enzyme. Likewise, the price of these enzymes will probably be somewhat higher than that of the well established enzymes with a long history of use and production on a large scale; an optimistic guess being for example roughly of the order of $10/kg. Accordingly, the enzyme potential in terms of quantity and value on a traditional enzyme market for the five examples selected is roughly as indicated in Figure 14. These numbers indicate the very important aspect of enzymatic synthesis that the potential of most of the enzymes involved is by far too small to justify their development and production - within the framework of the traditional enzyme market it will hardly be worthwhile to consider development of enzymes for synthesis of compounds of a production volume much smaller than 6-APA. Only the very big products as caprolactam and the best selling amino acids can, in themselves, justify development of enzymes for their production by enzymatic synthesis.

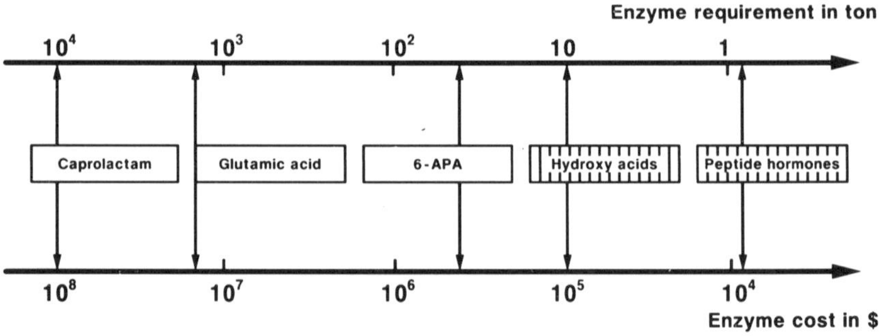

Fig. 14. Enzyme potential of selected enzymatic synthesis.

One may well argue that development of enzymes for organic synthesis would be more attractive if the enzymes could be applied for preparation of several rather than a single product. This, indeed, might very well be the case in some instances, for example, in the case of the nitrilases which exhibit an attractive, broad substrate specificity. In most instances, however, it is likely that economic considerations based on, for instance, the price of raw materials, requirements to reactor sizes, construction costs, etc. will create a strong pressure on the enzymatic process in regard to its efficiency, yield and other factors which, all in all, most probably will necessitate development of highly specialized enzymes for every single large scale enzymatic synthesis.

Another, indeed more attractive way of overcoming the barriers of the enzyme producers towards development of enzymes for organic synthesis could be based on a change in business philosophy of the enzyme producers and the organic chemical industry. Arrangements involving, for instance, some kind of more close collaboration between the enzyme producer and the chemical industry would thus most probably radically change the rather gloomy picture of the future of enzymatic synthesis implied in Figure 14. Presumably, this is what is going to happen, to the benefit of both industries and to enzymatic synthesis.

REFERENCES

1 Asano, Y., Y. Tani, and H. Yamada: 'A New Enzyme 'Nitrile Hydratase' which degrades Acetonitrile in Combination with Amidase.' Agric. Biol. Chem. **44**, 2251-2252 (1980).

2 Aunstrup, K.: 'Enzyme Technology, Problems and Possibilities.' Proceedings from the Third European Congress on Biotechnology, Munich (1984) (in press).

3 Boesten, W.H.J. and L.R.M. Meyer-Hoffmann: 'Process of preparing L and D alpha-amino acids by Enzyme Treatment of DL-alpha-amino acid amide.' US Patent 4,080,259 (1978).

4 Boesten, W.H.J.: 'Process for the Enzymatic Resolution of DL-phenyl glycine amide into its optically active antipodes.' US Patent 3,971,700 (1976).

5 Demain, A.L., R.B. Walton, J.F. Newkirk, and I.M. Miller: 'Microbial Degradation of Cephalosporin C.' Nature **199**, 909-910 (1963).

6 Jallageas, J.-C., A. Arnaud, and P. Galzy: 'Bioconversions of Nitriles and Their Applications.' In: Advances in Biochemical Engineering, **14**, 1-32 (1980).

7 Kirby, A.J.: 'Enzyme Catalysis.' In: Comprehensive Organic Chemistry (E. Haslam, ed.), 5, 389-455 (1979).

8 Walton, R.B.: '6-Aminopenicillanic Acid: Inhibition of Destruction of Cephalosporin C by Bacteria'. Science **141**, 1438-1439 (1964).

9 Zaks, A. and A.M. Klibanov: 'Enzymatic Catalysis in Organic Media at 100 °C.' Science **224**, 1249-1251 (1984).

ENANTIOSELECTIVE SYNTHESIS OF BIOLOGICALLY ACTIVE CYCLOPENTANOIDS VIA ENZYME CATALYSED ASYMMETRIC REACTIONS

H.-J. Gais
Institut für Organische Chemie und Biochemie der Technischen Hochschule, Petersenstraße 22, D-6100 Darmstadt (FRG)

Abstract. The enantioselective and enantioconvergent synthesis of the enantiomerically pure building blocks 8 and ent-8 for the total synthesis of natural and unnatural cyclopentanoids using the meso-concept and an efficient enzyme catalysed asymmetric transformation is described. 8 serves as a common intermediate for the asymmetric total synthesis of the macrolides brefeldin A 1 and 7-epi-brefeldin A 2 as well as the new prostaglandin building block 23.

1. Introduction

Because of their interesting structures and important biological activities and regulatory functions, cyclopentanoid natural and unnatural products such as the macrolides brefeldin A 1 (1) and 7-epi-brefeldin A 2 (2), the cell hormone prostacyclin 3 (3) and its potent analogue 6a-carba-prostacyclin 4 (R=H) (4), the cell hormone prostaglandin E_2 5 (5), and the metabolites pentalenolactone E 6 (6) and pentalenolactone 7 (6) are rewarding synthetic targets (scheme 1).

M. P. Schneider (ed.), Enzymes as Catalysts in Organic Synthesis, 97–122.

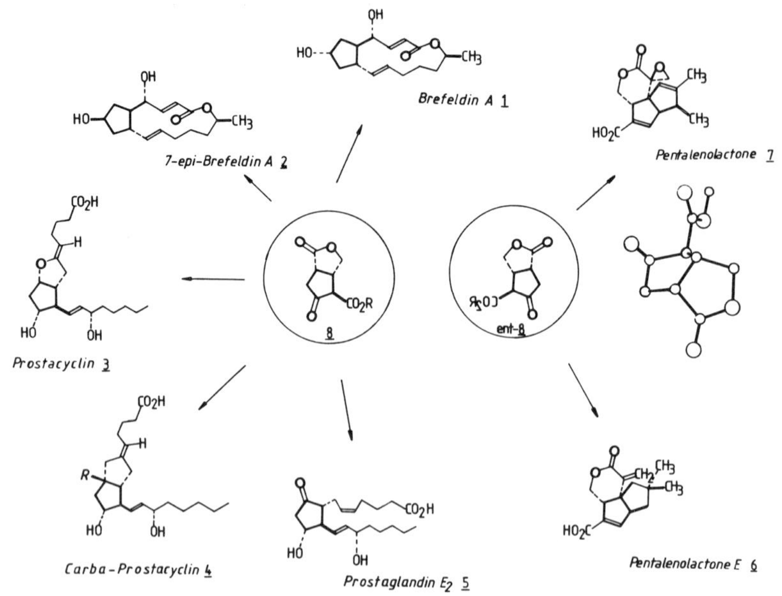

scheme 1.

2. Enantioselective and enantioconvergent synthesis of building blocks for the total synthesis of cyclopentanoids

A retrosynthetic analysis of the cyclopentanoids 1 to 7 led to the cyclopentanolactone 8 and its enantiomer ent-8 as key building blocks for their enantioselective total synthesis (scheme 1). The array of functional groups within and the topological bias of 8 and ent-8 (scheme 1) were chosen such as to provide a reliable platform to build upon in a highly chemo-, regio- and stereoselective fashion, and at the same time to attain the flexibility necessary for the enantioselective synthesis of 1 to 7 on routes not yet fully surveyable. The efficient creation of the common intermediates 8 and ent-8 enantiomerically pure in large quantities by asymmetric synthesis was the first problem to be addressed. The basic strategy for obtaining both enantiomers, 8 and ent-8 with only one step of asymmetric induction is the meso-concept outlined in scheme 2 (7).

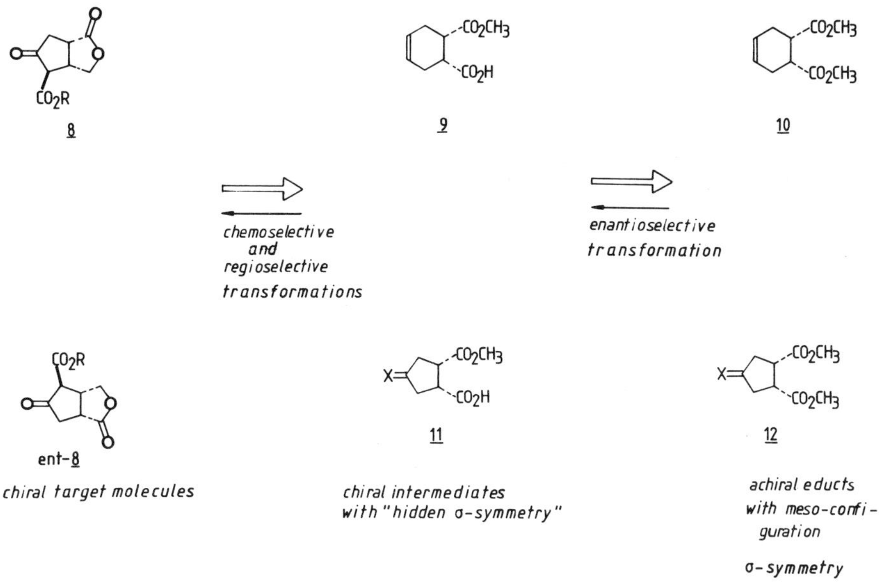

scheme 2.

A retrosynthetic analysis of the chiral target molecules 8 and ent-8 guided by the principles of the meso-concept led to the chiral intermediate 9 (or 11) with "hidden σ-symmetry" and from there to the achiral educt 10 (or 12) with meso-configuration, i. e. σ-symmetry. Synthesis of 8 and ent-8 would have to commence with an enantioselective hydrolysis of the meso-diester 10 (or 12) to the chiral half ester 9 (or 11), preferentially catalysed by an enzyme, and to terminate by their chemo- and regioselective transformations to the enantiomeric target molecules. There are three important features of the meso-concept for obtaining both enantiomers of a chiral compound with just one step of asymmetric induction, which deserve to be stressed:

1) The sense of asymmetric induction is of almost no importance, as long as the resulting half ester, 9 or ent-9 (11 or ent-11) is enantiomerically pure.
2) Both enantiomers of the chiral target molecules, 8 and ent-8 may be obtained by suitable chemo- and regioselective transformations either from 9 or ent-9 (11 or ent-11) because of their "hidden σ-symmetry" (enantioselective synthesis).
3) Both enantiomers of the chiral intermediate, 9 and ent-9 (or 11 and ent-11), e. g. obtained by an enantiomer separation, may be converted to a single enantiomer of the chiral target molecules, 8 or ent-8, simply by reversing the order of chemo- and regioselective transformations (enantioconvergent synthesis) (8).

Prerequisites for a successful realisation of the meso-concept for the enantioselective synthesis of 8 and ent-8 were the development of an efficient and highly selective enzyme catalysed asymmetric hydrolysis of the meso-diester 10 (or 12) as well as chemo- and regioselective transformations of 9 or ent-9 (11 or ent-11) suitable for large scale preparations.
For the asymmetric hydrolysis, the most promising enzyme at the outset of our investigations seemed to be the commercially available pig liver esterase (PLE) (9), because it had already been shown, that PLE hydrolyses dimethyl 3-hydroxy-3-methylglutarate with high enantioselectivity to the corresponding chiral half ester (10), and that the salts of half esters are in general not hydrolysed any further by this enzyme (11). Indeed PLE proved to be a highly successful enzyme for the enantioselective hydrolysis of meso- and other prochiral diesters to chiral half esters as shown in the meantime by us (7,12) and others (13).
The enantioselective hydrolysis of the meso-diester 10 with PLE results in practically exclusive attack at the (1R)-ester group to afford the enantiomerically pure (1S,2R)-half ester 9 in almost quantitative yield and with an enantiomeric excess (ee) ≥ 98% (12) (scheme 3).

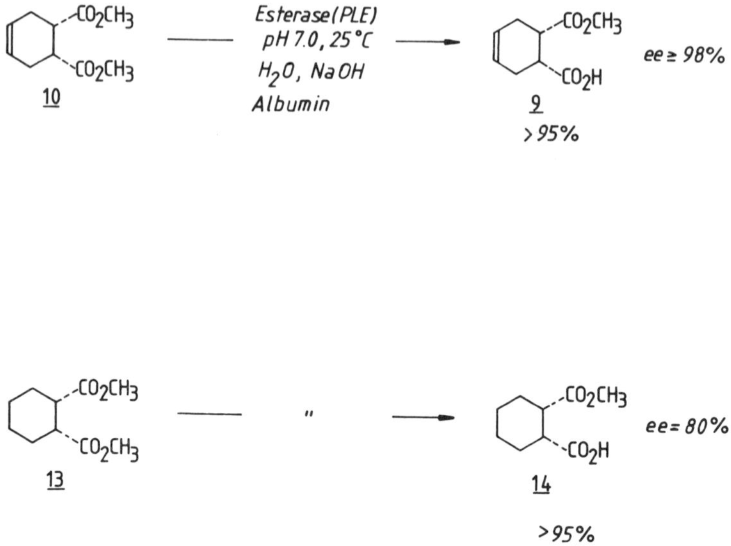

scheme 3.

For the synthesis of 1 mol 9, 9000 U PLE, 1000 ml phosphate buffer (pH 7.0), 500 ml 2 N sodium hydroxide, and a reaction time of 4 days at 25 °C are required. The reaction has been successfully run with almost the same results in regard to yield and ee on a 100 mol scale (14). The enzyme is stabilized by the addition of bovine serum albumin (BSA) and recovered by ultra filtration with only minor losses in activity. Interestingly the PLE catalysed asymmetric hydrolysis of the saturated meso-diester 13 to the chiral half ester 14 proceeds with the same sense of asymmetric induction, and with almost the same yield, but with a lower ee and rate (7,12) (scheme 3).
Enantiomer separation of rac-9 with (+)- and (-)-ephedrine enables large amounts of 9 and ent-9 to be obtained enantiomerically pure (ee≥98%) in yields higher than 90% (scheme 4).

Determination of Absolute Configuration and Enantiomeric Excess of 9 by X-Ray Analysis and ¹H-NMR of Ephedrine Salts

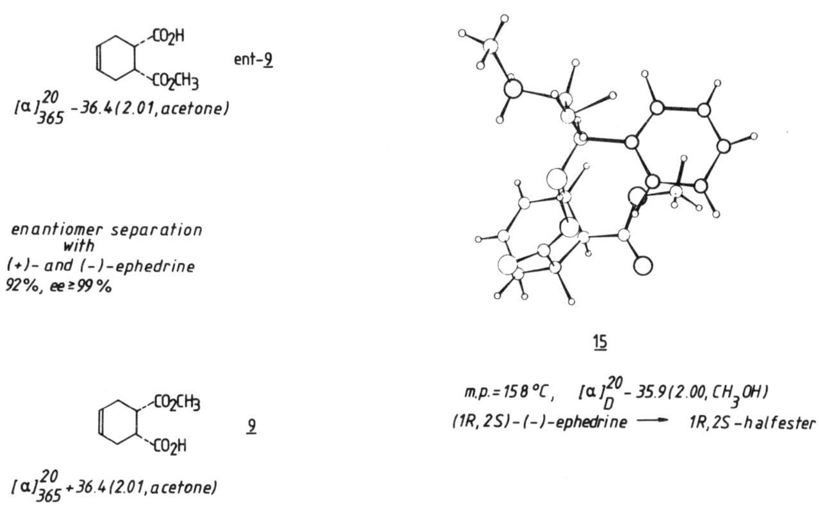

scheme 4.

The ee of 9 and ent-9 was determined from the diastereomeric excess (de) of the ephedrine salts (scheme 4), which in turn was determined by ¹H-NMR (300 MHz) spectroscopy from the different signals of the ester groups. Within the limits of detection (≥1%), admixture of the respective other diastereomer did not occur. The absolute configuration of 9 and ent-9 was established by X-ray structure analysis of 15, i. e. of the (-)-(1R,2S)-ephedrine salt of ent-9 (15).
The "hidden σ-symmetry" of 9 and ent-9 permits both their conversion into the lactones 16 or ent-16, respectively, and the synthesis of 16 and ent-16 from 9 by optional chemoselective reduction of the ester or

carboxy group to the hydroxymethyl group and subsequent lactonization (7) (scheme 5).

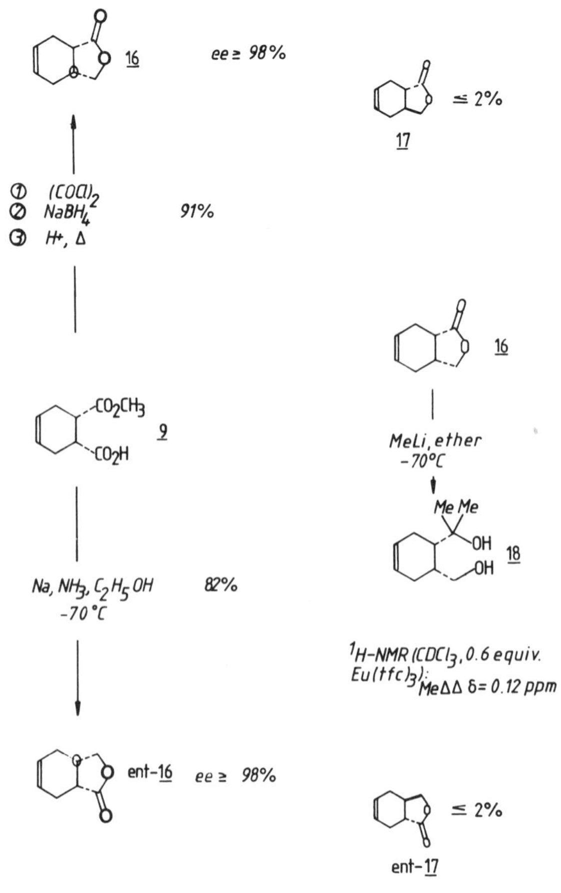

scheme 5.

For the ester group this proceeds by reduction with sodium and ethanol in liquid ammonia or with lithium triethylborohydride, and for the carboxy group via acid chloride and reduction with sodium borohydride in ethanol. Undesired epimerization to afford the trans lactones 17 and ent-17, respectively, can be avoided within the limits of detection (≥1%). For comparison, epimerization of 16 on the preparative scale with sodium hydroxide afforded in high yield ent-17, whose reduction to the known (1R,2R)-4-cyclohexen-1,2-dimethanol (16) allowed determination of the absolute configuration of 17 by a chemical correlation (7). The ee-values of 16 and ent-16 were determined indirectly on the dimethyldiols 18 and ent-18, obtained by treatment of the corresponding lactone with methyllithium (16), by ^{1}H-NMR (300 MHz) spectroscopy in the presence of the chiral shift reagent tris[3-(2,2,2-trifluoro-1-hydroxyethylidene)-d-camphorato]europium from the methyl signal to ≥98% (7). Within the limits of detection (≥1%), no admixture of the other enantiomer was detectable.

Oxidative ring cleavage of 16 with potassium permanganate in water in the presence of a phase transfer catalyst followed by esterfication of the corresponding mixture of dicarboxylic acids affords a 1:1-mixture of the isomeric lactone diesters 19 and 20 in high yield (scheme 6).

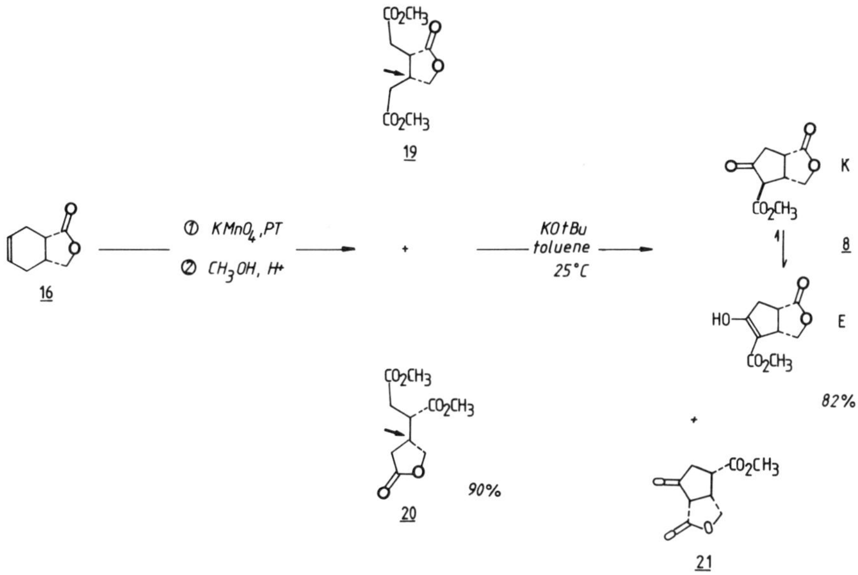

scheme 6.

As expected, both the pure diesters 19 and 20 and the mixture can be converted regioselectively with potassium tert-butoxide into the crystalline cyclopentanolactone 8, which exists almost exclusively in the enol form. The structure of 8 is proved by 400 MHz ^{1}H-NMR and 25 MHz ^{13}C-NMR (INADEQUATE technic) spectroscopy. As a side product the isomeric cyclopentanolactone 21 is formed in low yield (scheme 6).

Using the meso-concept, 8 and ent-8, the key building blocks for the enantioselective synthesis of the cyclopentanoids 1 to 7, are accessible rationally and enantiomerically pure by enantioselective and enantioconvergent synthesis from meso-tetrahydrophthalic anhydride (12).

3. Asymmetric synthesis of a novel enantiomerically pure prostaglandin building block

Their specific biological effects have again moved prostaglandins into the focal point of medical (3,5) and chemical (17) interest. 16.16-dimethyl prostaglandin E_2 5 ($R=CH_3$) effects cytoprotection of the gastro-intestinal tract of rats against aggresive agents (5), while the prostacyclin 3 is the most potent endogenous inhibitor of human blood platelet aggregation (3) (scheme 7).

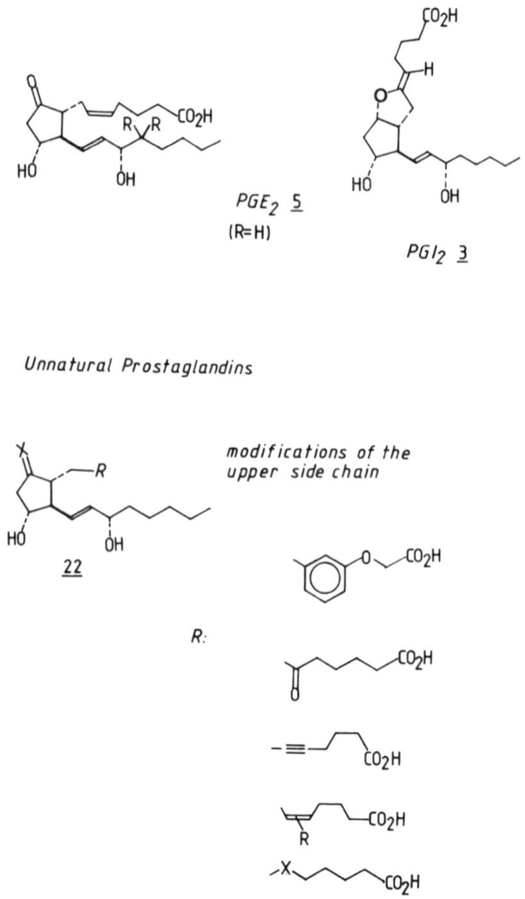

scheme 7.

We have developed an asymmetric synthesis of the novel, enantiomerically pure prostaglandin building block 23, which with alkyl-, alkenyl-, alkynyl- and arylmetal compounds 24 as well as phosphonates 25 permits attachment of various side chains R^1 and R^2 in a 1,4-addition, carbonyl-olefination strategy and, hence, should open a general enantioselective access to prostaglandins 22, especially those with modified upper side chains R^1, which are not easily synthesized by existing routes (17) (schemes 7 and 8).

1,4-Addition – Carbonylolefination – Strategy

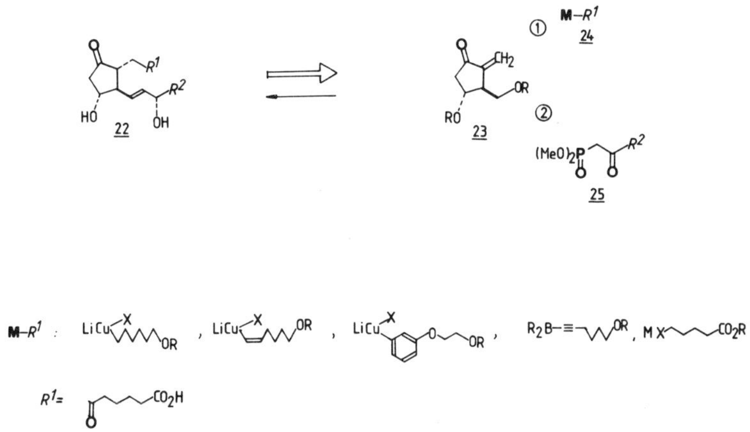

scheme 8.

The starting material for 23 is the enantiomerically pure keto ester 8, readily accessible by asymmetric synthesis (12). Stereoselective reduction of 8 with sodium borohydride affords the hydroxy ester 26 with 91% diastereoselectivity (ds), in addition to 4-epi-26. Recrystallization readily secures diastereomerically pure 26; 4-epi-26 can be isomerized to 26 using sodium methoxide in methanol. The stereoselective conversion of 8 to 26 via catalytic hydrogenation (85% ds) is preparatively even simpler (scheme 9). The relative configuration of 26 was determined by an X-ray structure analysis (15).

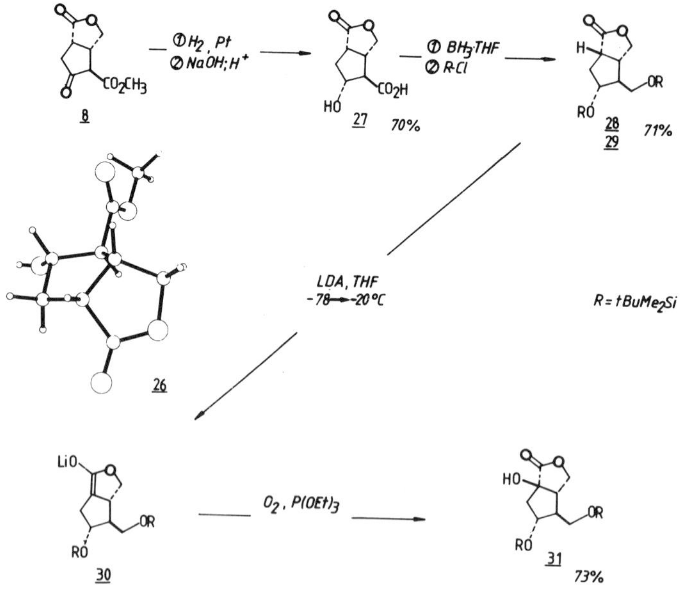

scheme 9.

Saponification of 26 to the carboxylic acid 27 and its chemoselective reduction with borane-tetrahydrofuran leads to the diol 28 in high yield, which is protected as the bissilyl ether 29. The decisive introduction of a hydroxy group at C-6a of 29 to give the hydroxylactone 31 is carried out by oxidation of the enolate 30, generated by metalation of 29 with lithium diisopropylamide (LDA) in tetrahydrofuran (THF), with the oxodiperoxymolybdenum-pyridine-hexamethylphosphoric triamide complex (MoOPH) (18) or with oxygen in the presence of triethylphosphite (19) (scheme 9).
With the reduction of 31 to the triol 32, the requirements are met for the final construction of the enone structural element in the synthesis of 23. 32 is oxidized with sodium metaperiodate to the stable, crystalline hydroxy ketone 33, which is then converted by methane sulfonyl chloride and triethylamine into the crystalline and enantiomerically pure novel methylenecyclopentanone 23 (20) (scheme 10).

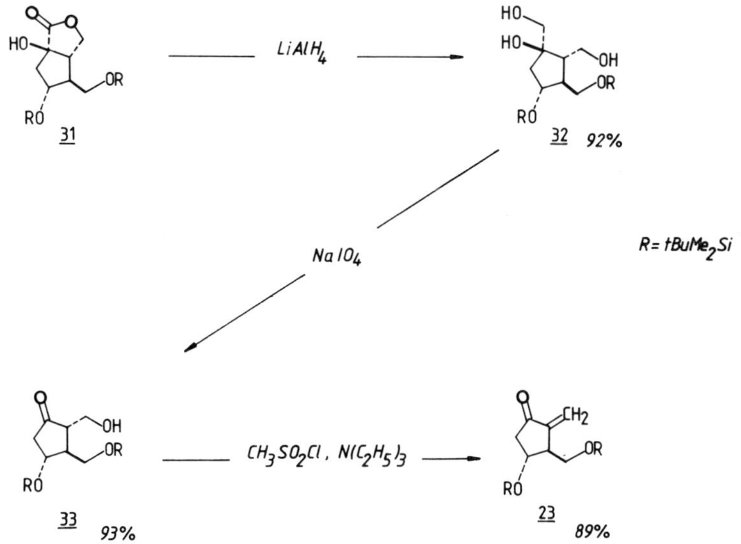

scheme 10.

A first demonstration of the synthetic potential of 23 is offered by 1.4-addition of the cuprate 24 (21), which affords the enantiomerically pure cyclopentanone derivative 34 (scheme 11).

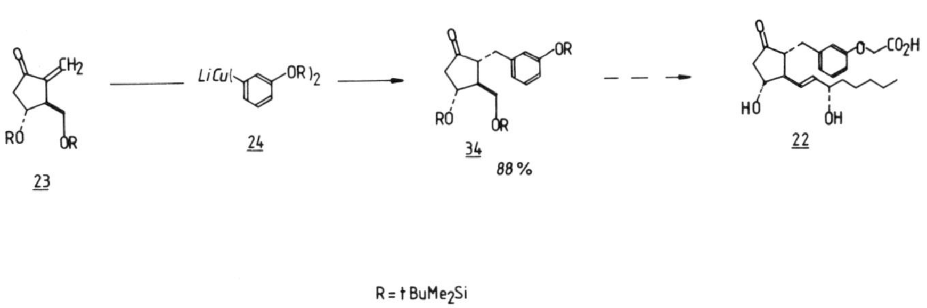

scheme 11.

Comparison of the relevant ^{1}H-NMR coupling constants of 34 ($J_{2,3}$=9.0, $J_{2,5ß}$=0.7 Hz) with those of 33 ($J_{2,3}$=10.0, $J_{2,5ß}$=1.0 Hz) and PGE_2 5 ($J_{8,12}$=11.6, $J_{8,10ß}$=1.2 Hz) proves the 2α-configuration, which is established during work-up with acetic acid and sodium acetate (thermodynamic control). 34 is an interesting precursor for 4,5,6-trinor-3,7-inter-m-phenylene-3-oxaprostaglandins E_1 22 (21), which are beeing clinically tested as selective bronchodilators for treatment of asthmatic attacks (22) (scheme 11).
The stereoselective synthesis of 38 from 28 via 35 and 36 demonstrates the method planned for the introduction of lower side chains to analogous diols, which are obtained by addition of 24 to 2 via appropriate transformations of the ring-carbonyl function and removal of the protecting groups R (scheme 12).

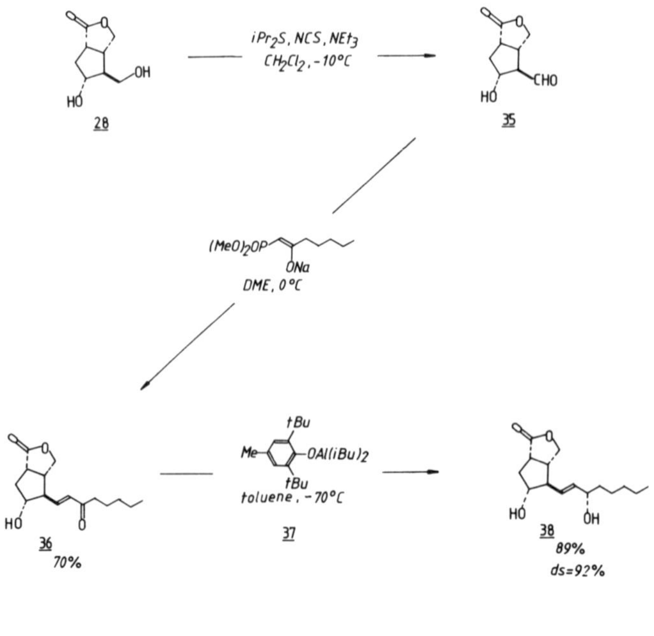

scheme 12.

Chemoselective oxidation of the diol 28 with N-chlorosuccinimide/diisopropylsulfide/triethylamine (23) yielded the labile hydroxyaldehyde 35, which is not isolated as such on a preparative scale, but immediately treated with the phosphonate salt to afford in 70% overall yield the enone 36. The stereoselective reduction of the enone 36 could be achieved with 92% ds and in high yield with 37 (24) (scheme 12).

4. Asymmetric total synthesis of the macrolides brefeldin A and 7-epi-brefeldin A

Because of their interesting structures, which resemble those of the prostaglandins, and unusally broad spectrum of interesting biological activity, the macrolides brefeldin A 1 and 7-epi-brefeldin A 2, biosynthesized by Curvularia lunata and other microorganism are attractive target molecules for chemical synthesis (25).
In 1981 we prepared the intermediate 66 (scheme 18) (7,26,27), formally the first asymmetric total synthesis of 1; later a second asymmetric total synthesis of 1 was described (25a). Here, we describe a novel asymmetric total synthesis of 1 and the first of 2, in which we employ besides a new strategy, new methods for the stereoselective construction of the C1-C4 structure element and for the macrolactonization (28).
Our synthetic strategy can be deduced from the synthons 39, 40 and 41; in their connection sequence, the corresponding building blocks are the cyclopentanolactone 43, the lithiosulfone 44, and the lithiopropiolate 42 (scheme 13).

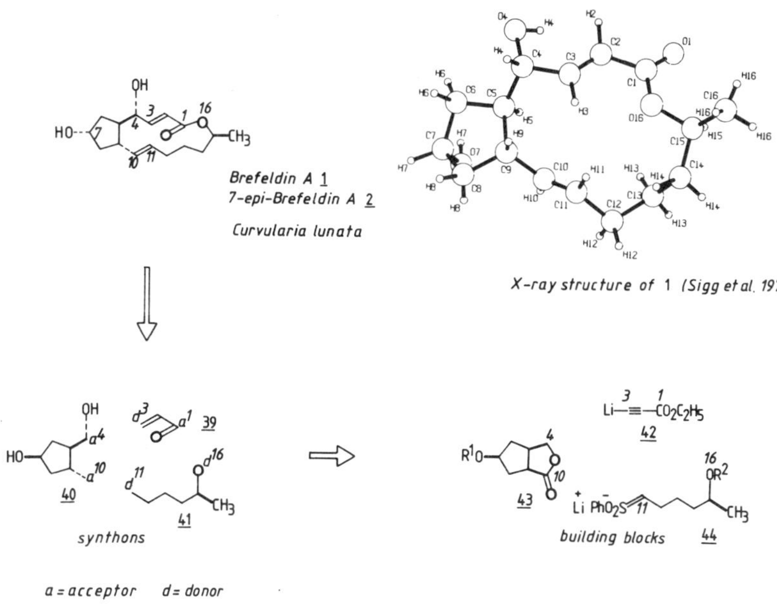

scheme 13.

A key element of our strategy for the asymmetric synthesis of 1 and 2 is the generation of the two E-configurated CC-double bonds stereoselectively from CC-triple bonds.
The synthesis of the cyclopentanoid building block 47 commences with the cleavage of the ß-keto ester 8 to afford the roof shaped bicyclic ketone

45 in almost quantitative yield, which after stereoselctive catalytic hydrogenation yields the ß-alcohol 46 with 98% ds, and which is subsequently protected as the methoxyethoxymethyl(MEM) ether 47 (scheme 14).

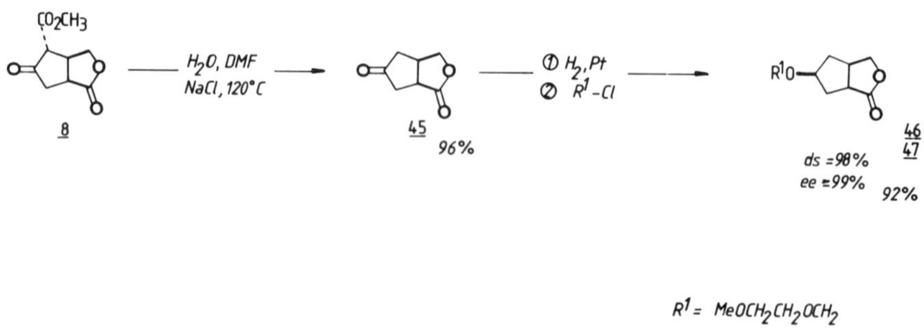

scheme 14.

An alternative route for the enantioselective synthesis of the cyclopentanoid building block 45 starts with the meso-diol 48, which is available on the preparative scale from meso-4-cyclohexen-1,2-dimethanol in five steps (7). Enantioselective, enzymatic oxidation of 48 with horse liver alcohol dehydrogenase (HLADH) and nicotineamide adenine dinucleotide (NAD^+) using flavin mononucleotide (FMN) coenzyme recycling (29) affords the hemiacetal 49, which is oxidized with pyridinium chlorochromate to the lactone 45. The cyclopentanoid building block 45 is obtained by this alternative, albeit less efficient route in 60% overall yield and with 45% ee (scheme 15). By recrystallization the ee can be raised to 95%. The remaining rac-45 is reduced to the meso-diol 48, which can be used again in the enzymatic oxidation. An analysis of the HLADH catalysed oxidation of 48 by the cubic model of Jones et al. (29) correctly predicts the absolute configuration of the enantiomer formed in excess (scheme 15, complexes 50S and 50R).

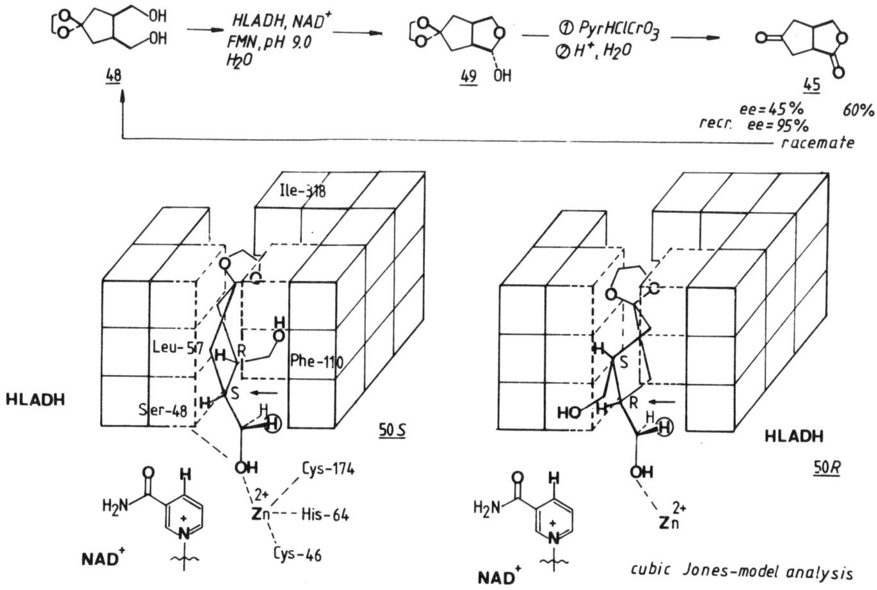

scheme 15.

The C11-C16 building block, the sulfone 61 was synthesized by three different routes. Two of them are using as chiral educts materials produced by microbial transformations: (S)-ethyl lactate and (S)-ethyl-3-hydroxybutanoate 51, which is easily prepared by bakers yeast reduction of ethyl acetoacetate (30), and the last starts with the unnatural amino acid D-glutamic acid (scheme 16).

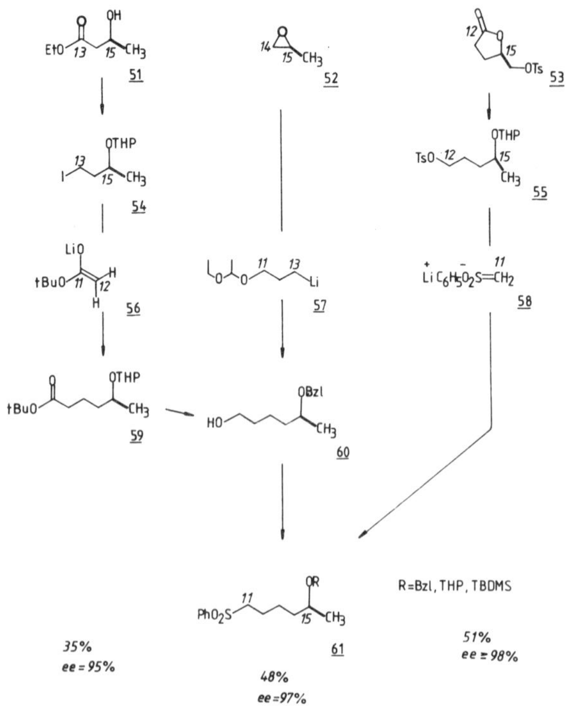

scheme 16.

Regioselective ring-opening of (S)-propylene oxide 52, obtained from (S)-ethyl lactate (31), with the organolithium compound 57 (32) leads to the alcohol 60 and from there via the corresponding tosylate and sulfide to the sulfone 61 in 48% overall yield and with 97% ee (28). The second route starts with the conversion of the lactone 53, which can be prepared from D-glutamic acid by known procedures enantiomerically pure (33), to the tosylate 55 by unexeptional transformations. Alkylation of the lithiosulfone 58 with 55 affords directly the sulfone 61 in 51% overall yield and with ≥98% ee. Chiral starting material for the third route to the sulfone 61 is the hydroxy ester 51, which is converted to the iodide 54. Alkylation of the enolate 56 with 54 generates the ester 59, which is further transformed to the sulfone 61 via the alcohol 60. In this case the overall yield is 35% and the ee 95% (34).
The C4-C10 building block 47 is linked smoothly in three steps to the C11-C16 building block 44 with the generation of the correct configuration at C-9 (scheme 17).

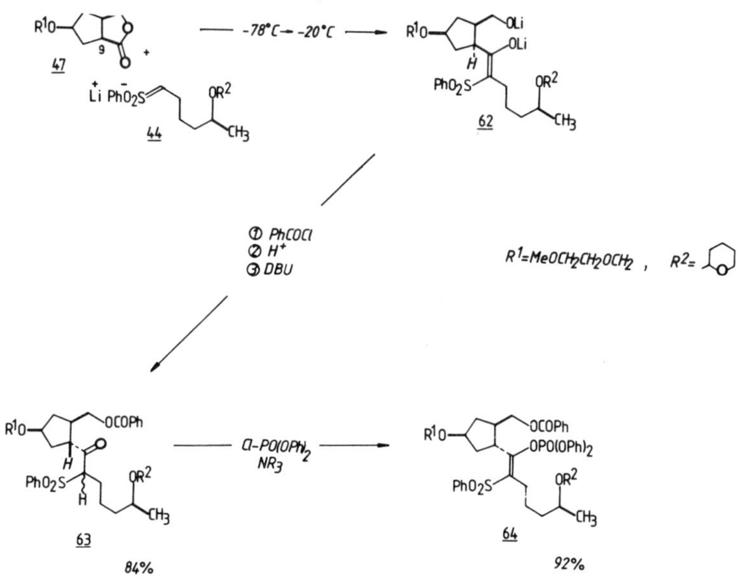

scheme 17.

Ring-opening of the lactone 47 with the lithiosulfone 44, either in the presence of an additional equivalent of 44 or of lithium diisopropylamide to give the dilithio compound 62 and chemoselective benzoylation; the initially formed ketosulfone 9-epi-63 is isomerized to the more stable ketosulfone 63 (63:9-epi-63 ≥ 30:1) with 1,8-diazabicyclo(5.4.0)-undec-7-ene (DBU) (scheme 17) and finally converted into the enol phosphate 64 in high yield. The enol phosphate 64 was reduced stereoselectively with concomitant release of the hydroxymethyl function to the (E)-alkene 66 via the alkyne 65 (35) (scheme 18).

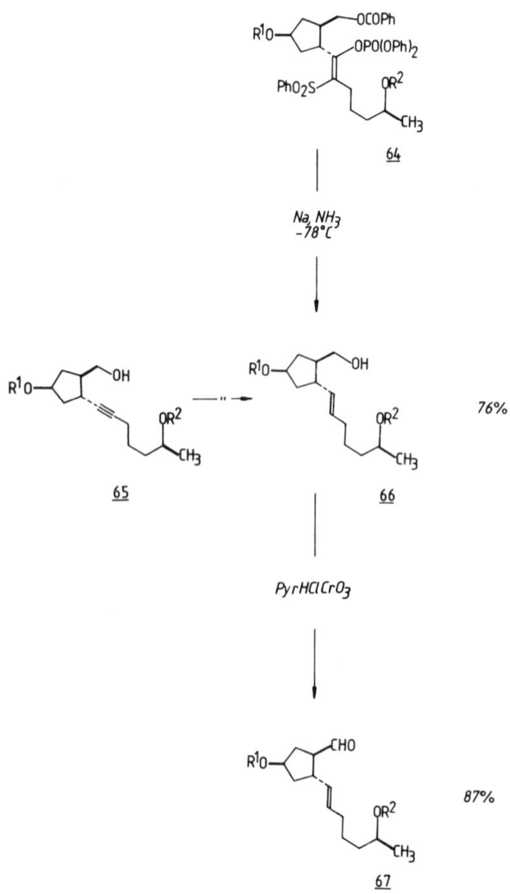

scheme 18.

Oxidation of the alcohol 66 to the aldehyde 67 with pyridinium chlorochromate completes the requirements for the coupling of the still absent C1-C3 structure element 42. A novel, potent method for this was developed (36), because not only the macrolides brefeldin A 1 and 7-epi-brefeldin A 2, but also the quite interesting macrolides colletodiol (37), and cytochalasin B (38) as well as others contain an (E)-4-hydroxy-2-alkenoic acid structure element (scheme 19).

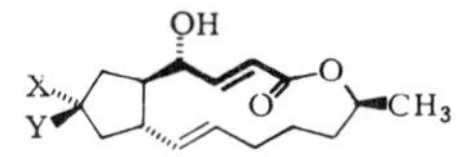

1, X = OH, Y = H
2, Y = OH, X = H

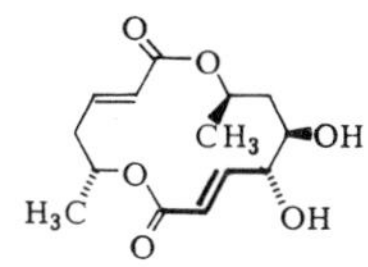

Colletodiol

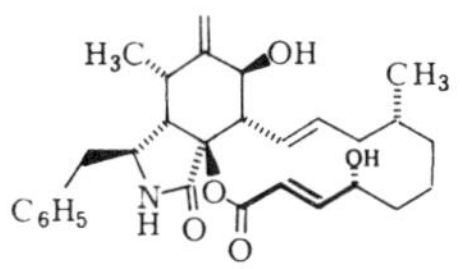

Cytochalasin B

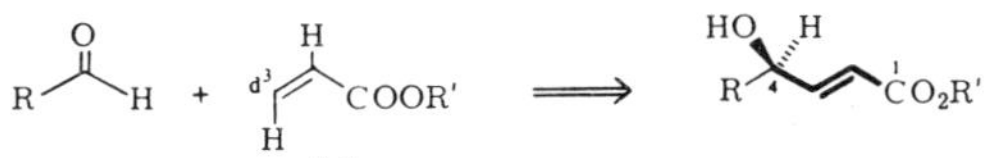

68

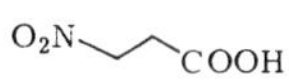

69a

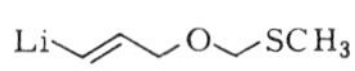

69b

Li–C≡C–$CO_2C_2H_5$

42

scheme 19.

For their direct synthesis from aldehydes and building blocks which correspond to the d^3-synthon 68, there are two methods; the d^3-reagents are either the acid 69a (39) or the vinyllithium compound 69b (40). We have developed an efficient procedure for the stereoselective synthesis of (Z)- and (E)-4-hydroxy-2-alkenoic acid esters from aldehydes and the d^3-reagent ethyl lithiopropiolate 42 (scheme 20) and its application to the (E/R)-stereoselective transformation of the aldehyde 67. Addition of the lithioalkyne 42, generated by deprotonation of ethyl propiolate with LDA in THF at -78°C (41), to the aldehyde 70 proceeds rapidly, even at -78°C, to yield the alkyne rac-72 (X=H) (scheme 20). The alkynes rac-72 are obtained by attaching the MEM, tetrahydropyranyl (THP) and tert-butyldimethylsilyl (TBDMS) groups to rac-72 (X≠H). The (Z)-alkenes rac-73 are accesible by chemo- and stereoselective hydrogenation of the alkynes rac-72 in yields of 95% and ds≥98%. Structural proof for rac-73 is provided by their ^{1}H-NMR spectra, which in addition to the signals for the Z-olefinic protons (J≈12 Hz) exibit the signal of H-4, which is shifted to low field by the anisotropic effect of the syn-orientated carbonyl group at δ = 5.00-5.20. Subsequent isomerization of the (Z)-alkenes rac-73 to the (E)-alkenes rac-71 is affected using 30 mol-% lithium ethyl thiolate in THF at 25°C. Within the limits of detection (≥1%), the (Z)-alkenes rac-73 were not found. Apart from the signals for the (E)-olefinic protons (J≈16 Hz), the ^{1}H-NMR spectra of rac-71 exibit the signal of H-4 at δ = 4.02-4.20, unaffected by the anti-orientated carbonyl group. Suitable protecting groups for the hydroxy function are the MEM, THP or TBDMS groups, but not the benzoyl (Bz) group since problems then arise in the isomerization step.

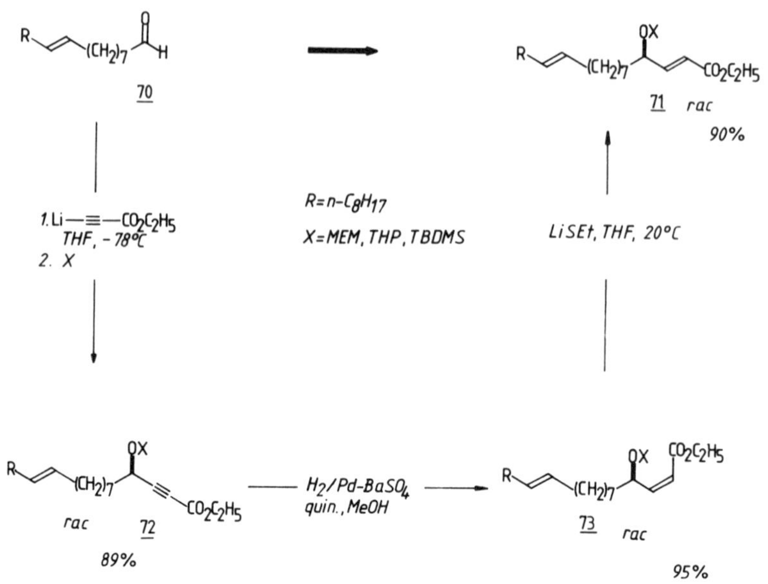

scheme 20.

The potential of this method is demonstrated by the conversion of the aldehyde 67 to the 7-epi-brefeldin A seco-acids 76 and 4-epi-76 (scheme 21).

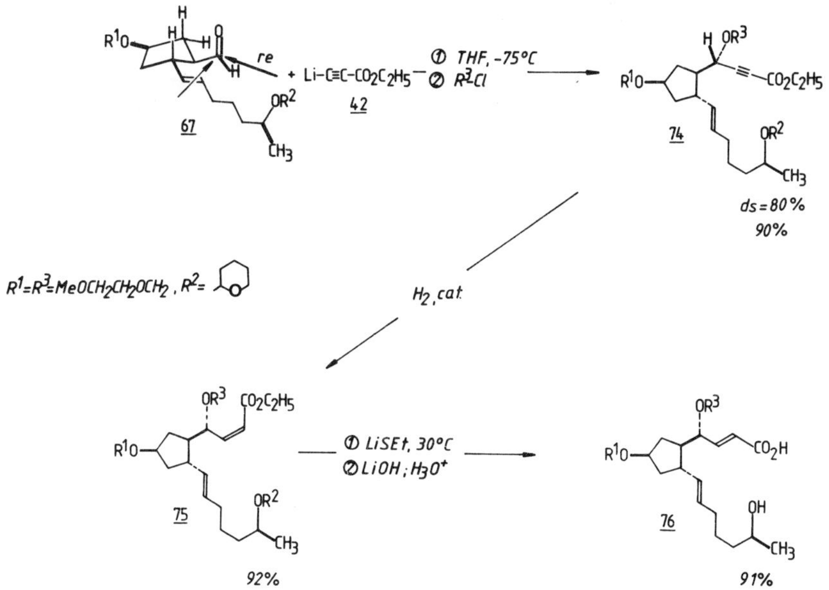

scheme 21.

The re-selective addition of 42 to the aldehyde 67 furnishes as expected the alkyne 74 with the correct configuration at C-4 in high yield with 80% ds, besides 4-epi-74, and thus completes the construction of the carbon framework of the target molecules brefeldin A 1 and 7-epi-brefeldin A 2. The diastereoselectivity of the addition step was determined by ^{1}H-NMR spectroscopy and by chromatographic separation of the isomers 74 and 4-epi-74. The configurations were ultimately assigned by conversion of the major diastereomer into 2 and 1, respectively. The alkynes 74 and 4-epi-74 were converted as described above into 76 and 4-epi-76 in excellent yield (scheme 21).

For the final lactonization of the seco-acids 76 and 4-epi-76 we have devised a new, efficient method (42) (scheme 22).

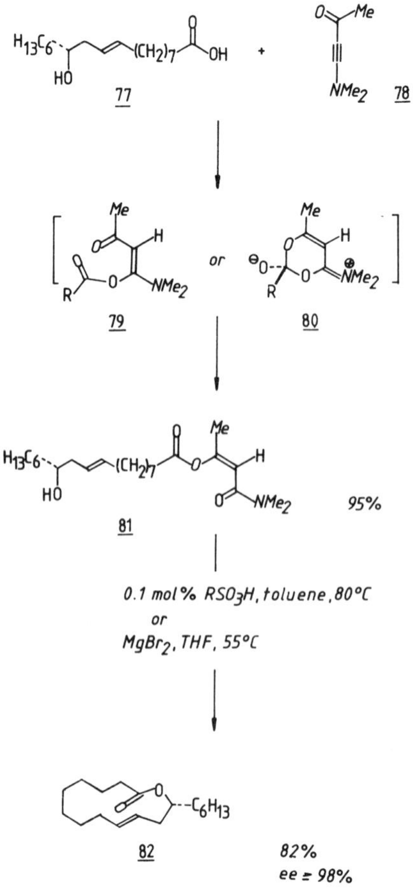

scheme 22.

Enantiomerically pure ricinelaidic acid 77 is converted with equimolar amounts of the push-pull alkyne 78 (43) via the primary adduct 79, which even at low temperatures rearranges rapidly in either a concerted or stepwise (via 80) fashion (44), in almost quantitative yield to the stable enol ester 81. 81 lactonizes under high dilution conditions smoothly in the presence of either 0.1 mol-% 10-camphersulfonic acid in toluene at 80°C or magnesium bromide in THF at 55°C to the enantiomerically pure 13-membered ricinelaidic acid lactone 82.
Application of this method to the lactonization of the seco-acids 76 and 4-epi-76 shows scheme 23.

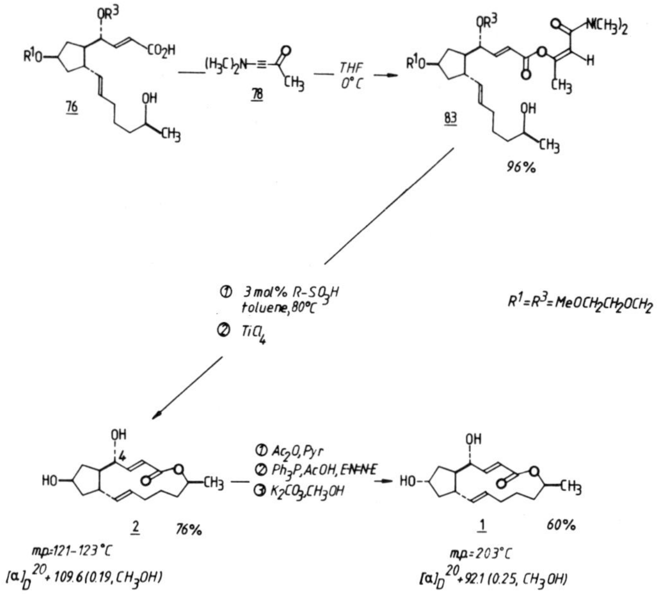

scheme 23.

Chemoselective reaction of 76 and 4-epi-76 with the alkyne 78 gives the stable enol esters 83 and 4-epi-83 and their H^+-catalysed cyclization the macrolides 84 and 4-epi-84 in good yields. No difference in the lactonization rates of 83 and 4-epi-83 was noticed.
84 and 4-epi-84 are cleaved with titanium tetrachloride to yield a 4:1-mixture of 2 and 4-epi-2, which are separated readily by columne chromatographie on silica gel. 2 is converted to 1 by the sequence shown in scheme 23; selective acylation of 2 to the corresponding 4-acetate, inversion of configuration at C-7 according to Mitsunobu (45) and cleavage of the brefeldin A diacetate thus obtained led to 1 (28).
The macrolides brefeldin A 1 and 7-epi-brefeldin A 2 thus obtained are indistinguishable both chromatographically and spectroscopically (300 MHz ^{1}H-NMR) from 1 biosynthesized from Penicillium brefeldianum and 2 from Curvularia lunata.

Acknowledgements

I would like to express my sincere appreciation to my able coworkers, Drs. K. L. Lukas and T. Lied for their contributions.
I am particulary grateful to Prof. Dr. H. J. Lindner for the X-ray analyses and to Drs. A. von Wartburg and C. P. Gorst-Allmann for providing biosynthetic samples of brefeldin A and 7-epi-brefeldin A.
The financial support from the Deutsche Forschungsgemeinschaft and the Fonds der Chemischen Industrie is gratefully acknowledged.

References

(1) E. Harri, N. Loeffler, H. P. Sigg, C. Tamm, Helv. Chim. Acta 46 (1963) 1235.
(2) C. P. Gorst-Allman, P. S. Steyn, C. J. Rabie, J. Chem. Soc. Perkin Trans. I 1982, 2387.
(3) J. R. Vane, Angew. Chem. 95 (1983) 782; Angew. Chem. Int. Ed. Engl. 22 (1983) 741.
(4) P. A. Aristoff, J. Org. Chem. 46 (1981) 1954.
(5) S. Bergström, Angew. Chem. 95 (1983) 865; Angew. Chem. Int. Ed. Engl. 22 (1983) 858.
(6) L. A. Paquette, Top. Curr. Chem. 119 (1984) 1, and references cited therein.
(7) H.-J. Gais, Habilitationsschrift, Technische Hochschule Darmstadt 1981.
(8) K.-K. Chen, N. Cohen, J. P. De Noble, A. C. Specian, Jr., G. Saucy, J. Org. Chem. 41 (1976) 3497.
(9) E. Heymann, W. Junge, Eur. J. Biochem. 95 (1979) 509, 519; D. Farb, w. P. Jencks, Arch. Biochem. Biophy. 203 (1980) 214, 227, 236.
(10) F. C. Huang, L. F. H. Lee, R. S. D. Mittal, P. R. Ravikumar, J. A. Chan, C. J. Sih, E. Caspi, C. R. Eck, J. Am. Chem. Soc. 97 (1975) 4144.
(11) M. Levy, P. R. Ocken, Arch. Biochem. Biophy. 135 (1969) 259.
(12) H.-J. Gais, K. L. Lukas, Angew. Chem. 95 (1984) 140; Angew. Chem. Int. Ed. Engl. 23 (1984) 142.
(13) C.-S. Chen, Y. Fujimoto, C. J. Sih, J. Am. Chem. Soc. 103 (1981) 3580; S. Iriuchijima, G. Tsuchihashi, Agric. Biol. Chem. 46 (1982) 1907; P. Mohr, N. Waespe-Sarcevic, C. Tamm, K. Gawronska, J. K. Gawronski, Helv. Chim. Acta 66 (1983) 2501; M. Arita, K. Adachi, Y. Ito, H. Sawai, M. Ohno, J. Am. Chem. Soc. 105 (1983) 4049 and references cited therein; M. Ohno, Y. Ito, M. Arita, T. Shibata, K. Adachi, H. Sawai, Tetrahedron 40 (1984) 145; C. J. Francis, J. B. Jones, J. Chem. Soc. Chem. Commun. 1984, 579; G. Sabbioni, M. L. Shea, J. B. Jones, J. Chem. Soc. Chem. Commun. 1984, 236; Y.-F. Wang, C.-S. Chen, G. Girdaukas, C. J. Sih, J. Am. Chem. Soc. 106 (1984) 3695; Y.-F. Wang, C. J. Sih, Tetrahedron Lett. 1984, 4999

K. Laumen, M. Schneider, Tetrahedron Lett. 1985, 2073; F. Björkling, J. Boutelje, S. Gatenbeck, K. Hult, T. Norin, P. Szmulik, Tetrahedron 41 (1985) 1347.
(14) B. Riefling, W. Brümmer, H.-J. Gais, NATO Advanced Research Workshop, Enzymes as Catalysts in Organic Synthesis, Schloss Reisensburg, 16-22 June 1985.
(15) H. J. Lindner, unpublished results
(16) I. J. Jakovac, H. B. Goodbrand, K. P. Lok, J. B. Jones, J. Am. Chem. Soc. 104 (1982) 4659.
(17) S. M. Roberts, F. Scheinmann, New Synthetic Routes to Prostaglandins and Thromboxans, Academic Press London 1982.
(18) E. Vedejs, D. A. Engler, J. E. Telschow, J. Org. Chem. 43 (1978) 188.
(19) E. J. Corey, H. E. Ensley, J. Am. Chem. Soc. 97 (1975) 6908.
(20) H.-J. Gais, T. Lied, K. L. Lukas, Angew. Chem. 96 (1984) 495; Angew. Chem. Int. Ed. Engl. 23 (1984) 511.
(21) D. R. Morton, J. L. Thompson, J. Org. Chem. 43 (1978) 2102.
(22) N. A. Nelson, R. C. Kelly, R. A. Johnson, Chem. Eng. News 60 (1982) No. 33, p. 30.
(23) B. Fraser-Reid, L. Magdzinski, B. Molino, J. Am. Chem. Soc. 106 (1984) 731.
(24) S. Iguchi, H. Nakai, M. Hayashi, H. Yamamoto, K. Maruoka, Bull. Chem. Soc. Jpn 54 (1981) 3033.
(25) a) C. Le Drian, A. E. Greene, J. Am. Chem. Soc. 104 (1982) 5473; b) T. Kitahara, K. Mori, Tetrahedron 40 (1984) 2935; c) K. Nakatani, S. Isoe, Tetrahedron Lett. 1985, 2209.
(26) H.-J. Gais, T. Lied, K. L. Lukas, Chemiedozententagung, Tübingen, March 1981.
(27) T. Kitahara, K. Mori, M. Matsui, Tetrahedron Lett. 1979, 3021
(28) H.-J. Gais, T. Lied, Angew. Chem. 96 (1984) 143; Angew. Chem. Int. Ed. Eng. 23 (1984) 145.
(29) K. P. Lok, I. J. Jakovac, J. B. Jones, J. Am. Chem. Soc. 107 (1985) 2521.
(30) D. Seebach, M. A. Sutter, R. H. Weber, Org. Synth. 63 (1984) 1
(31) B. Seuring, D. Seebach, Helv. Chim. Acta 60 (1977) 1175.
(32) P. E. Eaton, G. F. Cooper, R. C. Johnson, R. H. Mueller, J. Org. Chem. 37 (1972) 1947.
(33) U. Ravid, R. M. Silverstein, L. R. Smith, Tetrahedron 34 (1978) 1449.
(34) T. Lied, Ph. D. thesis, Technische Hochschule Darmstadt 1983.
(35) P. A. Bartlett, F. R. Green, J. Am. Chem. Soc. 100 (1980) 7583; B. Lythgoe, I. Waterhouse, J. Chem. Soc. Perkin Trans. I 1979, 242.
(36) H.-J. Gais, Angew. Chem. 96 (1984) 142; Angew. Chem. Int. Ed. Engl. 23 (1984) 143.
(37) R. Amstutz, E. Hungerbühler, D. Seebach, Helv. Chim. Acta 64 (1981) 1796.
(38) M. Binder, C. Tamm, Angew. Chem. 85 (1973) 370; Angew. Chem. Int. Ed. Engl. 12 (1973) 370
(39) P. Bakuzis, M. L. F. Bakuzis, T. F. Weingartner, Tetrahedron Lett. 1978, 2371

(40) E. J. Corey, R. H. Wollenberg, Tetrahedron Lett. 1976, 4705.
(41) E. J. Corey, C. U. Kim, R. H. K. Chen, M. Takeda, J. Am. Chem. Soc. 94 (1972) 4395; J. L. Hermann, M. H. Berger, R. H. Schlessinger, ibid 101 (1979) 1544; M. M. Midland, A. Tramontano, J. R. Cable, J. Org. Chem. 45 (1980) 28.
(42) H.-J. Gais, Tetrahedron Lett. 1984, 273.
(43) H.-J. Gais, K. Hafner, M. Neuenschwander, Helv. Chim. Acta 52 (1969) 2641.
(44) R. B. Woodward, R. A. Olofson, Tetrahedron Suppl. 7 (1966) 415; M. Neuenschwander, U. Stämpfli, Chimia 33 (1979) 439.
(45) O. Mitsunobu, Synthesis 1981, 1.

CREATION OF NOVEL CHIRAL SYNTHONS WITH PIG LIVER ESTERASE: APPLICATION TO NATURAL PRODUCT SYNTHESIS AND THE SUBSTRATE RECOGNITION

Masaji Ohno, Susumu Kobayashi and Kunitomo Adachi
Faculty of Pharmaceutical Sciences, University of Tokyo
7-3-1 Hongo, Bunkyo-ku
Tokyo 113
Japan

ABSTRACT. Enantioselective synthesis of some important antibiotics has been investigated by combination of enzymatic and non-enzymatic procedures. The synthetic strategy developed here is based on symmetrization-asymmetrization concept. Thus, retrosynthesis was carried out to generate, from target molecules, simplified symmetric diesters (symmetrization), and then the symmetric diesters were subjected to asymmetric hydrolysis with pig liver esterase to create the corresponding chiral half-esters. The chiral half-esters were converted to the target molecules by non-enzymatic procedures. Various types of carbapenem antibiotics, negamycin, showdomycin, 6-azapseudouridine, cordycepin, aristeromycin, neplanocin A, and aminocyclitol of fortimicin were efficiently synthesized with the desired absolute configurations. The structure-stereospecificty relationships revealed by the present study give significant and useful information about the topography of the active site of PLE and allow us to propose the working active site model.

1. Introduction.

Since the end of 1977, we have shown that the enantioselective total synthesis of some useful antibiotics can be efficiently achieved by reasonable combination of enzymatic and non-enzymatic procedures. (1-18) Our synthetic strategy for such optically active natural products is designed by the following principle.

(1) Symmetrization: retrosynthesis is performed to generate, from the target molecule, a simplified symmetric diester in the prochiral or meso form as shown in Scheme 1.

(2) Asymmetrization: the symmetric diester is subjected to asymmetric hydrolysis with pig liver esterase (PLE) to generate the corresponding chiral half-ester (Enzymatic conversion of a σ-symmery substrate to a C_1-symmetry intermediate)

(3) Non-enzymatic procedures: the chiral half-ester is converted to the targtet molecule and related molecules by means of usual organic synthesis, including some new method developed by us. At the beginning

M. P. Schneider (ed.), Enzymes as Catalysts in Organic Synthesis, 123–142.

of our enzymatic approach to the natural product synthesis, we ourseleves suspected that high substrate-specificity of enzymes may be disadvantage or difficult to overcome, to poduce general synthetic methods useful for asymmeric synthesis. However, we could have demonstrated that pig liver esterase, for instance, has so broad substrate-specificity extremely useful for asymmetric synthesis and thus the enzymatic approach to the new chiral synthons developed here has opened up a new avenue for natural product synthesis.

Scheme 1 Retrosynthesis of Natural Products based on Symmetrization–asymmetrization Concept

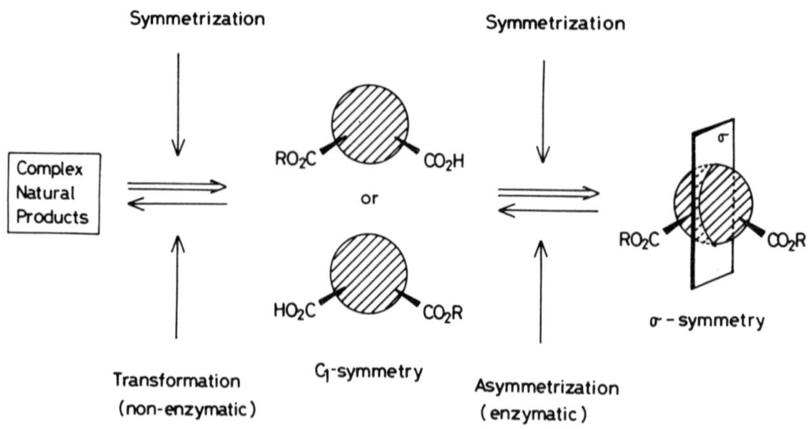

2. Natural Product Synthesis Based on Symmetrization-asymmetrization Concept.

2.1. Carbapenem Antibiotics

Naturally occurring β-lactam antibiotics belonging to the carbapenem group have been the subject of much synthetic study (19) because of their unique structures and interesting biological activity. They can be classified into three groups (trans, cis, and ene types) according to the structure of the side-chain of the β-lactam ring as shown in Scheme 2. The most important asymmetric carbon is C-5 with R-configuration common to all carbapenem antibiotics and essential for the biological activity. Therefore, it is a challenging problem for synthetic chemists how to introduce the asymmetric carbon with the desired configuration. We noticed that the problem of synthesizing the carbapenem antibiotics could be extremely simplified by considering a symmetric factor within the carbapenem nucleus. Three different approaches to the carbapenem nucleus were considered based on the symmetrization-asymmetrization concept as shown in Schemes 3, 5, and 6. First of all, 4-[(methoxycarbonyl)methyl]-2-azetidinone (**3S**) was considered to be a key intermediate for the trans- and ene-carbapenem nuclei. Thus, a chiral half-ester **4S** was considered to be the precursor for **3S** and finally β-amino dimethyl glutarate, a prochiral symmetric diester, was considered to be one of potential candidates as a substrate for PLE.

Scheme 2 Naturally Occurring Carbapenem Antibiotics

trans-Carbapenem

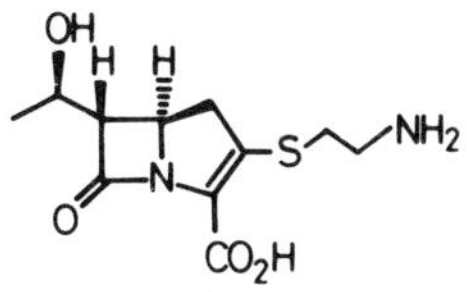

thienamycin

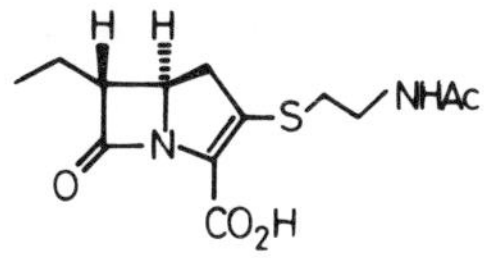

PS-5

cis-Carbapenem

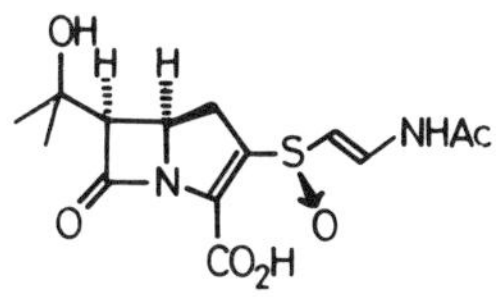

carpetimycin A

ene-Carbapenem

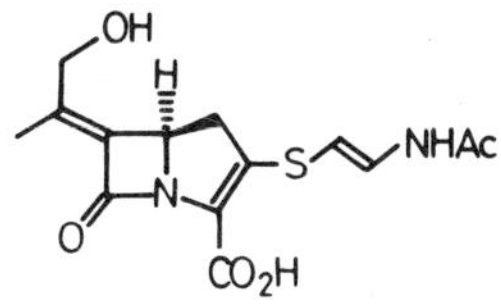

asparenomycin C

Scheme 3 Synthetic Strategy for trans-, cis-, and ene-Carbapenem Antibiotics based on Symmetrization-asymmetrization Concept (1)

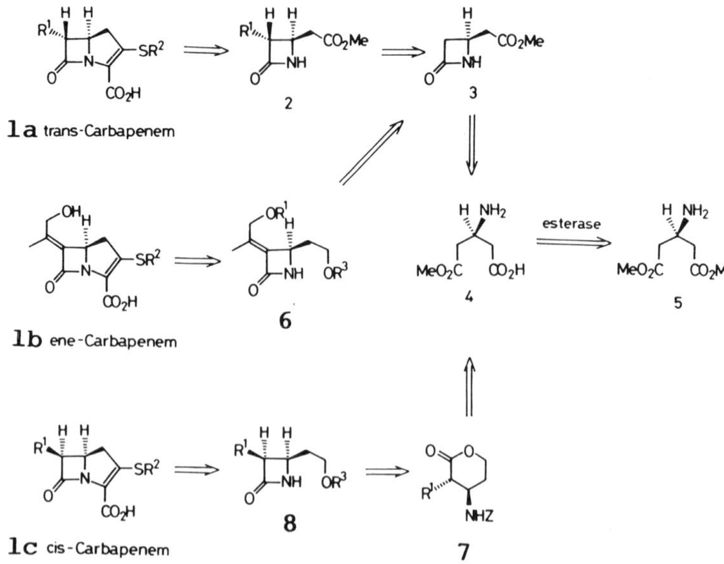

Scheme 4

Asymmetric Hydrolysis of Dimethyl β-Amino and β-Z-Aminoglutarates with PLE

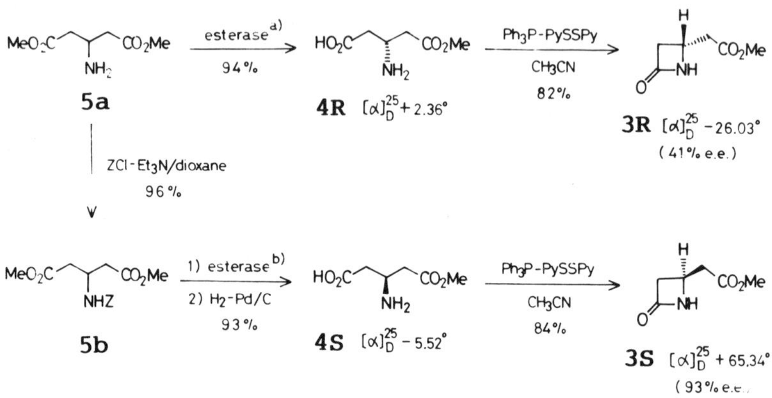

Conditions : 200 units pig liver esterase / 1 mmol substrate

a) pH 8.0, 0.05 M phosphate buffer, 25°C, 1.5 hr

b) pH 8.0, 0.05 M phosphate buffer-acetone, 25°C, 7 hr

Scheme 5

Synthetic Strategy for trans- and cis-Carbapenem Antibiotics
based on Symmetrization-asymmetrization Concept (2)

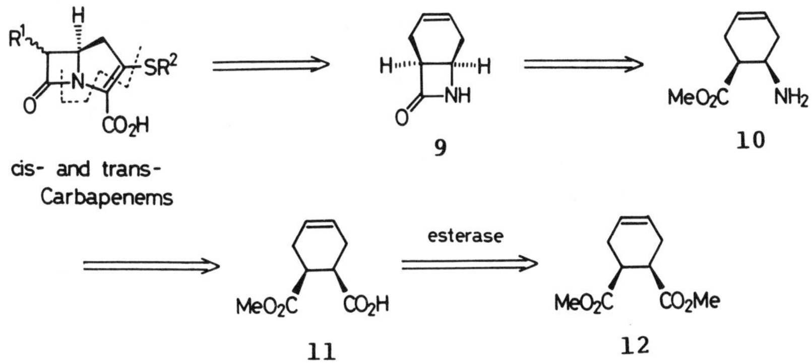

Scheme 6

Synthetic Strategy for Carbapenem Antibiotics
based on Symmetrization-asymmetrization Concept (3)

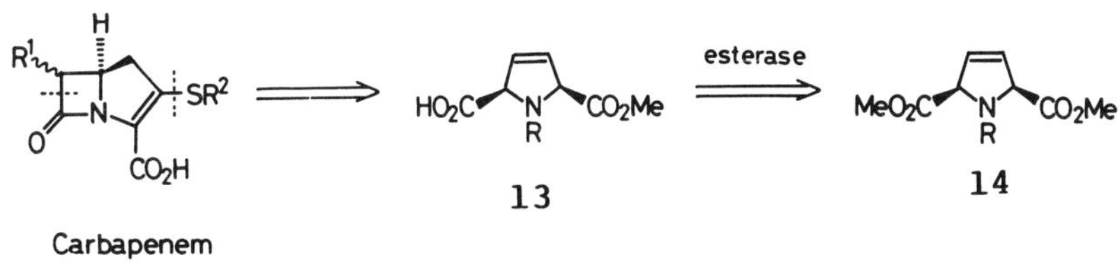

The chiral half-ester **4S** will be also a good synthon for cis-carbapenem, since thermodynamically more stable trans-δ-lactone **7** may be easily obtained from **4S** and correspond to thermodynamically less stable cis-β-lactam compound **8**. Experimentally, dimethyl β-aminoglutarate **5a** was first subjected to asymmetric hydrolysis with α-chymotrypsin, but the rate of hydrolysis was extremely slow. Therefore, we tried PLE which had been used first by Professor Sih in 1975 to hydrolyze β-hydroxy-β-methyl dimethyl glutarate. (20) PLE hydrolyzed **5a** very efficiently to give an optically active half-ester **4R**. The absolute configuration and optical purity of **3** were determined by comparison with an authentic sample after conversion to 2-azetidinone **3** with triphenylphosphine and 2,2'-dipyridyl disulphide in acetonitrile (2). The result showed that PLE cleaved the pro-S methyl ester group of **4** in a low optical yield (about 40% enantiomeric excess, ee). It was also shown that substrate **4** was partly hydrolyzed under the reaction conditions even in the absence of the enzyme. Therefore, the free amino group was protected by the benzyloxycarbonyl group (Z) to give **5b**. Surprisingly, incubation of **5b** with PLE under the same reaction conditions yielded the (3S)-half-ester **4S**. After hydrogenolysis of **4S**, and cyclization to the 2-azetidinone, (S)-4-[(methoxycarbonyl)methyl]-2-azetidinone was obtained in excellent chemical and optical yields. In the absence of the enzyme, almost no hydrolysis of **5b** took place. The results confirmed that PLE stereospecifically cleaved the pro-R methyl ester group of **5b**. It is worth noting that the protection of the amino group at the prochiral centre in **5a** with ZCl reversed the absolute configuration of the product obtained with PLE by selective hydrolysis of one of the enantiotopic (methoxycarbonyl)methyl groups (Scheme 4). This remarkable result gives us significant insights into the topography of the active site of PLE and we naturally became interested in the structure-stereospecificity relationships. The dependence of the absolute configuration of the hydrolyzed products on the protective groups is shown in Table 1. tert-Butyloxycarbonyl-, benzoyl-, and benzylamino derivatives of **5a** afforded the half-ester with the S configuration more selectively, but the acetyl derivative preferentially afforded the half-ester with R configuration. These findings are not only synthetically useful but also important in applying a working hypothesis of the topography of the active site of PLE. All kinds of carbapenem antibiotics were synthesized starting from **5S** (8 for trans, 9 for cis, and 11 for ene).

The second approach starting with dimethyl cis-1,2-cyclohex-4-enedicarboxylate (**12**) was most beautifully realized to afford the chiral half-ester **11** efficiently converted to the key intermediate **9** through **10** as shown in Scheme 5. The racemic compound **9** was already transformed to a cis-carbapenem (**21**). The third approach (Scheme 6) was investigate with a saturated diester of the tetrahydropyrrole ring system rather easily available, and the half-ester was obtained in reasonable chemical and optical yields as shown in Scheme 7.

Among the three approaches to carbapenem antibiotics, the total elegancy of the synthesis may depend upon the availability of the starting symmetric diesters, optical purity of the chiral half-esters obtained by the enzyme-catalyzed hydrolysis, and the efficiency of the conver-

Table I

Dependence of the Absolute Structures of Hydrolyzed Products on the N-Protective Groups

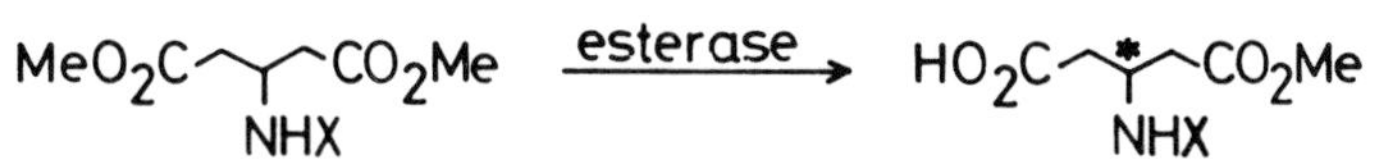

X	Chem. Yield(%)	Opt. Yield(%)	Config.
H	94	41	*R*
Ac	81	93	*R*
Bzl	94	50	*S*
Bz	60	72	*S*
Z	93	93	*S*
Boc	93	90	*S*

Scheme 7 Symmetrization–asymmetrization Concept

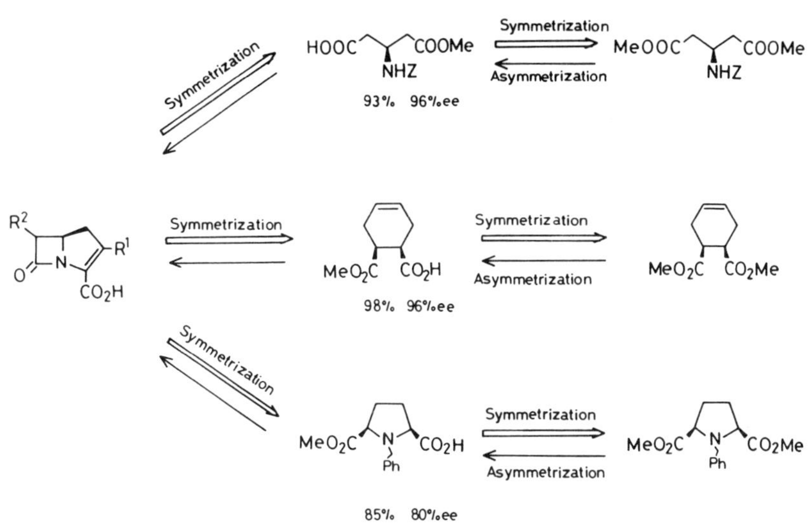

sion to the target molecule from the chiral half-esters. An important feature of the symmetrization-asymmetrization concept is able to design symmetric diester substrates in multiple forms from one target molecule and able to make choice of them according to the purpose of the synthesis.

2.2. Negamycin, a Peptide Antibiotic

Negamycin, 2- (3R,5R)-3,6-diamino-5-hydroxyhexanoyl -1-methylhydrazinoacetic acid, has also attracted a great deal of synthetic study, because it possesses a strong inhibitory activity against Gram-negative bacteria, including Pseudomonous (22). As shown in Scheme 8, a combination of enzymatic and non-enzymatic procedures was taken as our synthetic strategy and (+)-negamycin was synthesized in a completely stereocontrolled manner using iodocyclocarbamation as a key step as shown in Scheme 9 (5,13). (Compounds **18** and **21** are the desired intermediates with required stereochemistry.)

Scheme 8 Synthetic Strategy for Negamycin

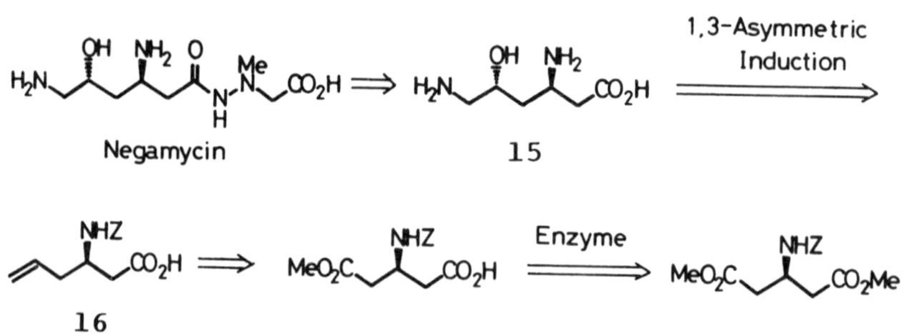

Scheme 9 Iodocyclocarbamation

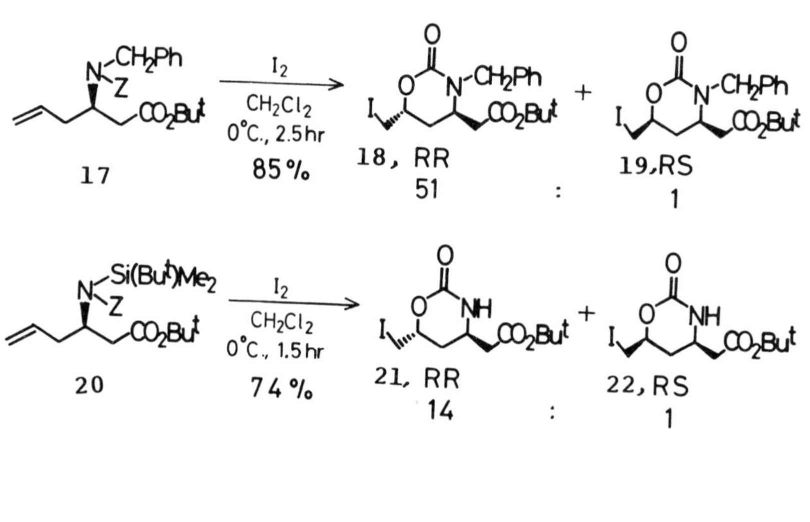

2.3. C- and N-Nucleosides (Showdomycin, 6-Azapseudouridine, Cordycepin) and Carbocyclic Nucleosides (Neplanomycin A and Aristeromycin)

As an extension of our enzymatic approach to natural product synthesis, sugar moiety of various nucleosides were considered to be a good target to demonstrate our concept. For instance, Neplanocin A (**23**) isolated in 1981 (23) from Actinoplanacea ampllariella is a novel carbocyclic anologue of adenosine with a cyclopentene moiety and it exhibits remarkable antitumor activity against L1210 leukemia in mice. All of the cyclopentane nucleosides previously synthesized were obtained in racemic forms, and the chiral carbocyclic moiety seems to be not easily accessible by conventional synthetic means or by the partial degradation of neplanocin A. The retrosynthesis of **23** based on the symmetrization-asymmetrization concept is shown in Scheme 10. After some attempts, a symmetrically consituted diester **24** was found to be a good substrate for PLE, although such a tricyclic and rigid meso compounds had never been subjected to enzyme-mediated hydrolysis. It should be mentioned here that only the unsaturated substrates **24**, **25** and **26** were efficiently hydrolyzed with PLE and the saturated exo **27** and endo **28** compounds were found to be hardly hydrolyzed with PLE, showing synthetically not useful, and the unsaturated meso compound **29** gave the half-ester with low optical purity.(Scheme 11, 12) Thus, various nucleosides were synthesized in optically active form as shown in Scheme 13, including neplanocin A and aristeromycin from **24**, cordycepin from **26**, L- and D-ribose, (+)-showdomycin, (-)-6-azapseudouridine from **25**.(3,4,6,10,12,14)

Scheme 10 Synthetic Strategy for (-)Neplanocin A by Chemicoenzymatic Approach

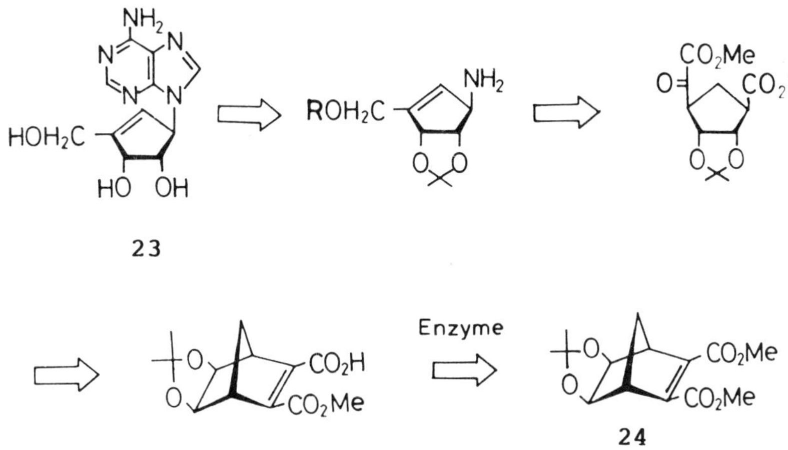

23

24

Scheme 11

Asymmetric Hydrolysis of Bicyclic diesters with Pig Liver Esterase

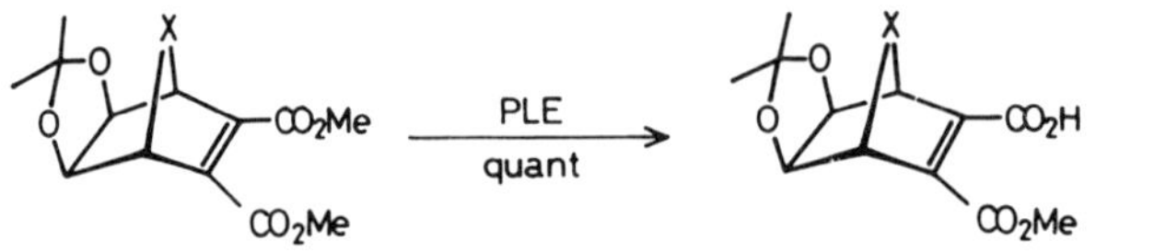

25 X=O
24 X=CH2

3,X=O
4,X=CH2 77~88% e.e.

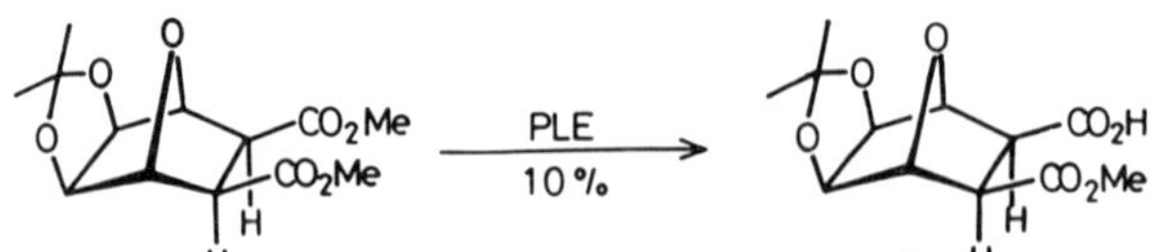

27

6
e.e. not determined

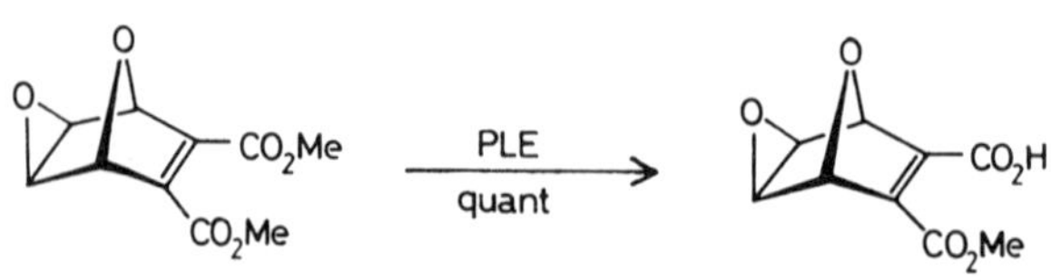

26

8 ~80%ee

Scheme 12

Asymmetric Hydrolysis of Bicyclic diesters with Pig Liver Esterase

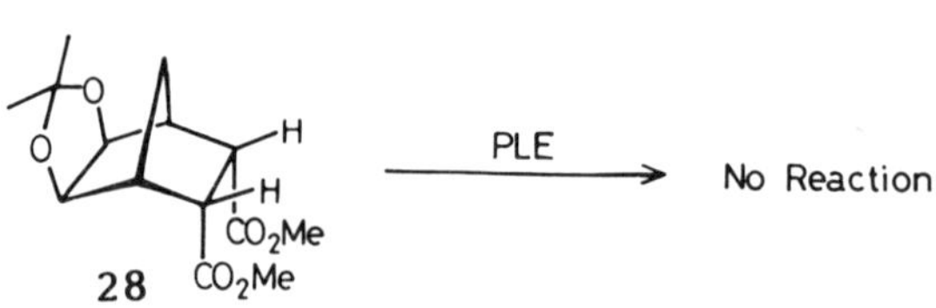

28

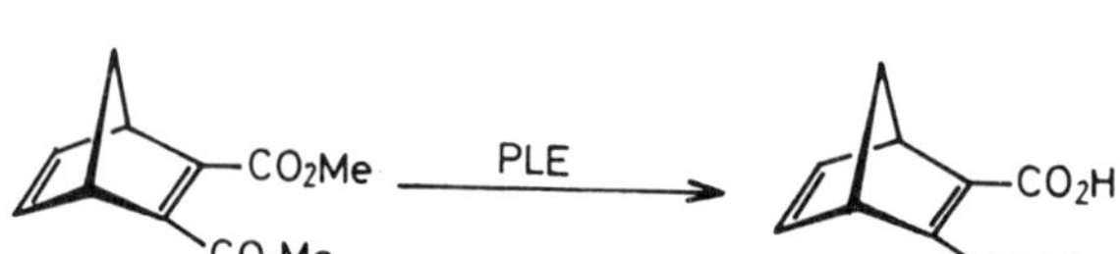

29

Scheme 13

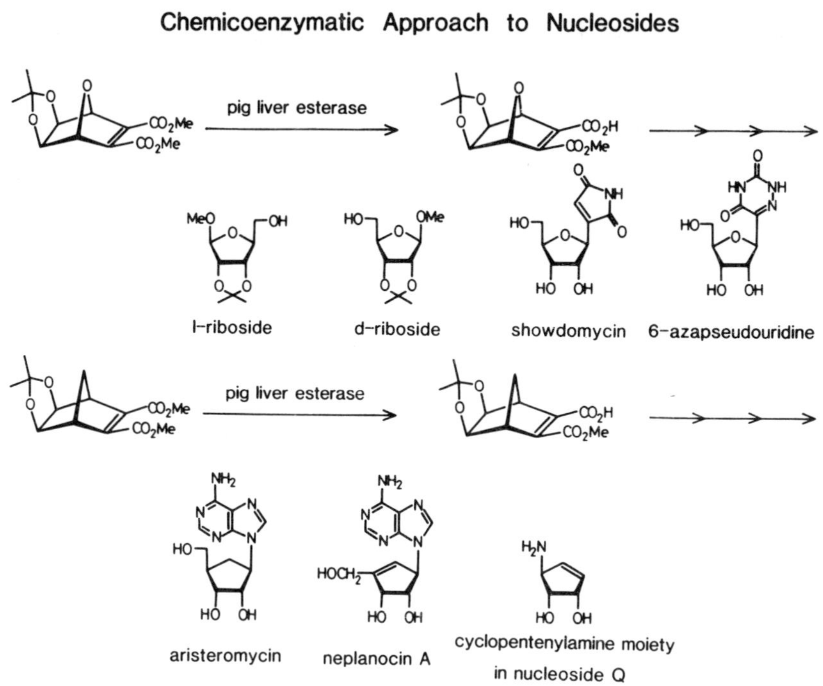

2.4. Natural Products Containing Chiral Cyclohexane Moiety

There are so many natural products containg chiral cycloalkane ring systems. We thought that creation of such ring compounds is also possible by the chemicoenzymatic approach. Thus, as shown in Scheme 14, a symmetric unsaturated diester **30** was treated with PLE to afford the chiral half-ester **31** in excellent chemical (98%) and optical (96% ee) yields. Various cyclohexane-derivatives were synthesized from **31** and all the derivatives are considered to be very useful chiral synthons for the synthesis of aminosugar and macrolide antibiotics (Scheme 15).

Scheme 14

Absolute Structure of Half-ester Prepared by Asymmetric Hydrolysis with Pig Liver Esterase (PLE)

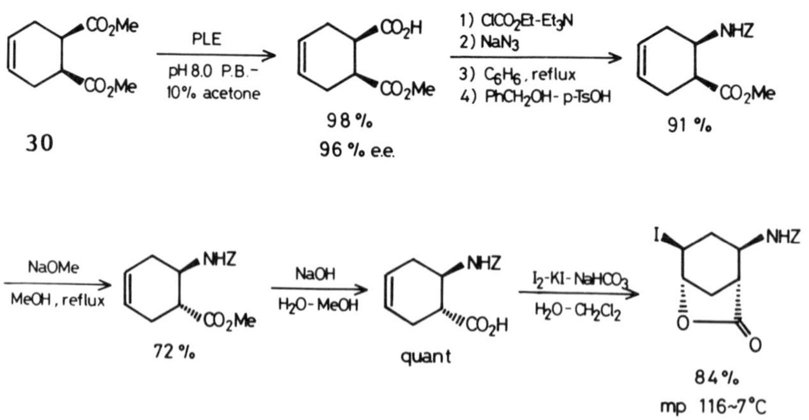

30

Scheme 15

Preparation of All Stereoisomers

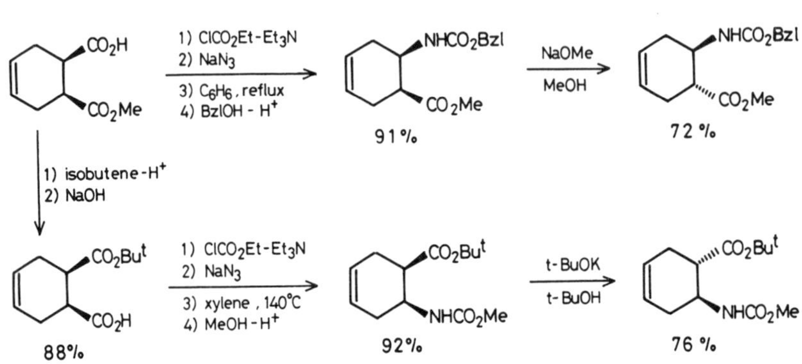

3. The Substrate Recognition of PLE

Our studies with a wide variety of symmetrically constituted diesters have shown that asymmetric hydrolysis with PLE is synthetically useful for the creation of new chiral half-esters. Furthermore, the structure-stereospecificity relationships revealed by the present study give significant and useful information about the topography of the active site of PLE and allow us to propose an active site model. For acyclic systems, the distance between the prochiral center and the hydrolyzed ester group seems to be very important as shown in Scheme 16. The dimethyl 2-monosubstituted malonate seems to be asymmetrically hydrolyzed but the resulted half-ester seems to be easily isomerized to show inconsistent optical rotations. 2,2-Disubstituted molonate may be synthetically useful.(24) 3-Substituted glutarates were found most important for the synthetic purpose, but further elongation (n=3 in Scheme 16) of the chain affords the half-ester with very low optical purity. Among the 3-substituted aminoglutarates, even the absolute configurations of the products may vary in response to small structural changes in the substrates.(Table I)

Tricyclic compounds **24, 25** and **26** are all hydrolyzed to afford synthetically useful half-esters. Topography of the active site of PLE may be most directly reflected from the interaction with such rigid systems. The active site seems to be hydrophobic and a flat region which can accommodate six-membered ring easily seems to be located at the place near from the catalytic site or serine moiety of PLE. Hydrophobic nature of the active site of PLE is shown by dimethylglutarates having much longer chains at 3-position. The orientation of the substrate in the active site is improved when hydrophobic groups on the diester bind to the hydrophobic site on the enzyme. This gives an increased optical yield of the half-ester as shown in Scheme 17 and 18. Cohen's pioneering model or 4-subunit theory proposed for α-chymotrypsin in 1969 is still very instructive when we think of the topography of the active site of an esterase.(25, 26) Of course, we know that it may be impossible to extrapolate perfectly mechanistic details from such small synthetic substrates to the large peptides like PLE, but such extraporation seems relatively reasonable with chymotrpsin.(27) Most of our results can be explained by assuming a working model of the active site of PLE as shown in Scheme 19, although there are some subtle results hard to explain. The model is useful not only for the explanation and prediction of the absolute configuration of the major chiral half-ester, but also for the design of better substrates.(Scheme 19)

The flat region can accommodate six-membered rings very easily and may contain carbonyl group(s) capable of hydrogen bonding or hydrophobic region capable of π-πinteraction. The pocket site (am site) is a cavity that lies at a variable angle to the catalytic site, and makes a small angle with (but is almost parallel to) the flat retion; it may also contain carbonyl group(s) capable of hydrogen bonding. The catalytic site is bounded by a steep wall that makes an angle (θ) of about 60° with the flat region; the serine residue is located somewhere in the catalytic site at a distance of 1.5-3.0 Å from the centre of the flat region. The part of the active site drawn with continuous lines repre-

Scheme 16 Preferred Distance between the Prochiral Center and the Ester Group

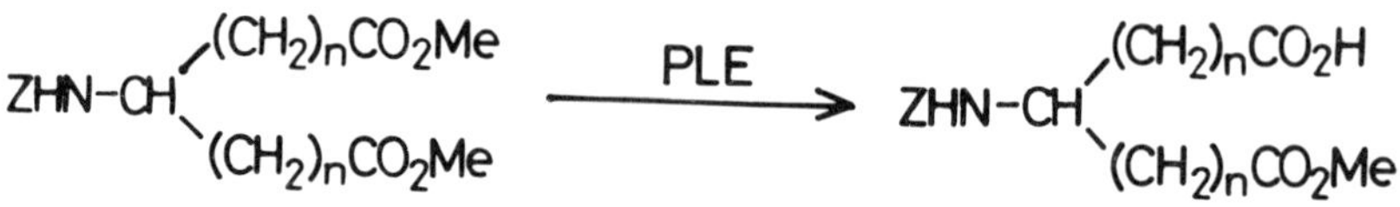

	e.e. (%)
n = 0	?
n = 1	93
n ≥ 2	~ 0

Scheme 17 Hydrophobic Nature of the Binding Site of PLE

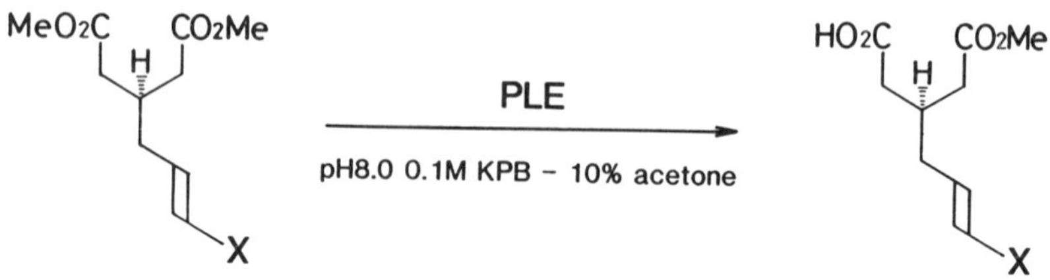

X	Yield (%)	e.e. (%)
CH_2OH	99	-19
CH_2OTHP	95	-74
Ph	97	-88

Scheme 18

Asymmetric Hydrolysis of Meso or Prochiral Esters with PLE

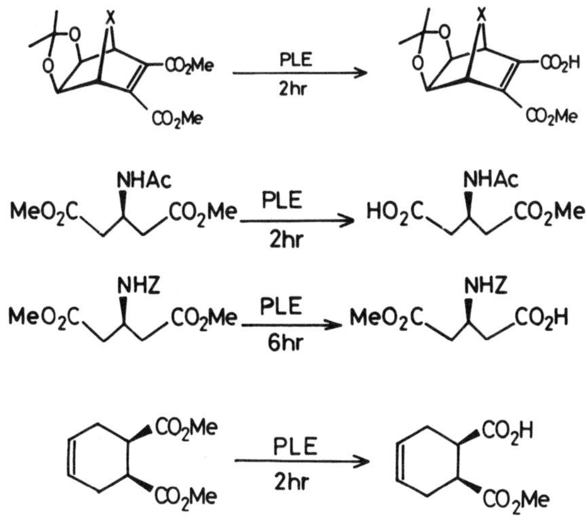

	C.Y.(%)	O.Y.(%ee)
X=O	96	80
X=CH_2	99	84
	81	93
	93	93
	98	96

Scheme 19 Active Site Model for PLE

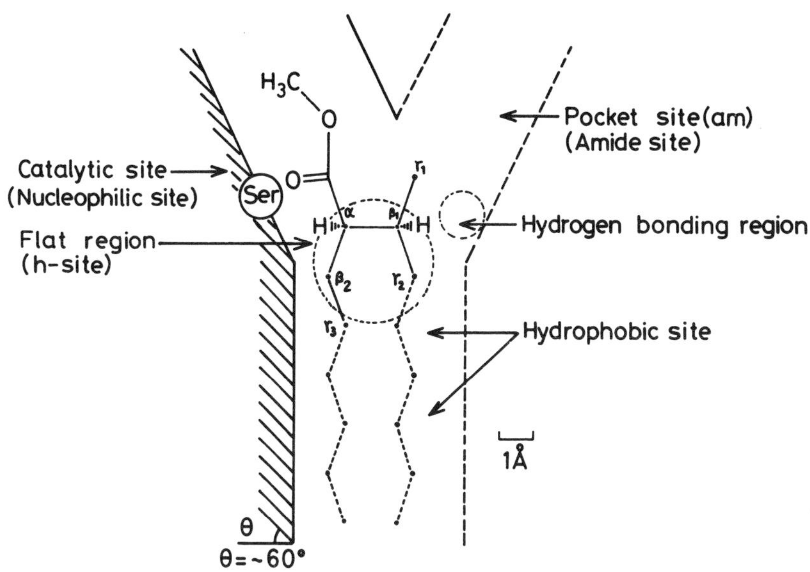

Scheme 20 The preferred (with the double circle) and less favorable Orientation for the Asymmetric Hydrolysis of 5b with PLE

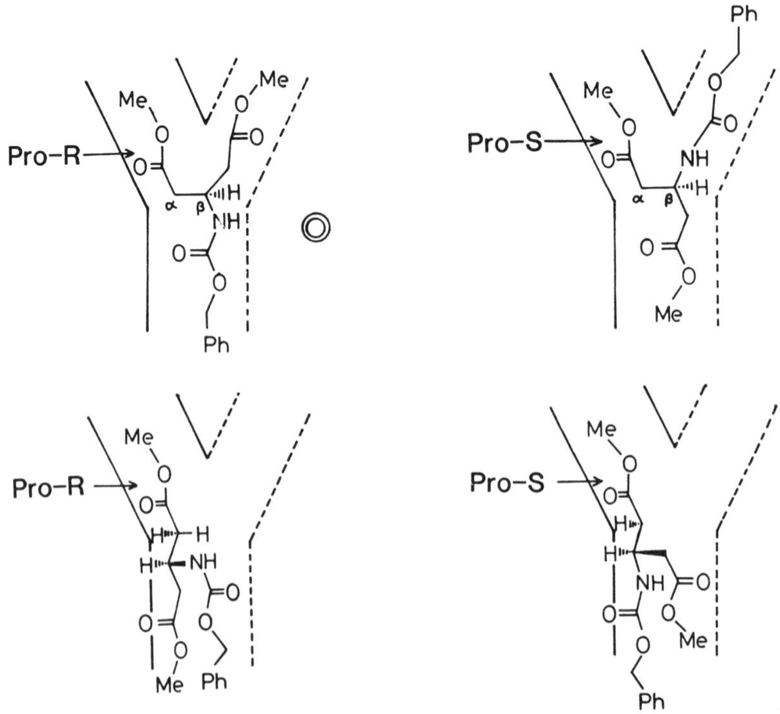

Scheme 21 The preferred (with the double circle) and less favorable Orientations for the Asymmetric Hydrolysis of N-acetyl 5a with PLE

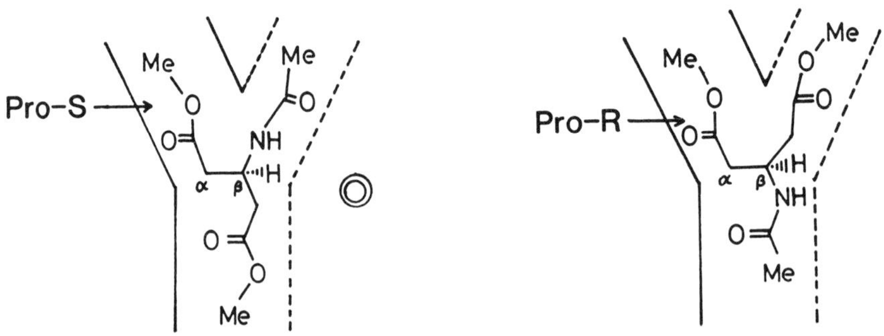

Scheme 22 Asymmetric Hydrolysis of 24 or 25 with PLE. Diagrams show the different orientations of the substrate in the active site: the preferred orientation is on the top-left.

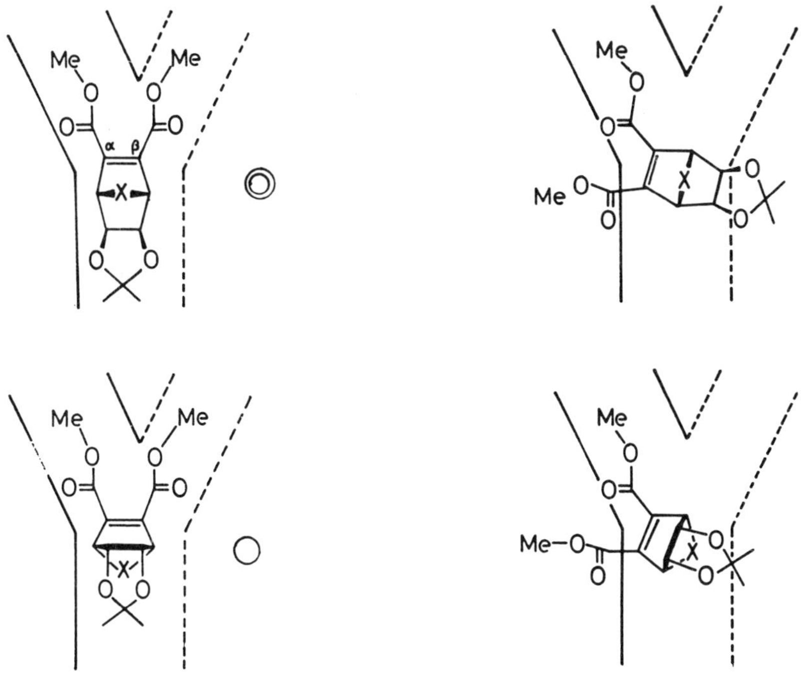

Scheme 23 Asymmetric Hydrolysis of 30 with PLE.

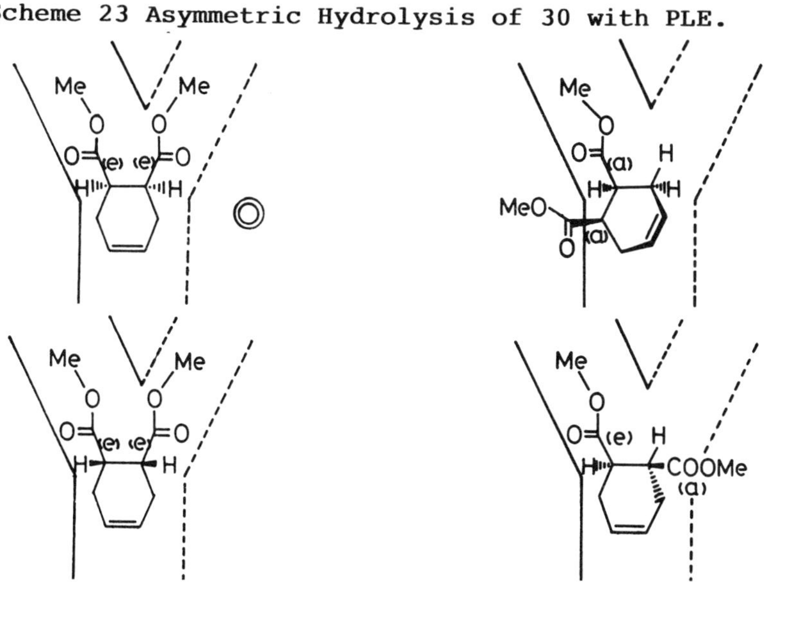

sents an open pocket which is readily accessible to substrates. The orientation of the substrate in the active site is improved when hydrophobic groups on the diester bind to the hydrophobic site on the enzyme. This may give an increased optical yield of the half-ester.
Typical examples of the accommodation of prochiral or meso substrates (shown in Scheme 18) in the active site are illustrated in Schemes **20**, **21**, **22** and **23**.

4. Conclusions

The enzymatic approach to the new chiral synthons described here has opened up a new avenue to natural product synthesis. More significantly, we have developed the symmetrization-assymmetrization concept in which a wide variety of diester-substrates possessing σ-symmetry undergo efficient asymmetric hydrolysis with PLE. It is also worthy of note that the optical yields of the chiral half-esters vary in response to even small synthetic change in the substrates and even the absolute configurations of some chiral half-esters may be synthetically controlled. Study on the enzyme by Jencks and his co-workers disclosed very interesting but complicated property of PLE, and the functional significance of the different forms of PLE still remains as mysterious as the physiological function of the enzyme itself.(28) We hope that the rapidly increased use of PLE as an important catalyst in organic synthesis will give also a great impact for such structural study of PLE again.

Acknowledgements- We would like to express our cordial thanks to our collabrators, Drs Y. Ito, Y. Wang, T. Iimori, K. Okano, T. Izawa, M. Arita, H. Nakai, H. Sawai, and T. Murakami, and Messrs T. Shibata, M. Kurihara, M. Nakada, K. Kamiyama, T. Tsuri, H. Yamashita, Y. Takahashi, T. Isobe and M. Shimada, and Miss H. Kakuda, for their great contribution at the various stages of our chemicoenzymatic study, and Dr. M. Yoshioka for the stimulating discussion.

REFERENCES AND NOTES

1. M. Ohno, S. Kobayashi, T. Iimori, Y-F. Wang, and T. Izawa, J. Am. Chem. Soc., **103**, 2405 (1981).
2. S. Kobayashi, T. Iimori, T. Izawa, and M. Ohno, J. Am. Chem. Soc., **103**, 2406 (1981).
3. Y. Ito, M. Arita, K. Adachi, T. Shibata, H. Sawai, and M. Ohno, Nucleic Acid Research, Symposium Series, No.10, 45 (1981).
4. Y. Ito, T. Shibata, M. Arita, H. Sawai, and M. Ohno, J. Am. Chem. Soc., **103**, 6739 (1981).
5. Y-F. Wang, T. Izawa, S. Kobayashi, and M. Ohno, J. Am. Chem. Soc., **104**, 6465 (1982).
6. M. Arita, K. Adachi, Y. Ito, H. Sawai, and M. Ohno, Nucleic Acids Research, Symposium Series, No.11, 13 (1982).
7. H. Kotani, Y. Kuze, S. Uchida, T. Miyabe, T. Iimori, K. Okano, S. Kobayashi, and M. Ohno, Agric. Biol. Chem., **47**, 1363 (1983).
8. K. Okano, T. Izawa, and M. Ohno, Tetrahedron Lett., **24**, 217 (1983).
9. T. Iimori, Y. Takahashi, T. Izawa, S. Kobayashi, and M. Ohno, J. Am. Chem. Soc., **105**, 1659 (1983).
10. M. Arita, K. Adachi, Y. Ito, H. sawai, and M. Ohno, J. Am. Chem. Soc., **105**, 4049 (1983).
11. K. Okano, Y. Kyotani, H. Ishihama, S.Kobayashi, and M. Ohno, J. Am. Chem. Soc., **105**, 7186 (1983).
12. M. Arita, K. Adachi, H. Sawai, and M. Ohno, Nucleic Acids Research, Symposium Series, No.12, 25 (1983).
13. M. Ohno, S. Kobayashi, T. Izawa, and Y-F. Wang, J. Chem. Soc. Jpn., Chem. & Ind. Chem., **12**, 99 (1983).
14. M. Ohno, Y. Ito, M. Arita, T. Shibata, K. Adachi, and H. Sawai, Tetrahedron Symposia in print, **40**, 145 (1984).
15. S. Kobayashi, K. Kamiyama, T. Iimori, and M. Ohno, Tetrahedron Lett., **25**, 2557 (1984).
16. M. Ohno, "Enzymes in Organic Synthesis", Ciba Foundation Symposium III, p.171 (1985).
17. M. Ohno, "Enantioselective Synthesis of Some Carbapenem-, Amino-sugar-, and Nucleoside-Antibiotics by Enzymatic Approach", Proceedings of the Vth International Conference on Organic Synthesis, Freiburg, in press (1985).
18. M. Ohno, "Enantioselective Synthesis of Carbocyclic Nucleosides by Enzymatic Approach and the Anticancer Activities", Proceedings of the 6th International Round Table "Nucleosides, Nucleotides and Biological Applications", La Grande Motte, in press (1985).
19. (a) D. B. R. Johnson, S. M. Schmitt, F. A. Bouffard, and B. G. Christensen, J. Am. Chem. Soc., **100**, 313 (1978). (b) J. J. Tufariello, G. E. Lee, P. A. Senaratne, and M. Al-Nuri, Tetrahedron Lett., 4349 (1979). (c) R. J. Ponsford, and R. Southgate, J. Chem. Soc., Chem. Commun., 845 (1979). (d) T. Kametani, S. Huang, S. Yokohama, Y. Suzuki, and M. Ihara, J. Am. Chem. Soc., **102**, 2060 (1980). (e) L. Cama, and B. G. Christensen, Tetrahedron Lett., **21**, 2013 (1980). (e) D. G. Melillo, I. Shinkai, T. Liu, K. Ryan, and M. Sletzinger, Tetrahedron Lett., **21**, 2783 (1980). (f) T. N. Salzmann, T. W. Ratcliffe, and B. G. Christensen, Tetrahedron Lett., **21**, 1193 (1980).

20. F. Huang, L. F. H. Lee, R. S. D. Mittal, P. R. Ravikumar, J. A. Chan, C. J. Sih, E. Caspi, and C. R. Eck, J. Am. Chem. Soc., **97**, 4144 (1975).
21. J. H. Bateson, R. I. Hickling, P. M. Roberts, T. C. Smale, and R. Southgate, J. Am. Chem. Soc., 1084 (1980).
22. (a) S. Shibahara, S. Kondo, K. Maeda, H. Umezawa, M. Ohno, J. Am. Chem. Soc., **94**, 4353 (1972). (b) S. Kondo, S. Shibahara, S. Takahashi, K. Maeda, H. Umezawa, M. Ohno, J. Am. Chem. Soc., **93**, 6305 (1971).
23. S. Yaginuma, N. Muto, M. Tsujino, Y. Sudate, M. Hayashi, M. Otani, J. Antibiot., **34**, 359 (1981).
24. P. Mohr, N. Waespe-Sarcevic, C. Tamm, K. Gawronska, J. K. Gawronski, Helv. Chim. Acta., **66**, 2501 (1983).
25. S. G. Cohen, Trans N. Y. Acad. Sci., **31**, 705 (1969).
26. J. B. Jones, Application of Biochemical Systems in Organic Chemistry, I. p. 133, John Wiley and Sons, New York (1976).
27. C. Walsh, Enzymatic Reaction Mechanisms, p. 60-62, W. H. Freeman and Company, San Francisco (1979).
28. D. Farb, W. P. Jencks, Arch. Biochem. Biophys., **203**, 214 (1980).

APPLICATION OF MICROBIAL TRANSFORMATIONS IN THE TOTAL SYNTHESIS OF NATURAL PRODUCTS

Dee W. Brooks
Abbott Laboratories
Pharmaceutical Products Division
Abbott Park, Illinois 60064
U. S. A.

ABSTRACT. Asymmetric microbial reduction of 2,2-disubstituted-1,3-cycloalkanediones using common baker's yeast effected an efficient monoreduction to provide ketol products of high enantiomeric excess. The scope and versatility of this method for a variety of substrates was examined and the integration of selected microbial products in synthetic strategies to various natural products including the trichothecene mycotoxins was illustrated. Yeast reduction of ß-ketoester systems was studied with respect to the effect of remote substituents on the enantioselectivity. Appropriate substrate design provided a useful [R] or [S] 1,3,5-trialkoxypentane synthon of high enantiomeric excess which was applied in synthetic studies of polyene macrolides.

1. INTRODUCTION

The discovery and development of enzyme catalysed processes to effect asymmetric transformations which are integrated as an important part of the strategy and design of synthetic schemes for various natural products is an interesting area of investigation. Increasing interest in enzymatic transformations as a means to prepare optically active intermediates for enantiospecific syntheses is evident from many examples in the recent literature.(1) Three general sources of optically active compounds provide intermediates for synthesis, as follows; 1. resolution of racemates by preparation and separation of diastereomeric derivatives, 2. chemical modification of biosynthetic products such as carbohydrates, terpenes, and aminoacids, and 3. asymmetric reactions with synthetic reagents or enzymes. These sources for the most part complement eachother and each can have merits or disadvantages in a given situation. Aspects of asymmetric synthetic transformations in the context of natural product targets is currently a popular area of research in organic chemistry. Integration of enzymatic catalysis as part of a synthetic strategy to provide enantiospecificity will be illustrated using a simple

M. P. Schneider (ed.), Enzymes as Catalysts in Organic Synthesis, 143–169.

microbial reduction. Common Baker's yeast (Saccharomyces cerevisiae) is a particularly versatile and easy to use microorganism for asymmetric reductions due to the presence of a variety and/or relaxed specificity of dehydrogenase enzymes.(2) Emphasis will be placed on the design and manipulation of the substrates for microbial reduction as this can be readily accomplished by organic chemists rather than the alternate approach of screening strains of microorganisms for reduction of a given substrate. Examples of substrate structural factors which control the enantioselectivity and stereoselectivity of selected microbial reductions will be described.

2. Microbial Reduction of 2,2-Disubstituted-1,3-Cycloalkanediones

We chose to embark on a study of microbial reduction of 2,2-disubstituted-1,3-cycloalkanediones by Baker's yeast hoping to effect an asymmetric monoreduction of these systems to provide a net prochiral distinction and the generation of two asymmetric centers. The resulting products would serve as useful intermediates in proposed schemes for the synthesis of selected natural products. The ability of certain enzymes to make prochiral distinctions is based on the formation of a preferred enzyme-substate complex which leads to a favored catalytic reaction for one of two enantiotopic homomorphic groups. Previous pioneering work in this context had fully substantiated the potential of microbial reductions of 2,2-disubstituted-1,3-cyclopentanediones with a variety of microorganisms particularly in application to steroid syntheses.(3)

2.1 Yeast Reduction of 2,2-Disubstituted-1,3-Cyclopentanediones

We first examined the reduction of the series 2-propyl-(**1**), 2-allyl-(**6**), and 2-propynyl-2-methyl-1,3-cyclopentandione(**11**) with baker's yeast.(4) A typical reaction procedure is described as follows: To a solution of 1L of water, 150g of D-glucose, and 4g of yeast extract, warmed at 40°C, was added 100g of dry active baker's yeast and the mixture was vigorously stirred at 35°C for 30min, after which 5-10g of dione was added slowly dropwise over 30min. The mixture was stirred at 25-30°C for 24h and then continuously extracted with CH_2Cl_2 for 48h to provide a crude product which was purified by chromatography on silica gel. The results of the microbial reduction are summarized in Figure 1 and compared with the ketol products formed by reduction with 1 eqiv of $NaBH_4$. It is interesting to notice the similar trend of decreased stereoselectivity in both the microbial and borohydride reductions in the series, propyl > allyl > propynyl but the greater sensitivity of the microbial system to 2-substituent differences.

Figure 1. Yeast Reduction of 2,2-Disubstituted-1,3-cyclopentanediones

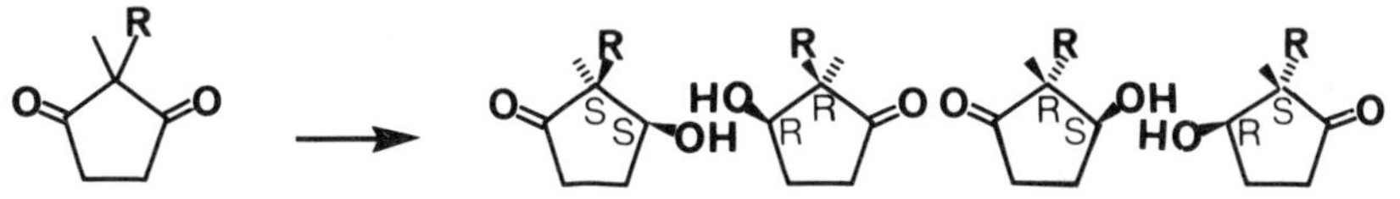

1 R=$CH_2CH_2CH_3$	**2**	**3**	**4**	**5**
6 R=$CH_2CH{=}CH_2$	**7**	**8**	**9**	**10**
11 R=$CH_2C{\equiv}CH$	**12**	**13**	**14**	**15**
16 R=$CH_2C(CH_3){=}CH_2$	**17**	**18**	**19**	**20**

entry	dione	reduction method	ketol products (% composition)	yield%	recovered dione%
1	**1**	yeast	**2**(100)	60	30
2	**1**	$NaBH_4$	**2**,**3**;**4**,**5**(85,15)	70	10
3	**6**	yeast	**7**;**9**(90,10)	75	15
4	**6**	$NaBH_4$	**7**,**8**;**9**,**10**(75,25)	75	5
5	**11**	yeast	**12**;**14**(67,33)	60	25
6	**11**	$NaBH_4$	**12**,**13**;**14**,**15**(67,33)	60	10
7	**16**	yeast	**17**(100)	75	10
8	**16**	$NaBH_4$	**17**,**18**;**19**,**20**(85,15)	77	5

The enantiomeric composition of the chiral ketols was determined by analysis of the ^1H and ^{19}F NMR spectra of the corresponding (+)-α-(trifluoromethyl)phenylacetates (MTPA) (5) using the racemic ketols as a reference. In all cases the microbial ketol products were greater than 98% enantiomerically pure.(6) The absolute configuration of the chiral ketol **7** was established by correlation with the known lactone **21** (7) which we prepared from **7** by ozonolysis followed by Jones oxidation. Ketols **7** and **12** were correlated to **2** by catalytic hydrogenation. Treatment of ketol **7** with p-toluenesulfonyl chloride in pyridine followed by heating in the presence of excess KNO_2 and subsequent aqueous acidic workup gave the epimeric ketol **10** which was catalytically hydrogenated to **5**. The alkynyl ketol **14** was hydrogenated to give ketol **4**. The absolute configuration of ketol **17** was established by further elaboration to **28** as shown in Figure 2. A preference to form the [2S,3S] diastereomer and consistent [S] configuration at the hydroxy group was evident. We were later informed of similar results in a previous study by Schwarz and coworkers (8) with other microorganisms.

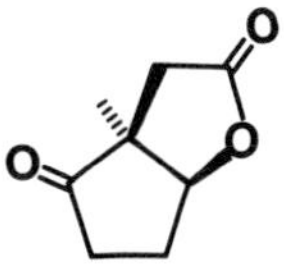

21

The utility of these readily available optically pure intermediates for the synthesis of natural products was demonstrated in the preparation of [R]-5-methyl-bicyclo[3.3.0]oct-1-ene-3,6-dione **28** from ketol **7** as shown in Figure 2. This preparation was subsequently improved by using the modified dione substrate **16** which upon reduction with yeast provided only the [2S,3S]-ketol **17** in good yield which was readily transformed to **28**. The alkynyl ketol diastereomers **12** and **14** were separated by chromatography and **14** was converted to the [S]-antipode **36**. The racemic dione **28,36** was elegantly used by Trost and Curran as a key intermediate in their synthesis of coriolin (**29**).(9) The microbial strategy provides ready access to both optically pure antipodes **28** and **36**.

Figure 2. Synthesis of 5-methyl- bicyclo[3.3.0]oct-1-ene-3,6-dione

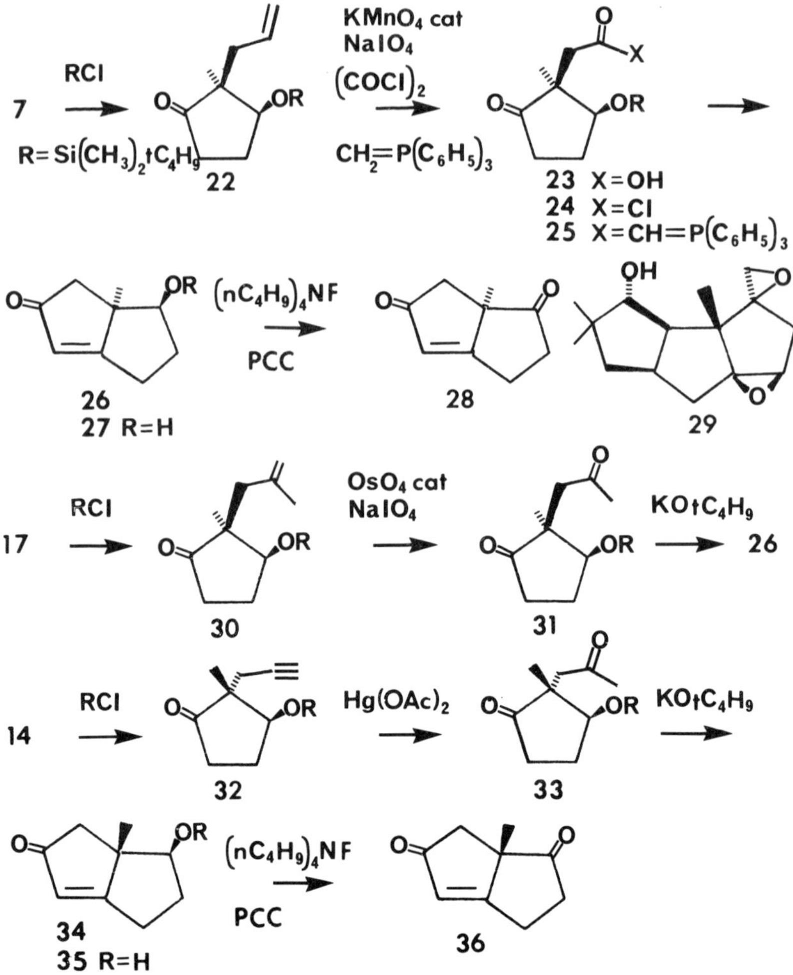

2.2 Total Synthesis of the Trichothecene Mycotoxin Anguidine

Our major initiative for the microbial study of the cyclopentanedione system was to provide a suitable optically pure precursor to develop a synthetic strategy for the total synthesis of trichothecene mycotoxins.(10) This group of natural products is of considerable interest due to the wide variety of biological activities they express depending on the respective functional groups present on their unique tricyclic skeleton.(11) Representative members of this group of natural products are shown in Figure 3. Since anguidine (40) exhibited antitumor activity, we thought it important to develop an enantioselective synthetic approach which could provide intermediates for structure-activity analysis as the target molecule was progressively assembled.

Figure 3. Representative Trichothecene Mycotoxins

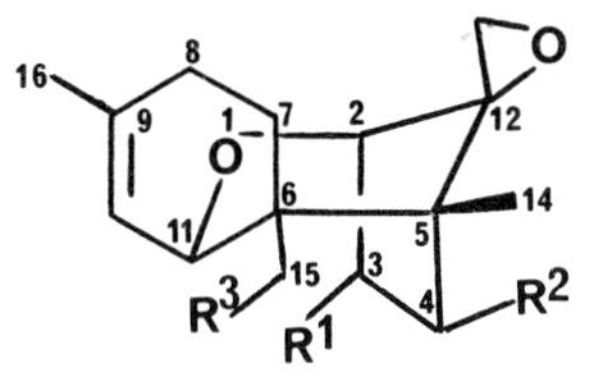

		R^1	R^2	R^3
TRICHODERMOL	(37)	H	OH	H
VERRUCAROL	(38)	H	OH	OH
CALONECTRIN	(39)	OH	H	OH
ANGUIDINE	(40)	OH	OAc	OAc

A scheme for assembling the trichothecene skeleton was conceived which involved the addition of an A-ring unit to a fully functionalized C-ring unit, followed by an intramolecular cyclization providing the B-ring and thus completing the requisite skeleton.

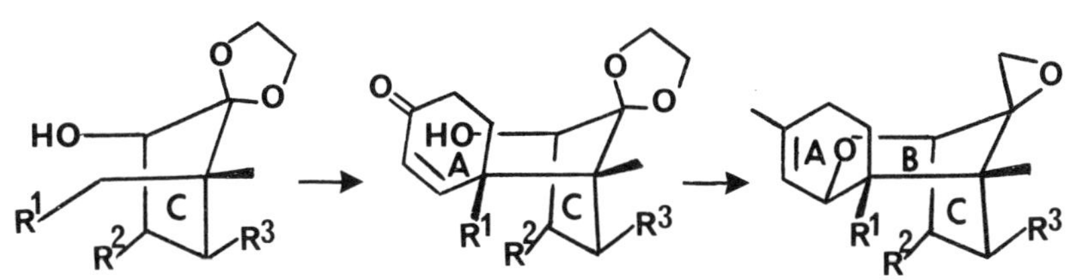

The optically pure ketol **7** served as the key starting material possessing the correct absolute configuration at the quaternary carbon (C5 of anguidine) and appropriately functionalized to facilitate efficient elaboration to a C-ring intermediate **49** as outlined in Figure 4.

Figure 4. Preparation of a C-Ring Intermediate for the Synthesis of Anguidine

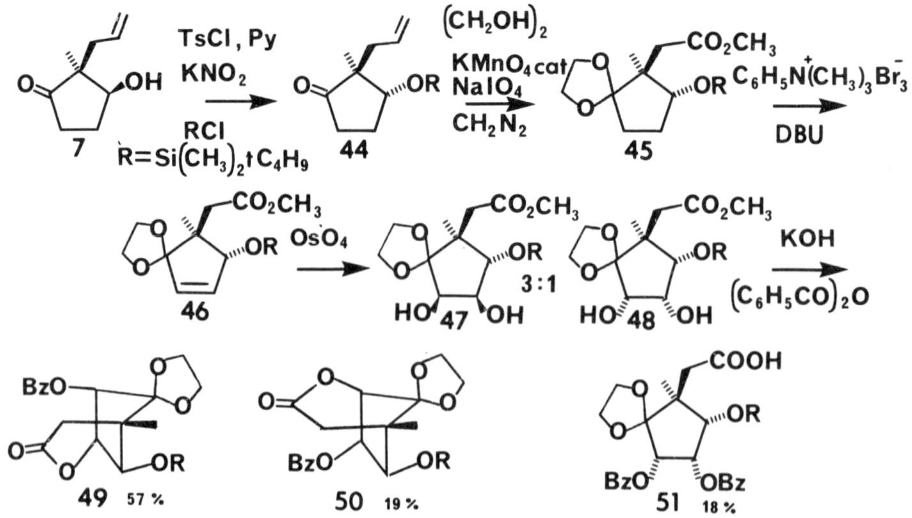

The next task involved the stereoselective intoduction of the A-ring which was successfully accomplished by a Robinson annelation sequence as shown in Figure 5. Reduction of the A+C intermediate gave the tetraol **56** and we reached the point for an interesting determination of preferred modes for intramolecular acid-catalyzed rearrangements. Four possible modes of tetrahydropyran (B-ring) formation by hydroxyl quenching of an intermediate allylic carbocation **57** were evident as shown in Figure 6. Crude analysis of conformational considerations predicted that cyclization leading to the trichothecene skeleton (path a) would be preferred. Brief treatment of the tetraol **56** with p-toluenesulfonic acid (catalytic amount) in dichloromethane gave a mixture of the desired trichothecene **58** and the unexpected [2.2.2] allyl ether **62**, resulting from cyclization of the C15 hydroxyl group. The undesired cyclization was circumvented by preparing the triacetate **63** which was selectively hydrolzed to provide the diol **64** which cyclized upon acid treatment to give only the trichothecene **65**. The synthesis of anguidine was completed in a straightforward manner guided by preceeding synthetic studies.(11a,e) This synthesis highlights the integration of a simple prochiral distinction effected by microbial reduction to provide an optically pure intermediate as a useful element of synthetic strategy in natural product syntheses.

Figure 5. Total Synthesis of Anguidine

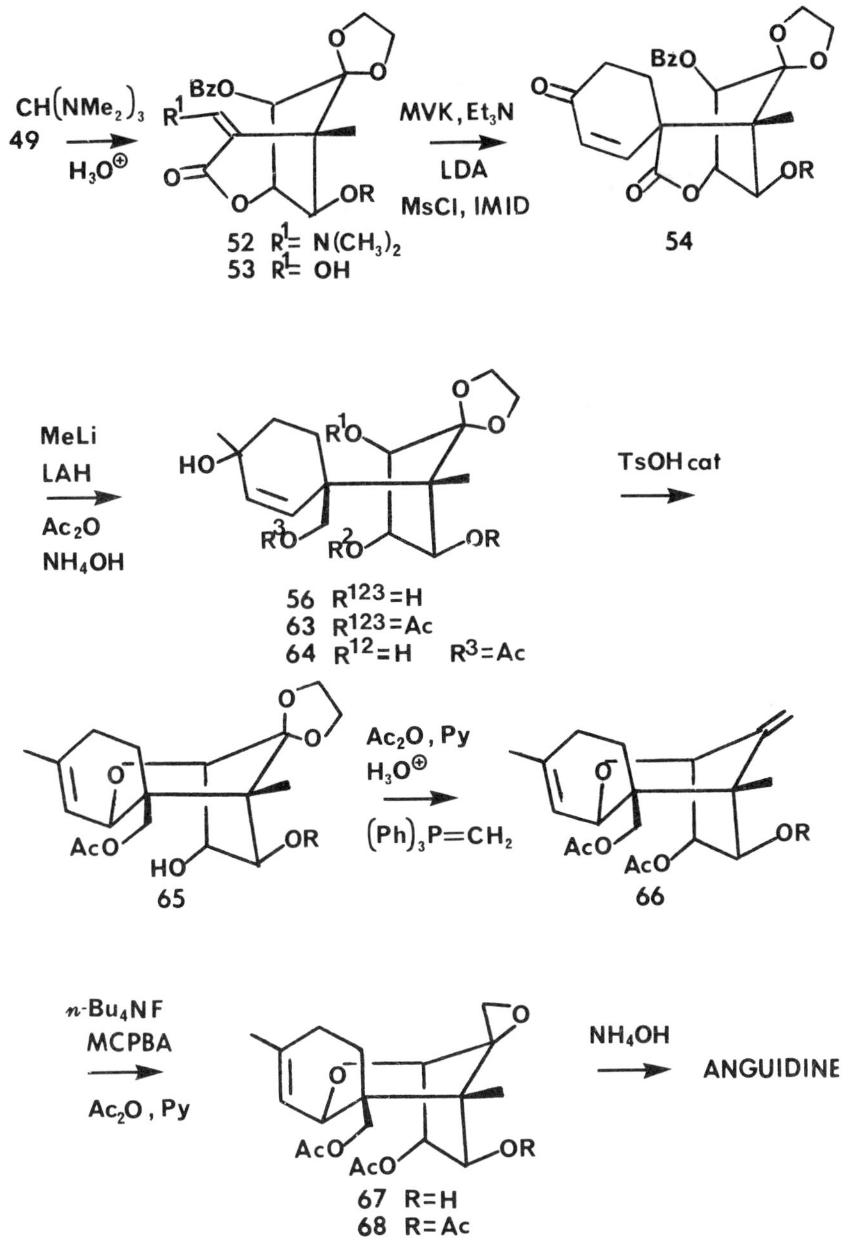

Figure 6. Possible Cyclization Modes for B-Ring Formation

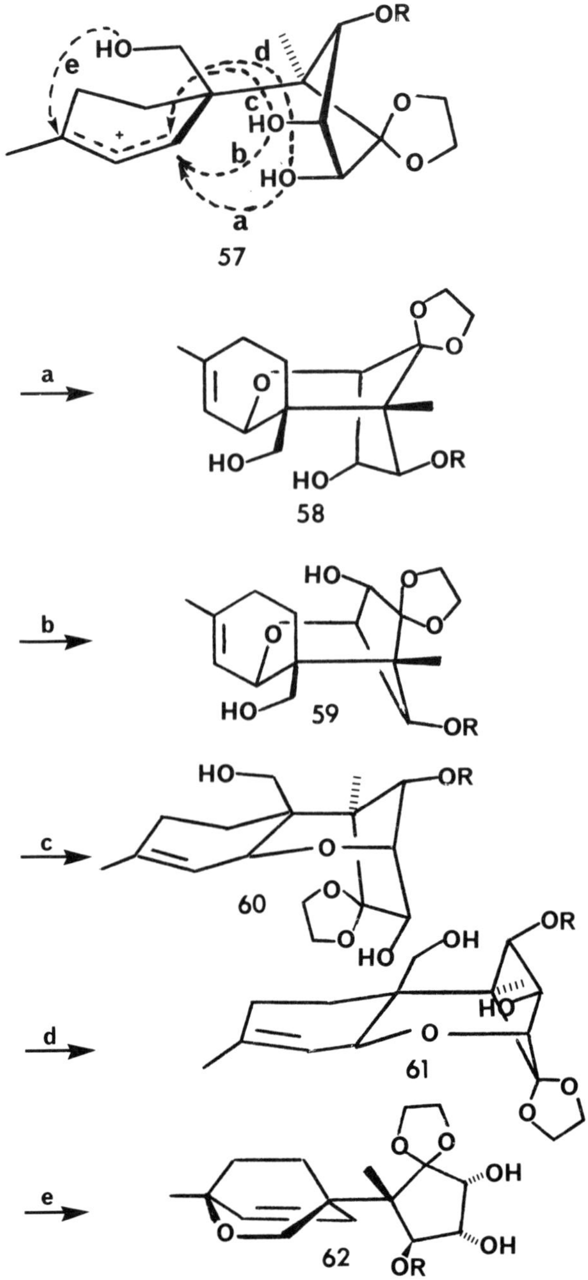

2.3 Yeast Reduction of 2,2-Disubstituted-1,3-Cyclohexanediones

In order to evaluate the generality of this strategy, we further investigated the yeast reduction of an analogous series of substituted cyclohexanediones.(12) The results of the microbial reductions compared to the products of reduction with 1 equiv. of $NaBH_4$ are shown in Figure 7. In all cases the microbial ketol products were found to have >98% enantiomeric purity. Unlike the cyclopentanedione system, the enzymatic versus borohydride reduction, proceeded with different stereoselectivity with respect to the distinction of the 2-substituents. The major diastereomer found in the cyclopentanoid system was [2S,3S] whereas the major diastereomer in the cyclohexanoid system had the [2R,3S] configuration.

The determination of absolute configuration for the chiral cyclohexanoids was established as follows. Ring expansion of the cyclopentanoid mixture of **2,4** (67,33%) of known absolute configuration, with diazomethane and $AlCl_3$ catalysis gave the cycloheptanoids **90;91** (67,33%). Similar ring expansion of the cyclohexanoid mixture **71;73** (45,55%) with 1eq of CH_2N_2 gave **90;91** (42,58%). The allyl mixture **76;78** (45,55%) was correlated to **71;73** (45,55%) by catalytic hydrogenation. The propynyl mixture **81;83** (27,73%) was also correlated to **71;73** (25,75%) in this manner. It is interesting to recognize the asymmetric consistency of the microbial reduction in this series to provide only the [S]-hydroxy configuration.

Figure 7. Reduction of Prochiral 2,2-Disubstituted-1,3-Cyclohexanediones with Bakers' Yeast and $NaBH_4$

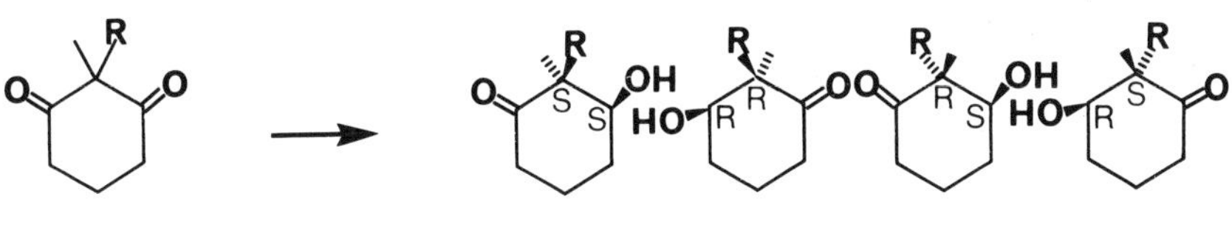

70 $R=CH_2CH_2CH_3$	**71**	**72**	**73**	**74**
75 $R=CH_2CH=CH_2$	**76**	**77**	**78**	**79**
80 $R=CH_2C\equiv CH$	**81**	**82**	**83**	**84**
85 $R=(CH_2)_3C(CH_3)=CH_2$	**86**	**87**	**88**	**89**

entry	dione	reduction method	ketol products (% composition)	yield%	recovered dione%
1	**70**	yeast	**71;73**(22,78)	80	15
2	**70**	$NaBH_4$	**71,72;73,74**(98,2)	70	10
3	**75**	yeast	**76;78**(45,55)	80	15
4	**75**	$NaBH_4$	**76,77;78,79**(83,16)	70	10
5	**80**	yeast	**81;83**(27,73)	75	20
6	**80**	$NaBH_4$	**81,82;83,84**(64,36)	70	15
7	**85**	yeast	**86;88**(24,76)	75	10
8	**85**	$NaBH_4$	**86,87;88,89**(95,5)	70	15

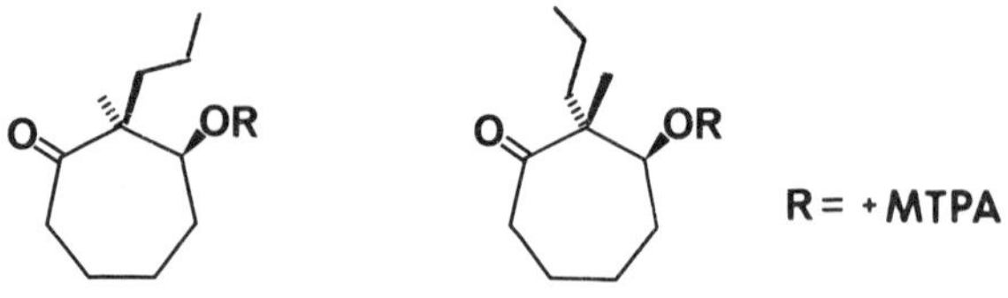

90 **91**

To further demonstrate the application of this methodology as a valuable element of synthetic design, we prepared the chiral cyclohexanoid **98** which closely resembles a racemic intermediate **92** used in a total synthesis of the diterpenoid zoapatanol (**93**).(13) Reduction of dione **85** with yeast provided a mixture of two chiral ketols **86** and **88** (see Figure 7). The absolute structures were assigned by comparison of the rotation and ^{13}C NMR spectra of the ketols and the ^{1}H NMR spectra of the corresponding (+)-MTPA esters with those of the previously assigned cyclohexanoids.

Inspection of zoapatanol (**93**) reveals that the requisite cyclohexanoid precursor should have the [2S,3R] configuration. A series of standard chemical transformations were performed to convert the microbial products **86** and **88** to the requisite [2S,3R] intermediate **97** as outlined in Figure 8. The ketols **86** and **88** were readily separated by chromatography on silica gel. A hydroxyl inversion procedure transformed **86** to the desired configuration providing **94**. The hydroxyl group of **88** was protected as a silyl ether and reduction with $NaBH_4$ gave a mixture of diastereomers **95**;**96** (25,75%) which were separated by chromatography on silica gel. Conversion of **96** to the benzoate, desilylation (Bu_4NF) and oxidation (PCC) provided **97**. Ketalization of **97** followed by hydroboration/oxidation provided **98**, a compound similar to the intermediate **92** used in Kane's synthesis (13) of zoapatanol (**93**). These procedures illustrate a combined microbial-synthetic scheme to access all four chiral products [2S,3S], [2R,3R], [2R,3S], and [2S,3R] from prochiral 2,2-disubstituted-1,3-diones.

Figure 8. Preparation of an Intermediate for the Synthesis of Zoapatanol

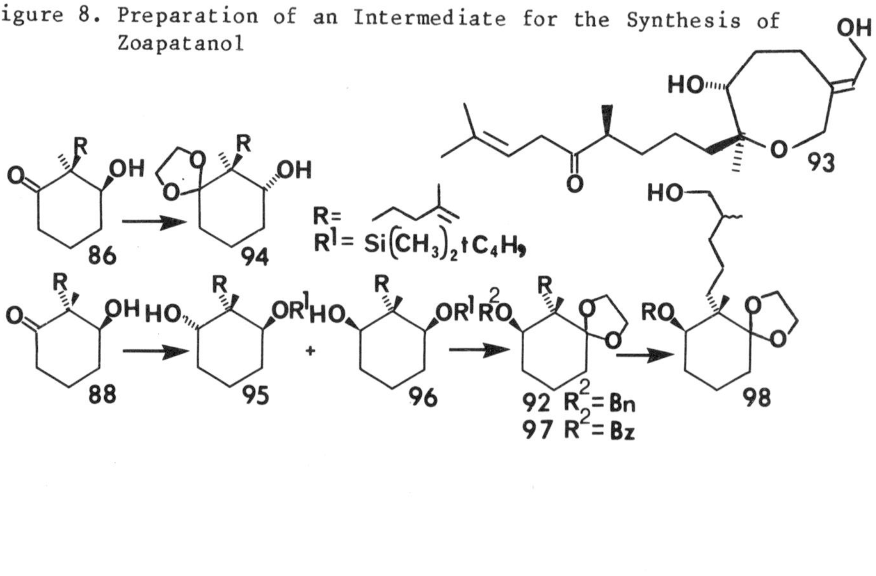

2.4 Yeast Reduction of 2,2-Disubstituted-1,3-Cycloheptanediones

Microbial reduction of a series of medium sized 7-, 8- and 9-membered diones was investigated (14) and the results are summarized in Figure 9. The enantiomeric composition of the chiral ketols produced by yeast reduction was determined by analysis and comparison of the ^{1}H NMR (470MHz) of the corresponding (+)-MTPA esters with those derived from the racemic ketols prepared by reduction with $NaBH_4$. In each case the singlet of the 2-methyl group and the methine proton alpha to the MTPA derivatized hydroxyl group were clearly resolved. In all cases the ketols generated by yeast reduction were found to have >98% enantiomeric purity. The assignment of structure and absolute configuration for the chiral cycloheptanoids was established by comparison of the ^{1}H NMR spectra of the propyl(+)-MTPA esters **90** and **91** of known absolute configuration which were previously reported with the same propyl(+)-MTPA esters derived from catalytic hydrogenation of the (+)-MTPA esters of **116**, **121**, and **123** obtained by yeast reduction. The structure of the microbial ketol products derived from diones **125** and **130** were not proven but assigned on the basis of comparison of their spectral data with that of the analogous (+)-MTPA esters of the correlated products. The medium size rings are not as efficiently reduced by bakers' yeast as the five or six membered diones. Substrates **120** and **125** provide synthetically useful conversions to novel chiral functionalized cycloheptanoids. In each case the ketols derived by yeast reduction were >98% enantiomerically pure regardless of the conversion efficiency. The completely opposite stereoselectivity in the yeast reduction of the propyldione **110** (entry 1) versus the allyldione **115** (entry 3) and the lack of stereoselectivity for the methallyldione **125** (entry 7) is very interesting. The stereoselectivity of the yeast reduction closely parallels that of the $NaBH_4$ reduction of these substrates. Microbial reduction of the cyclooctanedione **130** proceeded slowly to give a poor conversion to ketols **131** and **133**. No ketol product could be isolated from attempted yeast reduction of the cyclononanedione **135** under the conditions employed. Attempts to optimize the fermentation conditions to provide better conversions were not carried out. These results suggest that Bakers' yeast mediated reduction of prochiral 8- and 9-membered rings may not be a viable preparative method. Microbial reduction of these substrates with other microorganisms is another recourse.

Figure 9. Yeast Reduction of Medium Ring Cycloalkanediones

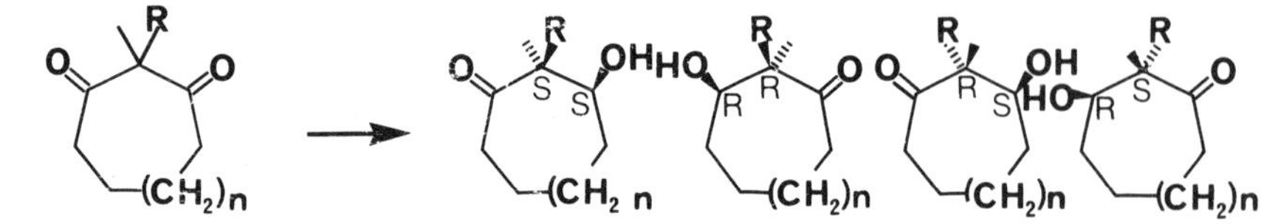

n	Dione	R				
n=1	**110**	$R=CH_2CH_2CH_3$	**111**	**112**	**113**	**114**
n=1	**115**	$R=CH_2CH=CH_2$	**116**	**117**	**118**	**119**
n=1	**120**	$R=CH_2C\equiv CH$	**121**	**122**	**123**	**124**
n=1	**125**	$R=CH_2C(CH_3)=CH_2$	**126**	**127**	**128**	**129**
n=2	**130**	$R=CH_2CH=CH_2$	**131**	**132**	**133**	**134**
n=3	**135**	$R=CH_2CH=CH_2$	**136**	**137**	**138**	**139**

Entry	Dione	Reduction Method	Ketol Products (% composition)	Yield%	Recovered dione %
1	**110**	yeast	**111**;**113**(2,98)	10	75
2	**110**	$NaBH_4$	**111**,**112**;**113**,**114**(2,98)	60	15
3	**115**	yeast	**116**(100)	20	60
4	**115**	$NaBH_4$	**116**,**117**;**118**,**119**(95,5)	65	10
5	**120**	yeast	**121**;**123**(71,29)	60	30
6	**120**	$NaBH_4$	**121**,**122**;**123**,**124**(7,93)	15	15
7	**125**	yeast	**126**;**128**(55,45)	40	50
8	**125**	$NaBH_4$	**126**,**127**;**128**,**129**(53,47)	65	10
9	**130**	yeast	**131**;**133**(82:18)	5	75
10	**130**	$NaBH_4$	**131**,**132**;**133**,**134**(88:12)	60	15
11	**135**	yeast	no ketol product	0	80
12	**135**	$NaBH_4$	**136**,**137**;**138**,**139**(85,15)	60	15

2.5 Substrate Versatility in Yeast Reduction of 1,3-Diones

With the effect of ring size investigated, we next examined the substrate versatility of the yeast reduction for various functionalized substituents at the 2-position.(15) For a series of 5- and 6-membered-1,3-diones containing ester and nitrile groups, microbial reduction proceeded as before to give good yields of optically pure ketols as shown in Figure 10.

Figure 10. Yeast Reduction of Functionalized Cycloalkanediones

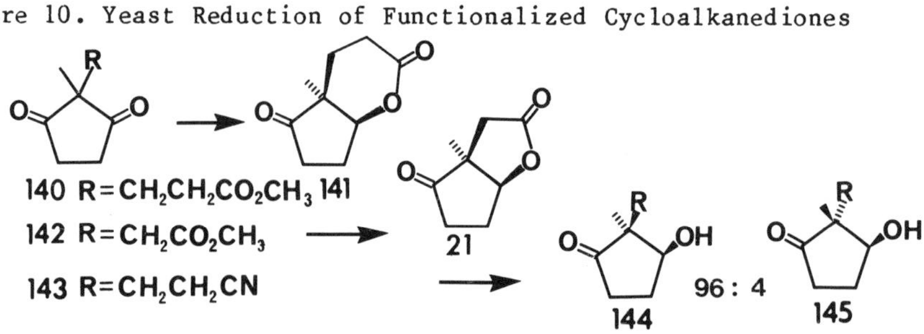

Figure 10. continued

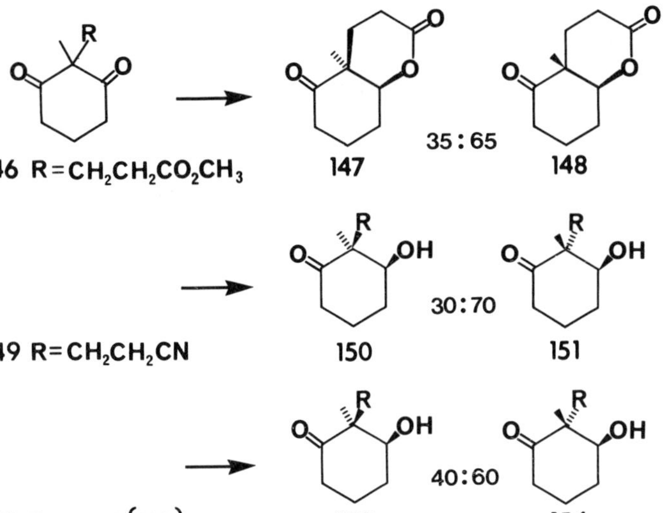

146 R=$CH_2CH_2CO_2CH_3$ → 147 35:65 148

149 R=CH_2CH_2CN → 150 30:70 151

152 R=$CH_2C(CH_3)CH_2$ → 153 40:60 154

Application of the geminal acylation reaction developed by Kuwajima and coworkers (16) provided access to several interesting 2,2-disubstituted-1,3-cyclopentanediones. The spirofused diones **155** and **157** and the 2-aryldione **159** were efficiently reduced by yeast to ketols with >98% enantiomeric purity.(17) Therefore this microbial reduction appears to be very tolerant to varied 2-substitution.

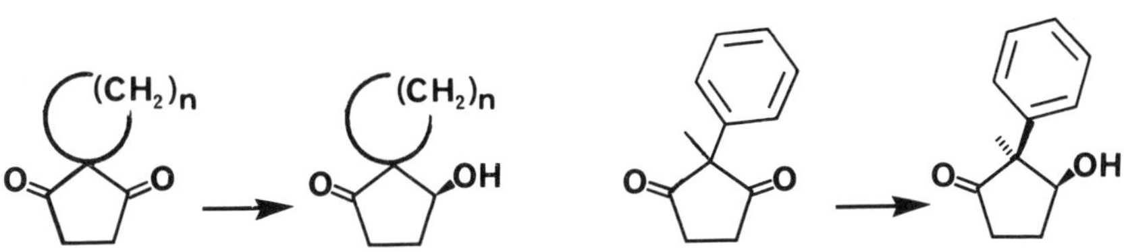

155 n=6 → 156; 157 n=7 → 158; 159 → 160

Yeast reduction of two acyclic substrates **161** and **164** also provided diastereomeric mixtures of ketols with >98% enantiomeric purity as shown.(18)

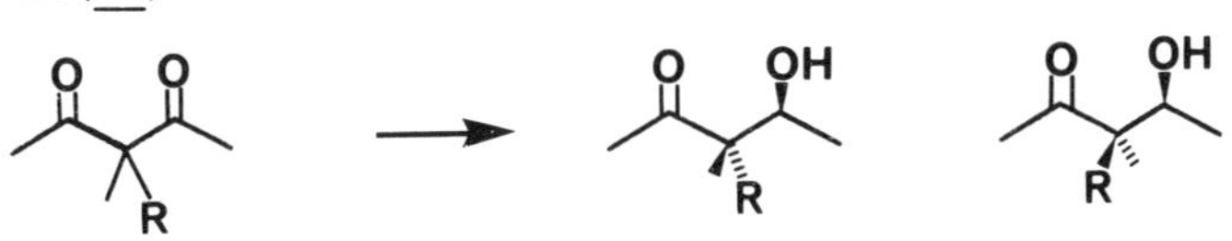

161 R=$CH_2CH{=}CH_2$ → 162 73 : 27 163

164 R=$CH_2C(CH_3){=}CH_2$ → 165 75 : 25 166

From these results we have demonstrated that yeast reduction of 2,2-disubstituted-1,3-diones is a useful and versatile method which can be integrated into synthetic strategies for enantiospecific syntheses of natural products. Methods to convert a given diastereomeric ketol derived from microbial reduction to its epimer or enantiomeric antipode have been illustrated. We have further observed that access to novel chiral acyclic intermediates like compound **B** are readily obtained by a Baeyer-Villiger oxidation of chiral ketols such as **A** with peracid followed by reduction of the resulting lactone. Further transformations of chiral compounds like **A** to incorporate them into synthetic schemes awaits the ingenuity of synthetic chemists.

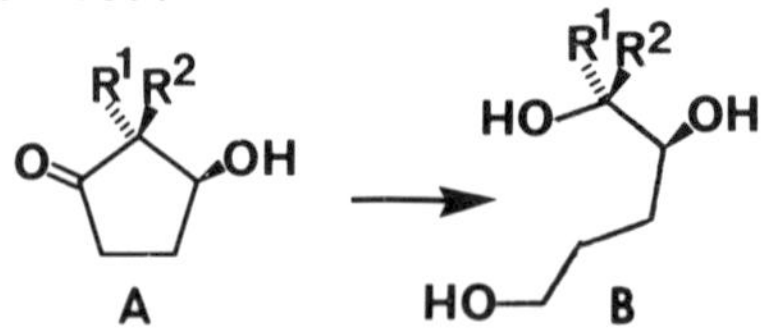

The strategy that has been presented could be further expanded by designing substrates with **n** prochiral elements such as compound X (n=7). Asymmetric reduction by yeast of one carbonyl could then generate a product with **n-1** chiral centers.

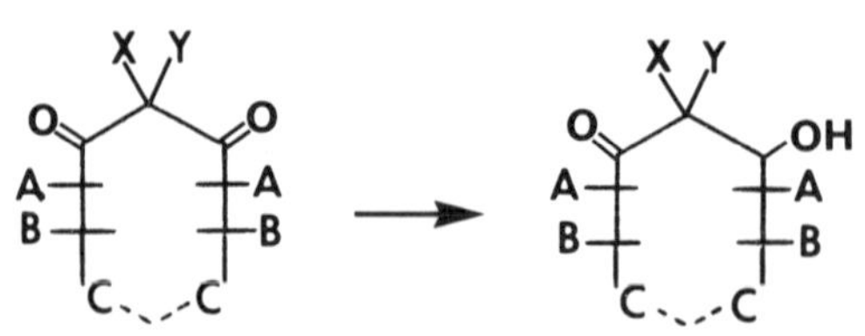

3. Microbial Reduction of Beta-Ketoesters

3.1 Remote Substituent Effects in Microbial Reduction of 3-Ketoglutarate and 3-Ketoadipate Esters

Recently several studies (<u>19</u>) have illustrated interesting results concerning the effect of relatively remote substituents from the carbonyl group on the enantiomeric outcome of yeast mediated reductions of ß-ketoesters. We investigated the yeast reduction of a series of prochiral 3-oxoglutarate and 3-oxoadipate esters and observed that simple differences in the remote ester functionality influenced the enantioselectivity observed.(<u>20</u>) The enantiomeric composition of the hydroxyesters was determined by analysis and comparison of the ^{19}F NMR (188MHz) spectra of the corresponding (+)-MTPA esters with those derived from the racemic ketols prepared by reduction with $NaBH_4$. In each case the diastereomeric CF_3 signals were clearly resolved and integrated to provide the results shown in Figure 11.

Figure 11. Yeast Mediated Reduction of Prochiral 3-Oxoglutarate Esters and 3-Oxoadipate Esters

C D

R= (+)-$OCC(CH_3)(CF_3)C_6H_5$

Entry	Substrate	Y	Z	Product	diastereomeric composition%	yield%
1	**171**	OCH_2CH_3	OCH_3	**178**	50:50	20
2	**172**	$OC(CH_3)_3$	$OCH_2C_6H_5$	**179**	50:50	20
3	**173**	$OC(CH_3)_3$	OCH_2CF_3	**180**	50:50	10
4	**174**	$OC(CH_3)_3$	OCH_3	**181**	62:38	36
5	**175**	OCH_2CH_3	SCH_2CH_3	**182**	65:40	10
6	**176**	OCH_3	$SC(CH_3)_3$	**183**	57:43	8
7	**177**	$OC(CH_3)_3$	SCH_2CH_3	**184**	72:28	32

E F

Entry	Substrate	Y	Z	Product	S:R	yield%
8	**185**	OCH_3	OCH_3	**191**	74:26	19
9	**186**	$OCH_2C_6H_5$	OCH_3	**192**	51:49	24
10	**187**	OCH_3	SCH_2CH_3	**193**	52:48	56
11	**188**	$OC(CH_3)_3$	OCH_3	**194**	65:35	26
12	**189**	OCH_3	$OC(CH_3)_3$	**195**	83:17	29
13	**190**	$OCH(CH_3)_3$	SCH_2CH_3	**196**	92:8	26

The isolated yield represents the amount of 3-hydroxyester isolated after a 24 h reaction time. A mass balance of about 80% was obtained in each case by the sum of 3-hydroxyester and recovered starting material. The absolute configuration of the predominate enantiomer in the products **194** and **196** was established as 3-[S] by a correlation sequence outlined in Figure 12. Chymotrypsin catalyzed hydrolysis of dimethyl 3-hydroxyglutarate provided the 3-**R** half-acid ester **198** (21) which was acetylated and then homologated by a standard Arndt-Eistert reaction. Reduction with $LiAlH_4$ and acetylation provided 3-[R]-1,3,6-hexanetriol triacetate (**200**); $\alpha = -8.0°$, c=2, $CHCl_3$, as a reference compound. The yeast products **194** and **196** were reduced with $LiAlH_4$ and acetylation provided the corresponding triacetates with $\alpha = +4.0°$ and + 11.4° respectively. The ^{19}F chemical shift of the CF_3 group in the predominate (+)MTBA diastereomer for products **191-196** was consistently upfield and this observation was used as the basis for assigning the 3-[S] configuration to the major enantiomer of the uncorrelated 3-hydroxyadipate product mixtures.

Figure 12. Correlation of 3-Hydroxyadipate Products

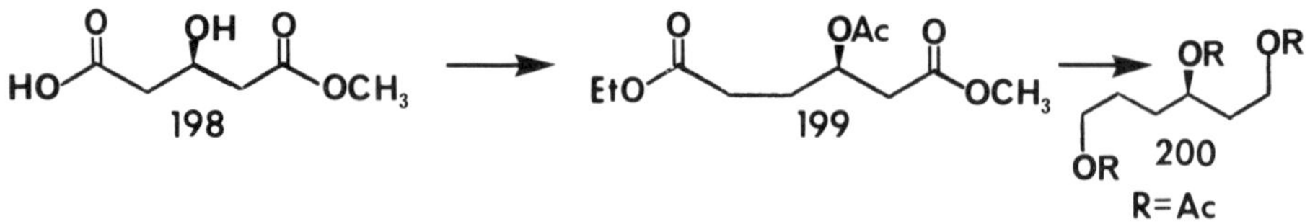

The results of the yeast reductions of prochiral 3-oxoglutarates **C** reveals that, in general, the enzyme reducing system does not readily differenciate small differences in the size of the ester group (entry 1), the presence of an aromatic ring (entry 2), or a trifluoromethyl group (entry 3). It is interesting to note that a distinction between an ethyl ester versus a ethanethiol ester (entry 5) is similar to that of a tert-butyl versus a methyl ester (entry 4). The best distinction of this set of substrates was provided by a ethanethiol ester versus a tert-butyl ester (entry 7).

The symmetry of this substrate system provides a good test of the ability of the yeast reducing system to distinguish differences in substituents remote from the carbonyl group. These results provide support for the concept of developing further refinements in adjusting groups Y and Z such that optical purities approaching 100% might be obtained. Also, if Y and Z are designed as in the case for entry 7, to allow selective reaction at either ester site specifically, an **R**- or **S**-3-hydroxyglutarate synthon from the same substrate would be available as a useful precursor in natural product synthesis.

For the 3-ketoadipate substrates **E**, the dimethyl ester (entry 8) might be considered as an starting point for comparison of the effect of changes in the ester groups. In some cases, entries 9, 10, and 11 the distinction became worse. Entries 11 and 12 show that the effect of the tert-butyl ester is greater when it is beta to the carbonyl. Entry 13 provides a synthetically useful 3-hydroxyadipate **196** of 84% ee that can be selectively reacted at either ester group. These results illustrate that simple changes in the ester group of ß-ketoesters can be used to control the enantioselectivity of yeast mediated reductions.

3.2 Yeast Reduction of 5-Alkoxy-3-oxopentanoate Esters

Beta-ketoesters appear to be a rather general substrate for yeast mediated reduction providing access to 3-hydroxyester products of variable optical purity. This methodology would be generally more useful and appealing to synthetic chemists if there was a simple means to control or optimize the enantioselectivity of microbial reductions. After examining the ability of bakers' yeast to distinguish differences in the ester groups of prochiral glutarate esters a logical extension work was to study yeast mediated reduction of 5-alkoxy-3-oxopentanoate esters where the choice of ester group and/or the 5-alkoxy group might be used to "fine tune" the enantioselectivity of the enzymatic reducing system at the carbonyl group to provide an optically pure 5-alkoxy-3-hydroxyester. A series of such substrates were examined and the results of this study clearly demonstrate the ability to optimize the enantioselective outcome of yeast mediated reductions of beta-ketoesters by simple changes of the ester group employed. Furthermore, the optically active 5-alkoxy-3-hydroxyester produced by microbial reduction can serve as a useful intermediate in asymmetric syntheses of natural products as either an [R]- or [S]-1,3,5-trihydroxypentane synthon, as will be briefly illustrated in the following discussion.

During our investigation two significant reports on a similar theme appeared. Hirama and coworkers studied 6-alkoxy-3-oxohexanoates and found that yeast mediated reduction of the potassium carboxylate salt provided complete enantioselectivity.(19a) Sih and coworkers discovered an amazing reversal of enantioselectivity from [S](65%) to [R](100%) in yeast mediated reduction of 4-chloro-3-oxopentanoates influenced by the size of a homologous series of C1 to C12 n-alkylesters.(19b) Also, evidence for the existence of at least two active oxidoreductases in yeast for the 4-chloro-3-oxopentanoate substrates was provided.

The results of yeast mediated reduction of the ß-ketoesters **201-211** to provide the corresponding 5-alkoxy-3-hydroxyesters **212-222** are summarized in Figure 13. Reduction of the ß-ketoesters with $NaBH_4$ provided the corresponding racemic hydroxyesters for comparison. The ß-hydroxy esters, both chiral and racemic, were then converted to the corresponding (-)-MTPA esters and the enantiomeric purity was determined by analysis of ^{1}H and ^{19}F NMR spectra.

The absolute configuration of the optically active products was determined by conversion of the yeast reduction product **215** to 3-[S]-silyloxyvalerolactone **224** (α = -10.2°) by the reaction sequence summarized in Figure 14. 3-[R]-silyloxyvalerolactone **226** (α = + 9.2°) was prepared from **225** with known [R] configuration, derived from chymotrypsin hydrolysis of dimethyl 3-hydroxyglutarate (21) and subsequent silyl ether protection. The microbial products were then more efficiently correlated by reduction with $LiBH_4$ to the chiral differentially protected 1,3,5-trihydroxy synthon **227**. By appropriate manipulation this chiral intermediate can be readily incorporated into enantioselective synthetic schemes, particularly for acetogenin type natural products, as a key [R] or [S] 1,3,5-trihydroxy or 3-hydroxyglutarate synthon.

The major enantiomer produced by the microbial reduction was consistently of the [S] configuration for the examples studied. No reversal of enantioselectivity was observed as reported by Sih and coworkers for 4-chloro-3-oxopentanoates.(19b) The effect of substrate concentration on the enantioselectivity of the yeast reduction of **206** was investigated at four concentrations; 0.056M,

0.028M (the standard concentration used for the results listed in Figure 13), 0.007M, and 0.0018M. The enantiomeric composition remained constant at 95[S]:5[R] +/-2%. The potassium carboxylate salt **201** with R=OK) was reduced in poor yield (<10%) to the corresponding hydroxyacid.

Figure 13. Yeast Reduction of 5-Alkoxy-3-oxopentanoates

Entry	Substrate	R	Enantiomeric Composition	Product	Yield%	Configuration
1	**201**	OCH_3	65:35	**212**	65	-
2	**202**	OCH_2CH_3	78:22	**213**	60	-
3	**203**	$OCH(CH_3)_2$	67:33	**214**	40	-
4	**204**	$OCH_2CH_2CH_2CH_3$	85:18	**215**	55	S
5	**205**	$SCH_2CH_2CH_2CH_3$	84:16	**216**	40	-
6	**206**	$OCH_2CH(CH_3)_2$	95:05	**217**	70	S
7	**207**	$SCH_2CH(CH_3)_2$	87:13	**218**	30	-
8	**208**	$OCH_2(CH_2)_3CH_3$	96:04	**219**	40	S
9	**209**	$OCH_2CH_2CH(CH_3)_2$	77:23	**220**	35	-
10	**210**	$OCH_2C(CH_3)_3$	98:02	**221**	35	S
11	**211**	$OCH_2(CH_2)_6CH_3$	-----	**222**	0	-

In this work the enantiomeric excess was influenced by small changes in the ester hydrocarbon group. Longer hydrocarbon chains appeared to give better enantioselectivity but conversions became poor for n-hexyl (10%) and n-octyl esters (0%). Branching as in i-propyl **203** and i-amyl **220** resulted in poor ee whereas branching as in i-butyl **206** and neopentyl **210** caused a dramatic increase in ee. A change of the electronic environment with thiol esters n-butyl S **205** or i-butyl S **207** had little effect compared to the corresponding O-esters.

This study clearly demonstrates the utility of adjusting substrate substituents in a rational manner to obtain improved enantiomeric excess. Using this approach we realized an increase in the enantiomeric selectivity of the yeast reduction from 65:35 for the methyl ester **201** to > 98:2 with the neopentyl ester **210**.

Figure 14. Correlation of 5-Benzyloxy-3-hydroxypentanoates

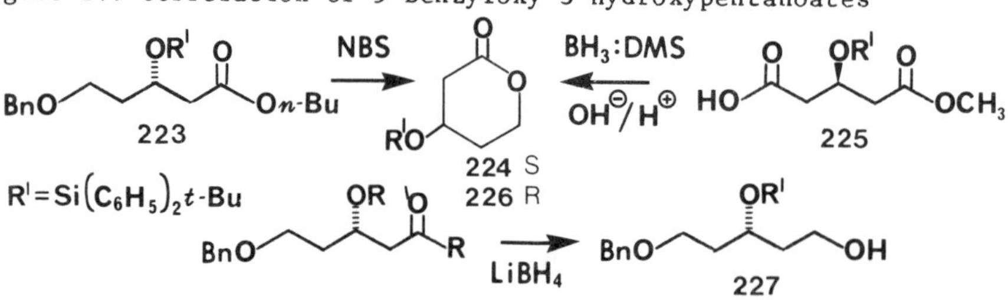

3.3 Synthetic Studies of Polyene Macrolides

The impetus for the study of microbial reduction of the ß-ketoester systems mentioned was to generate appropriate intermediates of high enantiomeric purity which could be incorporated into a general synthetic strategy for the polyene macrolides. The polyene macrolide antibiotics are endowed with a unique structure composed of a highly functionalized macrocyclic lactone which incorporates opposed lipophilic polyene and hydrophilic polyol carbon chains and in some cases contain a free carboxyl group and an aminoglycosyl residue. They were discovered as products of microbial fermentation of the large Streptomyces genus in the 1950's.(22) Early studies were hindered by the difficulties of purification at that time and by the inherent instability of these compounds to various conditions including heat, light, mild acid, and mild base. However, brilliant degradative studies elucidated the major structural features. The most widely studied member is amphotericin B (**230**), whose absolute structure has been fully established by X-ray crystallography.(23) The structures of other members such as pimaricin **231** are based mainly on chemical degradation and mass spectral studies.(24)

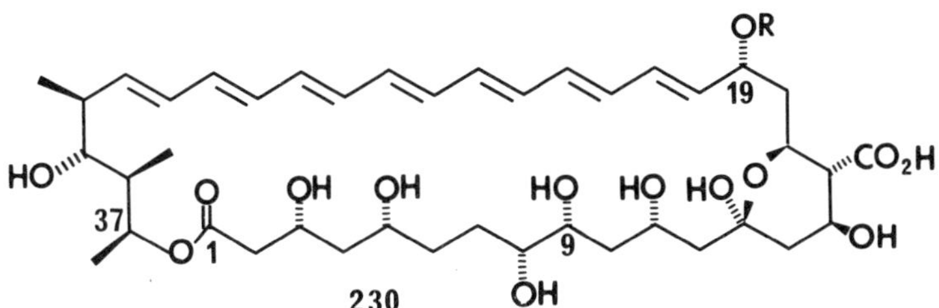

230

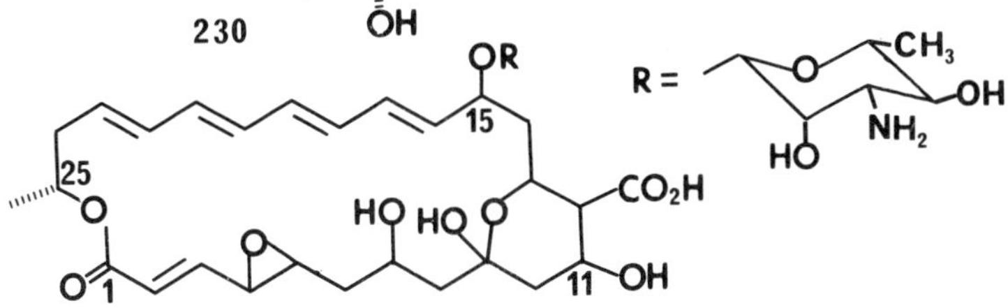

231

Several synthetic studies toward this complex group of natural products have been reported.(25) Indeed, they are the last remaining group of known naturally occuring macrolides which have eluded total synthesis. They provide a formidable challenge to presently available synthetic methods for the following reasons; 1) the reactive nature of the functionality present in these natural products, 2) the large size of the macrolide implies a lengthy synthetic process, and 3) the numerous chiral centers incorporated in one large cyclic compound. These factors obligate a highly convergent synthetic strategy with careful orchestration of specific and stereoselective reactions.

A convergent synthetic plan for polyene macrolides which we initiated is illustrated for amphotericin B (**230**) in Figure 15, where imaginary fragmentation of **230** provides five main components which formed the focus of our synthetic studies. We completed an enantioelective synthesis of fragment **C** from ethyl 3[S]-hydroxybutanoate.(25b) The preparation of fragment **B** was accomplished as outlined in Figure 16 from the chymotrypsin derived intermediate **225**.(26) Reduction of the carboxyl group and oxidation gave the aldehyde **232** which was treated with an allyl titanate complex to give a diastereomeric mixture of **234;235**(2:1)in good yield. Several other nucleophilic allyl species were studied but gave increasing amounts of lactone products or poorer yields. Oxidation of the mixture of alcohols followed by reduction with $NaBH_4$ in the presence of $CeCl_3$ gave a 10:1 ratio of **234;235** and the hydroxy group was protected as the silyl ether **237**. Hydroboration/oxidation provided the desired fragment **B** intermediate **238**. The structure **238** was proven by conversion to the tetraacetate **239** and observed to be the same tetraacetate by ^{13}C and optical rotation as that derived from a degradation sequence of amphotericin B (shown in Figure 18).

Figure 15. Imaginary Fragmentation of Amphotericin B

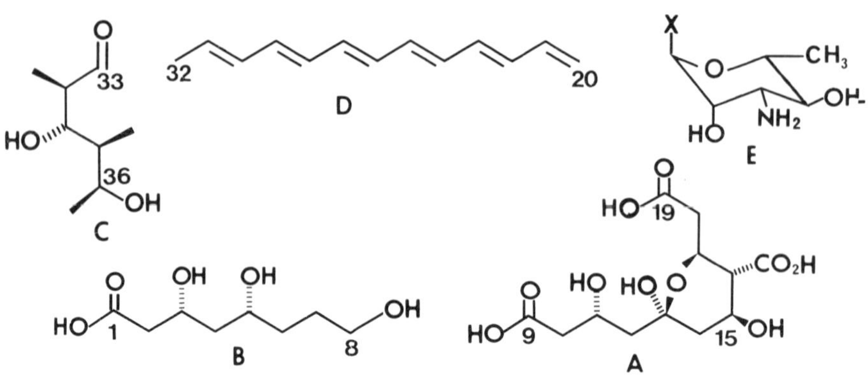

Figure 16. Preparation of Fragment **B**

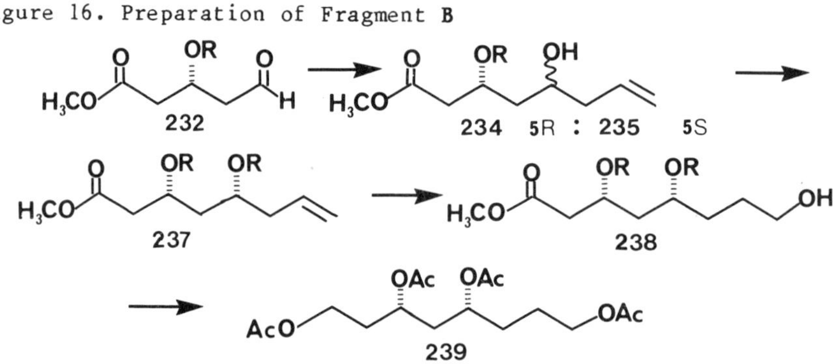

Fragment **A** is a two carbon extension of a C11-19 skeletal unit which is found in most of the polyene macrolides which have been isolated. Close examination of this key fragment reveals a duplicate retro-synthetic element, 3-hydroxyglutaric acid (**232**), when the carbon bonds C13-14 and C15-16 are fictiously oxidatively cleaved. An efficient strategy to synthesize fragment A exploiting this apparent symmetry would involve the combination of two chiral derivatives of **232** and a suitable acetate fragment representing C14-15, as shown.

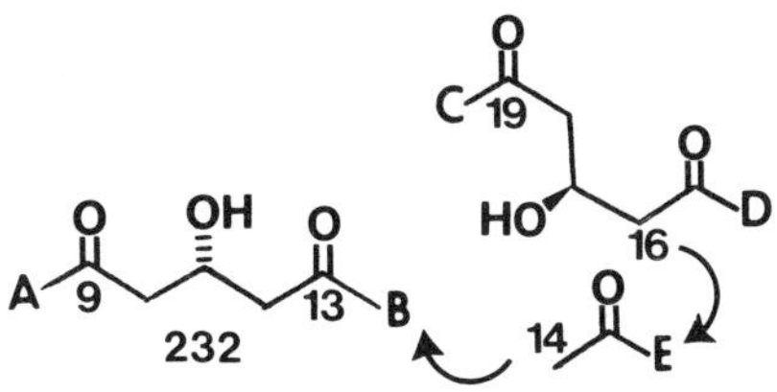

The synthesis of fragment **A** was approached as outlined in Figure 17. Chymotrypsin catalysed hydrolysis of dimethyl 3-hydroxyglutarate provided the 3[R]-hydroxymonoacid **198** which by a silylation protection sequence followed by selective deprotection of the silylester gave **225** which was smoothly C-acylated, by a mild biomimetic method,(27) to provide the two carbon extended unit **240**. Other methods of acylation involving strongly basic or acidic conditions do not proceed as well in this case. Our method of C-acylation provides a key solution to a high yield acetate-type extension of polyfunctionalized acyclic carbon chains. The carbonyl group was protected as an ethylene acetal, the benzyl ester was hydrogenated, reduction or the carboxyl group and protection of the hydroxyl group as a t-butyldimethylsilyl ether gave **241**. Reduction of the ester with $LiBH_4$ and subsequent acetylation provided **244**. The silyl ether was cleaved and the hydroxyl group was oxidized to provide the aldehyde **245** representing carbons 9 - 15 of amphotericin B.

The yeast product **217** served as a convenient 3[S]-hydroxyglutarate synthon which was converted to the corresponding p-nitrobenzoate and methanolysis provided the methyl ester **246**. Formation of the dianion of **246** with two equivalents of lithium diisopropylamide at -78°C and condensation with the aldehyde **245** provided a mixture of two diastereomeric adducts **247**;**248**(2:1). The assignment of structure is based on the known stereochemical preference for condensations of lithiodianions of 3-hydroxyesters.(28) In order to verify this assignment of structure the mixture of **247**:**248** was converted to the heptaacetate mixture **249** consisting of four diastereomers. We attempted to obtain a similar heptaacetate mixture **251**, consisting of two diastereomers, by a degradation sequence on amphotericin B (shown in Figure 18) in order to establish the structure of the synthetic material. However we were unable to isolate any pure material representing a heptaacetate **251** from amphotericin B.

Figure 17. Synthetic Scheme for Fragment A

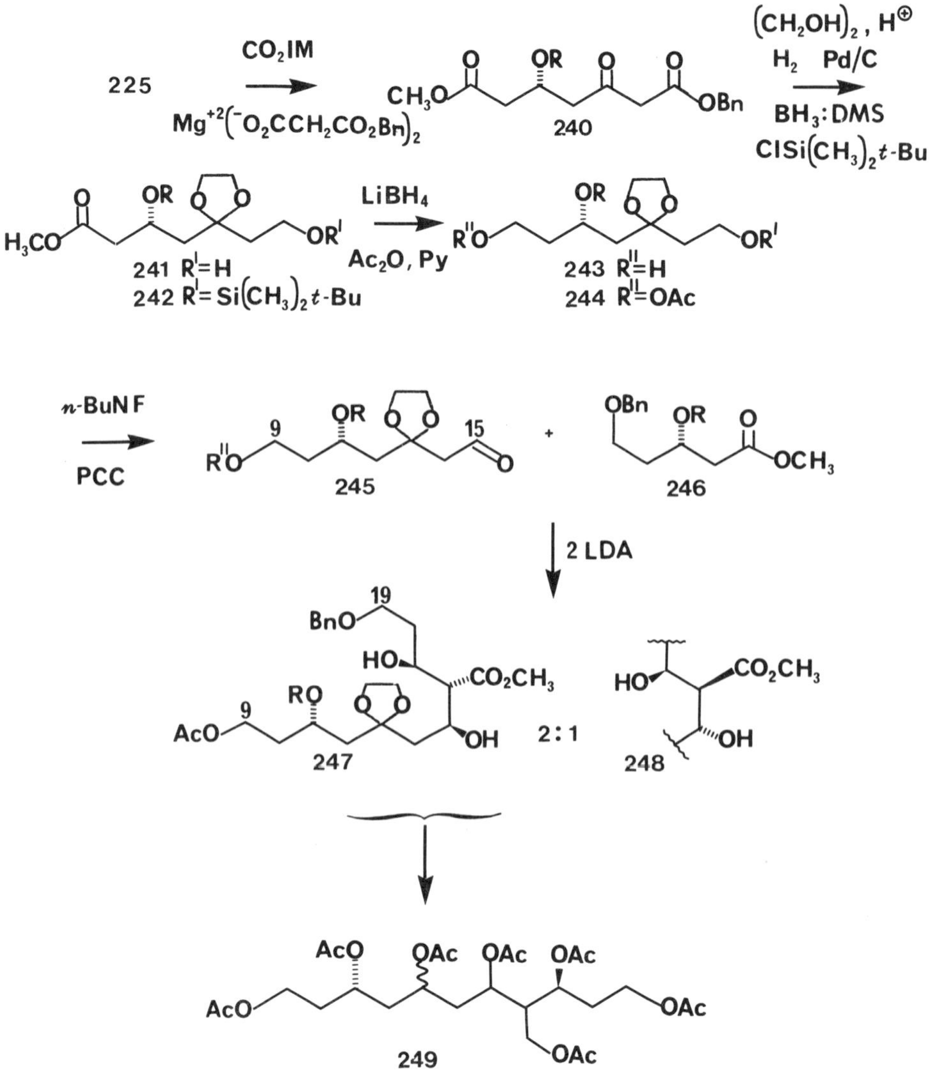

Figure 18. Degradation of Amphotericin B

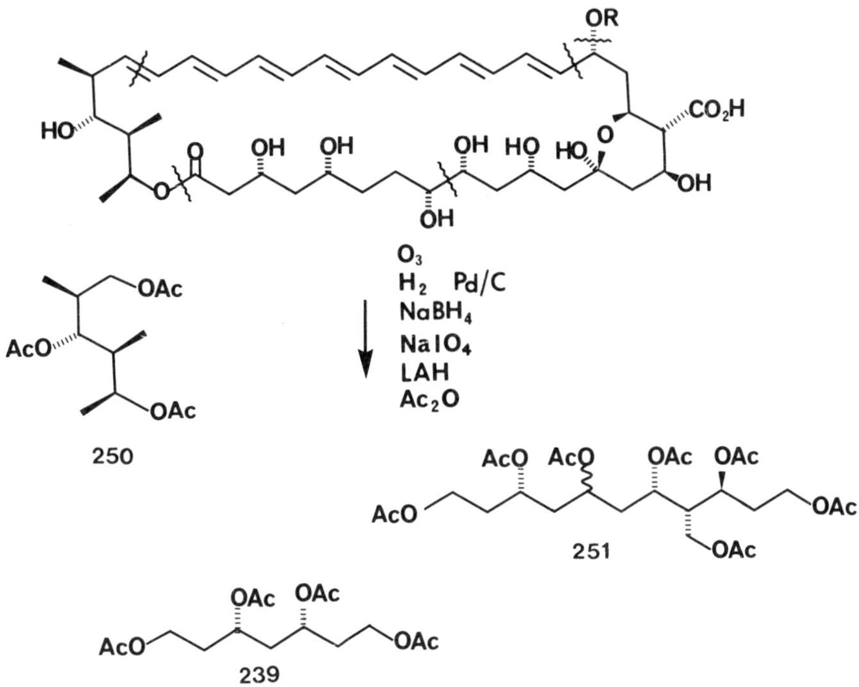

With several chiral fragments available we embarked on a synthetic assembly of the aglycone of pimaricin as shown in Figure 19.(29) The polyene ester unit **255** was prepared by a Wittig reaction of aldehyde **253** derived from 3[R]-hydroxybutanoate with the phosphonium salt **254**. The chiral unit **243** previously described (Figure 17) was oxidized to the aldehyde **256** and subjected to a Wittig-Horner condensation with the phosponate **257** to provide the ester **258** which was carefully hydrolysed to the acid **259**. Coupling of fragments **255** and **259** using dicyclohexylcarbodiimide in the presence of dimethylaminopyridine gave the desired ester **260**. The t-butyldimethylsilylether group was selectively cleaved with (HF)n pyridine and the hydroxyl group was oxidized to the unstable aldehyde **261**. Condensation of **261** with the dianion derived from **246** by treatment with two equivalents of LDA at -78°C gave a mixture of two diastereomers in a 2:1 ratio which were assigned as **262** by analysis of the 470MHz ^{1} NMR spectrum.

Figure 19. Synthetic Approach to Pimaricin

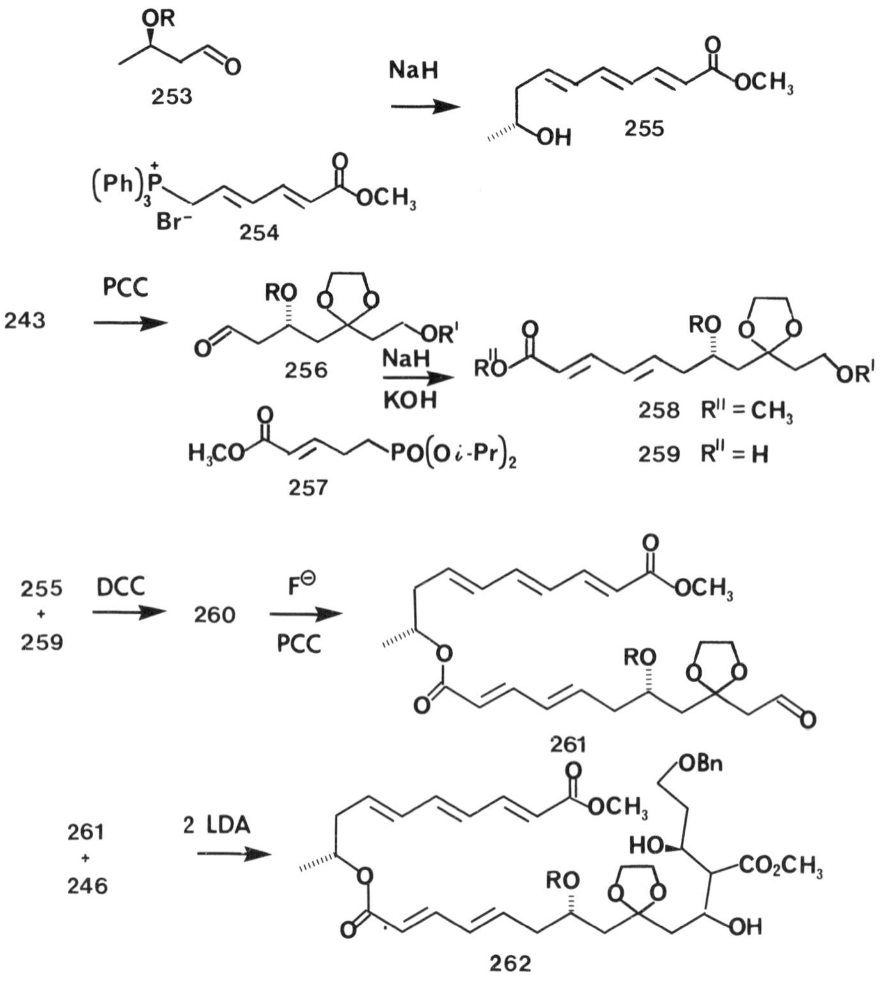

We have demonstrated a few applications of enzyme catalysed transformations as an integral part of the strategy for enantiospecific syntheses of natural products. Increasing numbers of additional applications and new developments are very likely to be forthcoming as more investigators are attracted to the exciting research endeavor of exploring the use of enzymes in organic synthesis.

Acknowledgements

I thank my collaborators; Drs. P. Grothaus, J. T. Palmer, R. Kellogg, H. Bevinakatti, C. Cooper; and H. Mazdiyasni, N. Castro de Lee, M. Wilson, and V. Camara for their contributions. Financial support of this research was provided by The Public Health Service (CA 31987), Showalter Trust Fund, Petroleum Research Fund, and Research Corporation. The excellent facilities and support offered by the Department of Chemistry, Purdue University, West Lafayette, Indiana, U.S.A. during the period of this research endeavor is appreciated.

References

1. For a general overview refer to the following: a. "Enzymes in Organic Synthesis", Ciba Foundation Symposium 111, Pitman: London, 1985; b. Whitesides, G. M.; Wong, C.-H. Aldrichimica Acta 1983, **16**, 27; c. Findeis, M. A.; Whitesides, G. M. Annu. Rep. Med. Chem. 1984, **19**, 263; d. "Applications of Biochemical Systems in Organic Chemistry", Jones, J. B.; Sih, C. J.; Perlman, D., Eds.; Wiley: New York, 1976, Part I and II.

2.a. Neuberg, C. Adv. Carbohydr. Chem. 1949, **4**, 75; b. MacLeod, R.; Prosser, H.; Finkentscher, L.; Lanyi, J.; Mosher, H. S. Biochemistry 1964, **3**, 838, c. Kieslich, K. Synthesis 1969, 147; d. Deol, B. S.; Ridley, D. D.; Simpson, G. W. Aust. J. Chem. 1976, **29**, 2459; e. Sih, C. J.; Chen, C.-S. Angew. Chem. Int. Ed. Engl. 1984, **23**, 570.

3.a. Kieslich, K. "Microbial Transformations of Non-Steroid Cyclic Compounds"; Thieme: Stuggart, 1976, p29 and references cited; b. Gibian, H.; Kieslich, K.; Koch, H. J.; Kosmol, H.; Rufor, C.; Schroder, E.; Vossing, R. Tetrahedron Lett. 1966, 2321.

4. Brooks, D.W.; Grothaus, P. G.; Irwin, W. L. J. Org. Chem. 1982, 47, 2820. The structures for compounds 3, 6, and 9 reported in this paper were incorrectly shown and should be the corresponding enantiomers with the [2R,3S] configuration.

5. Dale, J.A.; Dull, D.L.; Mosher, H. S. J. Org. Chem. 1969, **34**, 2543.

6. A control experiment established a practical limit of detection of 1.5% of a MTPA diastereomer in a mixture by ^{1}H NMR at 470 MHz.

7. Schwarz, S.; Carl, C.; Schick, H. Z. Chem. 1978, **18**, 401.

8. Schwarz,S.; Truckenbrodt, G.; Meyer, M.; Zepter, R.; Weber, G.; Carl, C.; Wentzke, M. J. Prakt. Chemie. 1981, **323**, 729

9.a. Trost, B. M.; Curran, D. P. J. Am. Chem. Soc. 1980, **102**, 5699; b. Tetrahedron Lett. 1981, 4929; c. J. Am. Chem. Soc. 1981, **103**, 7380.

10. Reviews: a. Tamm, C. Fortschr. Chem. Org. Naturst. 1974, **31**, 63; b. Bamberg, J. R.; Strong, F. M. In "Microbial Toxins"; Kadis, S., Ed.; Academic Press: New York, 1973; Vol.3, pp 207-292; c. Doyle, T. W.; Bradner, W.T. In "Anticancer Agents Based on Natural Product Models"; Cassidy, J.M.; Douros, J., Eds.; Academic Press: New York, 1980, Chapter 2; d. Isolation and structure of anguidine: Dawkins, A. W. J. Chem. Soc. C 1966, 116.
11. Several total syntheses of trichothecenes have been reported: a. Colvin, E. W.; Malchenko, S.; Raphael, R. A.; Roberts, J. S. J. Chem. Soc., Perkin I 1973, 1989; b. Still, W. C.; Tasi, M. Y. J. Am. Chem. Soc. 1980, **102**, 3654; c. Schlessinger, R. H.; Nugent, R. A. J. Am. Chem. Soc. 1982, **104**, 1116; d. Kraus, G. A.; Roth, B.; Frazier, K.; Shimagaki, M. J. Am. Chem. Soc. 1982, **104**, 1114; e. Trost, B. M.; McDougal, P. G. J. Am. Chem. Soc. 1982, **104**, 6110; f. Roush, W. R.; D'Ambra, T. E. J. Am. Chem. Soc. 1983, **105**, 1058; g. Brooks, D. W.; Grothaus, P. G.; Mazdiyasni, H. J. Am. Chem. Soc. 1983, **105**, 4472; h. Grothaus, P. G., Ph. D. Thesis, Purdue University, 1984. Other pertinent synthetic studies are referenced in the above reports.
12. Brooks, D. W.; Mazdiyasni, H.; Chakrabarti, S. Tetrahedron Lett. 1984, 1241.
13. Kane, V. V.; Doyle, D. L. Tetrahedron Letters 1981, 3027, 3031.
14. Brooks, D. W.; Mazdiyasni, H.; Sallay, P. J. Org. Chem. 1985, in press.
15. Mazdiyasni, H., M.Sc. Thesis, Purdue University, 1984.
16. Shimada, J.-I.; Hashimoto, K.; Kim, B. H.; Nakamura, E.; Kuwajima, I. J. Am. Chem. Soc. 1984, **106**, 1759.
17. Wilson, M., M.Sc. Thesis, Purdue University, 1985.
18. Camara, V., M.Sc. Thesis, Purdue University, 1985.
19.a. Hirama, M.; Shimizu, M.; Iwashita, M. J. Chem. Soc., Chem. Commun. 1983, 599; b. Zhou, B.; Gopalan, A. S.; Van Middlesworth, F.; Shieh, W.; Sih, C. J. J. Am. Chem. Soc. 1983, **105**, 5925.
20. Brooks, D. W.; Castro de Lee, N.; Peevey, R. Tetrahedron Lett. 1984, 4623.
21. Cohen, S. G.; Khedouri, E. J. Am. Chem. Soc. 1961, **83**, 4228.
22.a. For a historical account of the discovery of the first polyene macrolide see: Baldwin, R. S. " The Fungus Fighters ", Cornell University Press, 1981, Ithaca, N. Y.; b. For a review of the early isolation and structural studies see: Oroshnik, W.; Mebane, A. D. Fortschritte d. Chem. Org. Naturst. 1963, **21**, 17; c. A more recent review of this area is provided by: Hamilton-Miller, J. M. T. Bact. Rev. 1973, **37**, 166.
23. Gannis, T.; Aritabile, G.; Mechlinski, W.; Schaffner, C.P. J. Am. Chem. Soc. 1971, **93**, 4560.

24. Golding, B. T.; Rickards, R. W.; Meyer, W. E., Patrick, J. B. Tetrahedron Lett. 1966, 3551.
25.a. Masamune, S.; Kaiho, T.; Garvey, D.S. J. Am. Chem. Soc. 1982, **104**, 5521; b. Brooks, D. W.; Kellogg, R. P. Tetrahedron Lett. 1982, 4991; c. Brooks, D. W.; Palmer, J. T. Tetrahedron Lett. 1983, 3059; d. Masamune, S.; Ma, P.; Okumoto, H.; Ellingboe, J. W.; Ito, Y. J. Org. Chem. 1984, **49**, 2834; e. Boschelli, D.; Ellingboe, J.; Masamune, S. Tetrahedron Lett. 1984, 3395; Blanchette, M.; Choy, W.; Davis, J. T.; Essenfeld, A.; Masamune, S.; Roush, W. R.; Sakai, T. Tetrahedron Lett. 1984, 2183; f. Hirama, M.; Uei, M. Tetrahedron Lett. 1982, 5307; g. Nicolaou, K. C.; Uenishi, J. J. Chem. Soc. Chem. Commun. 1982, 1292; h. Georges, M.; Tam, T. F.; Fraser-Reid, B. ibid. 1984, 1122, 1123; i. Lipshutz, B. H.; Kozlowski, J. A. J. Org. Chem. 1984, **49**, 1147.
26. Kellogg, R. P., Ph. D. Thesis, Purdue University, 1985.
27. Brooks, D. W.; Lu. L. D. L.; Masamune, S. Angew. Chem. Int. Ed. Engl. 1979, **18**, 72-73.
28.a. Frater, G. Helv. Chim. Acta 1979, **62**, 2825; b. ibid. 1980, **63**, 1383.
29. Palmer, J. T., Ph. D. Thesis, Purdue University, 1985.

Approaches to Chiral Building Blocks for Natural Products Synthesis

Pierre Sinaÿ, Laboratoire de Biochimie Structurale, U.A. 499, U.E.R. de Sciences Fondamentales et Appliquées, 45046 Orléans, France

ABSTRACT. The main objective of this investigation is the stereocontrolled synthesis of higher sugars possessing multiple contiguous chiral centers. For this purpose, glycosyl-lithium reagents have been prepared from thiophenyl glycosides, anomeric sulfones, or glycosylstannanes. They react in a stereocontrolled manner with aldehydes to provide, after opening of the tetrahydropyran ring, the expected higher sugars.

INTRODUCTION

During the last decade, monosaccharides have frequently been used as natural sources of stereocenters for the total synthesis of "chiraly rich" natural products. In most of the work published to date appropriate manipulation of various cyclic pentoses and hexoses--favorite chiral templates--ultimately resulted into short cyclic segments which are starting materials for "off-template" chemistry. All things considered, the anomeric center of a monosaccharide is thus an "off-template" site which is only used to lock the molecule in the most desirable heterocyclic form thus allowing the principles of conformational analysis to operate for the domestication of the endocyclic chiral transformations (Scheme 1).

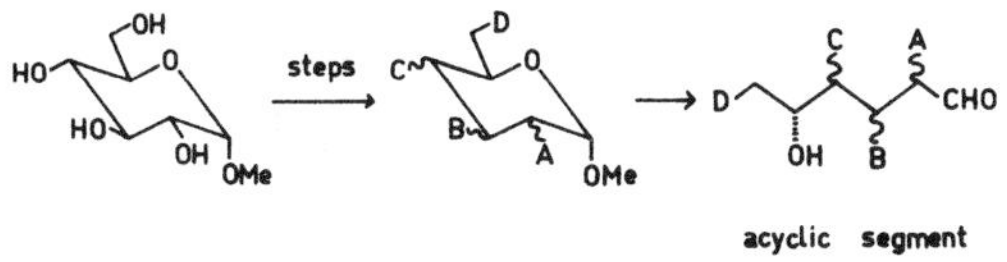

Compared with monosaccharides, disaccharides--as cellobiose, maltose or lactose--are unfortunately of no additional value in this respect, inasmuch as the chiral carbon backbone is interrupted by the interglycosidic oxygen atom, so that the impressive "super-chirality" of a disaccharide is not a bonus and cannot be directly translated into a "pseudo" higher carbon sugar. This is frustrating, more especially as the formation of the interglycosidic linkage is generally under efficient stereoelectronic control (Scheme 2).

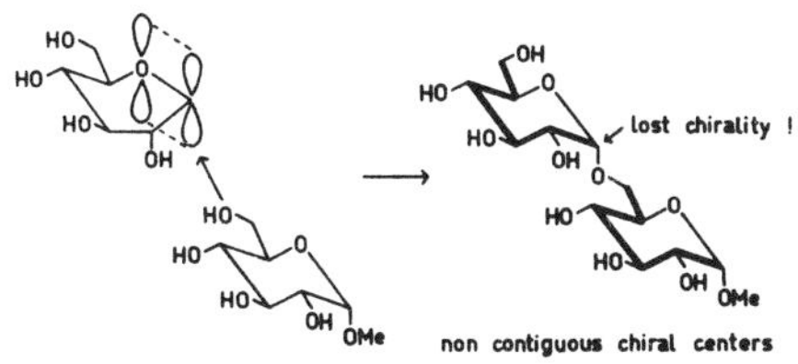

M. P. Schneider (ed.), Enzymes as Catalysts in Organic Synthesis, 171–176.

SYNTHESIS OF HIGHER SUGARS

We might therefore wonder how we can take advantage of the anomeric stereoselection (stereoelectronic control) to synthesize a higher carbon sugar--"hyper sugar"--possessing *multiple contiguous chiral centers.*

Before answering this question, let's first consider the biosynthesis of glucose. In the presence of three enzymes, D-glyceraldehyde 3-phosphate and dihydroxyacetone phosphate are first condensed (one enzyme) and the hexose converted (two enzymes) into glucose. Going now from this reality to bioscience fiction, it is possible to fictively "polymerise" dihydroxyacetone phosphate and get a nonose, a dodecose...and a glucose type hyper-sugar! (Scheme 3)

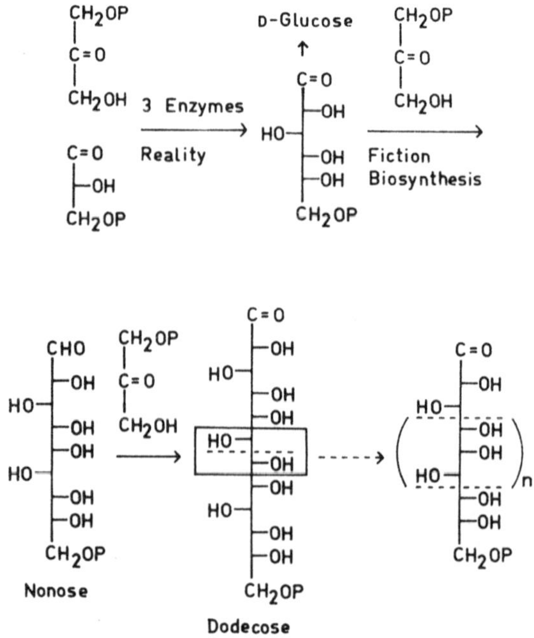

But such a process, operating in the world of lipids (synthesis of gutta percha from isopentenyl pyrophosphate, Scheme 4) is not a reality in the carbohydrate field. Nature doesn't care to elaborate higher-sugar.

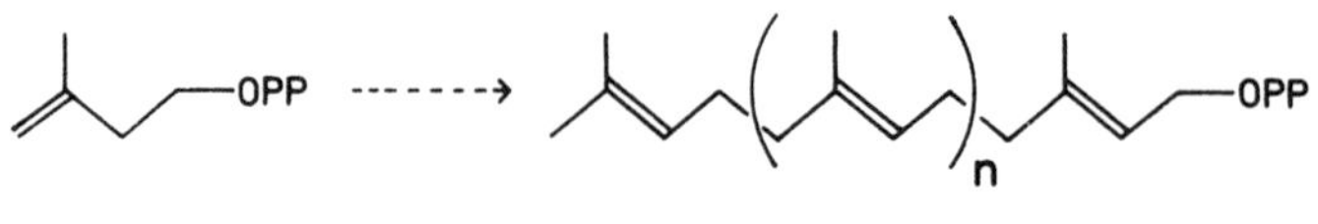

gutta percha

Let's now consider the ghostly dodecose (Scheme 3). It is a dimer of glucose which after all could be considered as such and thus elaborated by chemical processes shown in Scheme 5.

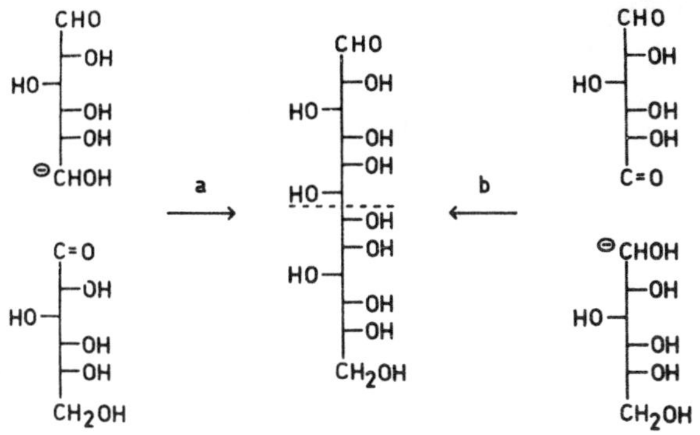

This is indeed the process b that we have materialized, using simple generations of reactive glycosyl-lithium derivatives recently developed in our laboratory (1). The concept is apparent from Scheme 6.

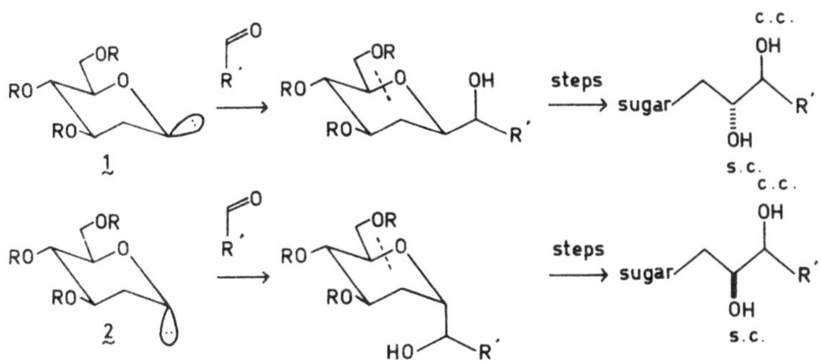

s.c. : stereoelectric control ; c.c. : classical control (cyclic Cram, Felkin)

R = CH_2Ph or Si$\!+$

The glycosyl-lithium reagents 1 and 2 have been generated from thiophenyl glycosides (1), anomeric sulfones (1) or glycosylstannanes (2) (Scheme 7).

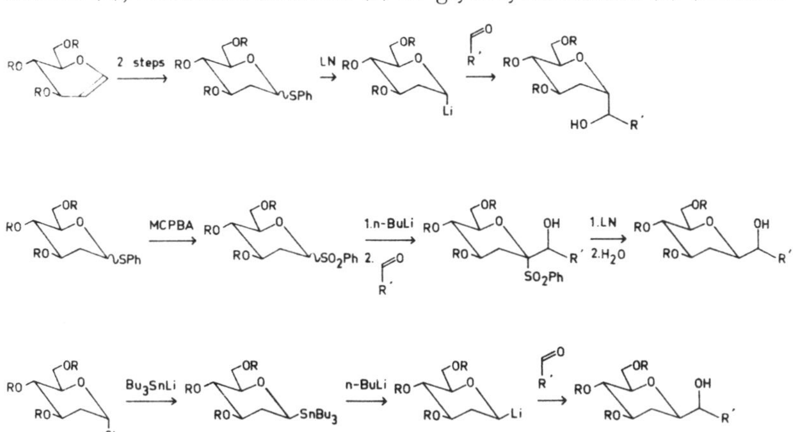

An application of this concept to the preparation of a dodecose via a "C-disaccharide" is shown in Scheme 8.

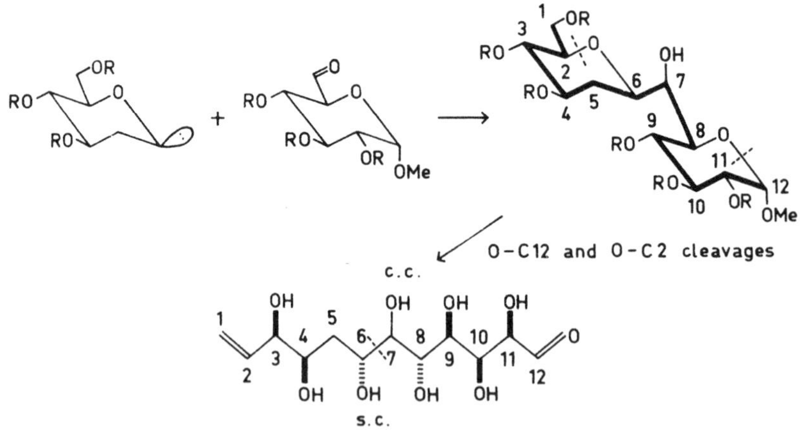

The C-6 chiral center has been created through stereoelectronic control (s.c.) and pathway b of Scheme 5 is thus mimicked. The cleavage of the O-C12 bond is straightforward (acid hydrolysis of a methyl glycoside) and the cleavage of the O-C2 bond has been trivially achieved along lines which are delineated in Scheme 9 on a model compound prepared in our laboratory (3) (Scheme 9).

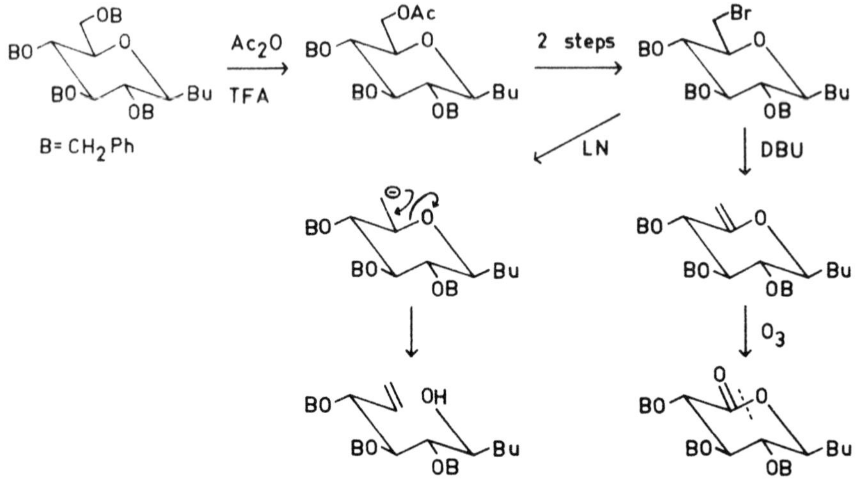

At this stage, we can summarize our approach as follows: the equivalent of 2-deoxy-glucose (D-glucal is a convenient source) has been converted into a higher-sugar, where one chiral center nicely results from stereoelectronic control, the contiguous one being established through an "off-template" classical control (Scheme 10).

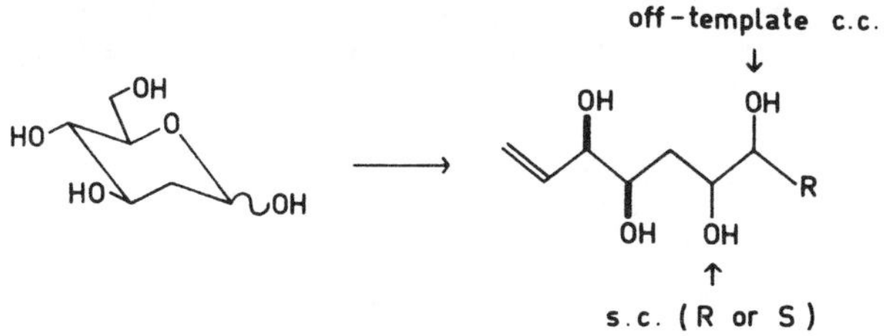

Combining this approach with another one recently developed by us (3), we were now able to synthesize a higher-sugar where the stereospecific formation of a vic-diol resulted from a double and very high stereoelectronic control (Scheme 11).

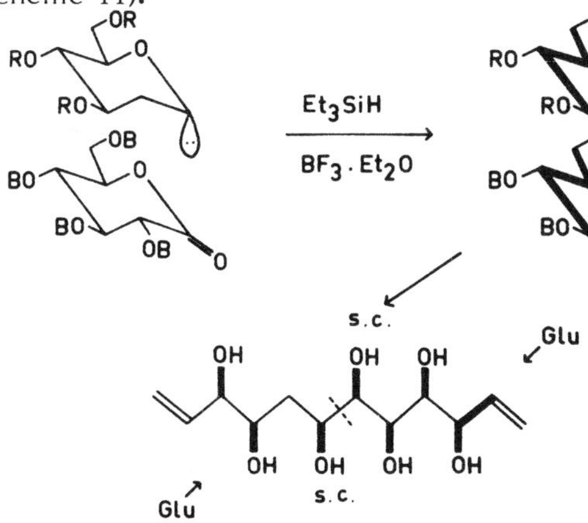

We recently reported (4) a stereoelectronically controlled synthesis of the first member to date of a true "C-disaccharide", that is the synthetic counterpart of α-methylgentiobiose in which a methylene group replaces the inter-unit-oxygen atom. The unsaturated precursor may now be selectively cis-hydroxylated to give, after manipulation, a polyhydroxy acid, which may be considered as the seco-acid of an artificial carbohydrate-type macrolide! (Scheme12)

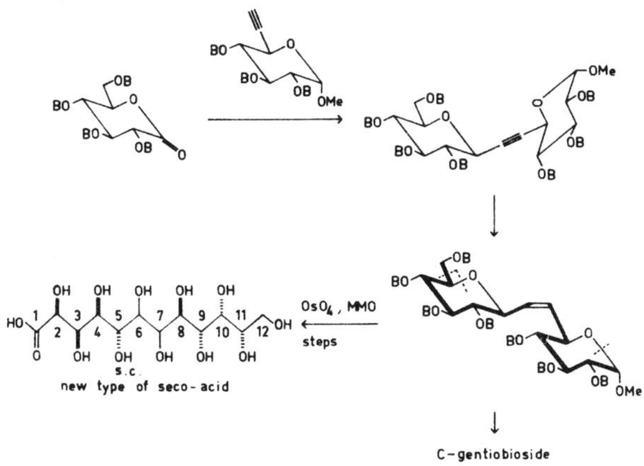

Let's conclude with the exotic preparation of a "palytoxin-type" polyhydroxylated molecule, which derives directly from the previously developed chemistry (Scheme 13).

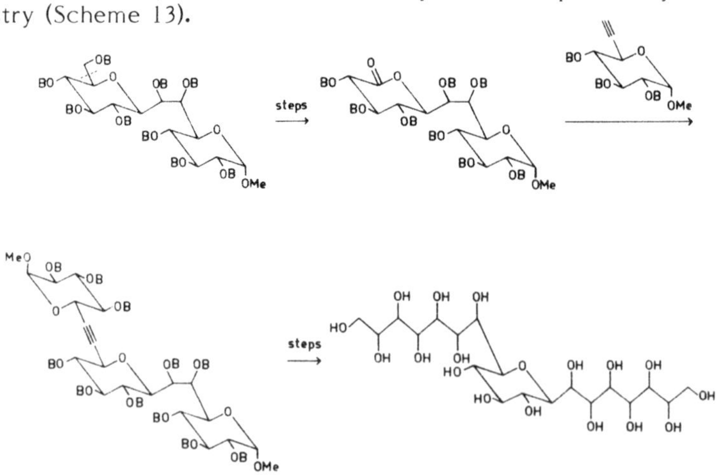

CONCLUSION

Stereocontrolled formation of C-C bond at the anomeric center of a monosaccharide opens the way to the synthesis of higher sugars via tetrahydropyran rings.

ACKNOWLEDGEMENTS

I would like to thank Dr. Jean-Marie Beau, Dr. Francesco Nicotra, Dr. Taku Chiba, Mr. Jean-Marc Lancelin and Mr. Dominique Rouzaud for their invaluable contributions to this programme, Mr. Jacques Esnault and Ms. Anne-Marie Noirot for their fine technical expertise, as well as the Centre National de Recherche Scientifique for their generous financial support.

REFERENCES

1. P. Sinaÿ, J.-M. Beau and J.-M. Lancelin, Organic Synthesis: An Interdisciplinary Challenge, Blackwell Scientific Publications, 1985, 307.
2. P. Lesimple, J.-M. Beau and P. Sinaÿ, J. Chem. Soc., Chem. Commun., (1985) 894.
3. J.-M. Lancelin, P. H. Amvan Zollo and P. Sinaÿ, Tetrahedron Lett., (1983) 4833.
4. D. Rouzaud and P. Sinaÿ, J. Chem. Soc., Chem. Commun., (1983) 1353.

SYNTHESIS OF ENANTIOMERICALLY PURE UNNATURAL COMPOUNDS VIA NON-BIOMIMETIC HOMOALDOL REACTIONS

Dieter Hoppe(*)
University of Göttingen, Department of Organic Chemistry
Tammannstr. 2
D-3400 Göttingen
Federal Republic of Germany

ABSTRACT. 2-Alkenyl N,N-diisopropylcarbamates, which are prepared from allylic alcohols, are easily deprotonated by n-butyllithium. The 1-oxy-allyl lithium compounds formed serve as homoenolate reagents and add to aldehydes and ketones with high γ-regioselectivity to yield enol carbamates of 4-oxy-alkanals or -alkanones. After exchange of the cation by titanium or aluminium reagents, the addition proceeds with high anti or syn diastereoselectivity. Altogether, the new reagents provide the first more general method for stereocontrolled homoaldol reactions. A report on asymmetric homoaldol reactions, including results of several research groups, is also given.

1. INTRODUCTION

Since our group was not involved in the development of enzymatic reactions, I am not able to contribute to the topic of the workshop. We learned during recent days that both - enzymatic and chemical synthesis - have their special merrits in the synthesis of enantiomerically pure compounds; so synthetic chemists should be familiar with both. I like to talk this morning, how chiral information, which is granted by enzymes, can be utilized and extended by new synthetic methods for carbon-carbon bond formation, which - so far - is not recogneized to occur in nature.

Aiming the synthesis of an enantiomerically pure target molecule, which carries several stereogenic centers, in principle, the synthetic chemist has the choice between two strategies. These often are not applied in plain form, but are combined (scheme 1):

1.The "chiral template approach". An enantiomerically pure, poly-stereogenic precursor, which is readily available from nature, for example glucose, is transformed to the target molecule. The synthetic steps include stereospecific manipulations of the stereocenters, like inversions, substitutions, chirality transfer, also sometimes removal of excess stereocenters, and reactions in the vicinity, which do not touch the configuration.

2. The second strategy, the "induced diastereoselective synthesis", starts from an enantiomerically pure precursor with one or few stereo-

M. P. Schneider (ed.), Enzymes as Catalysts in Organic Synthesis, 177–198.

centers. The further ones are formed from prochiral centers by diastereoface selecting reactions under the influence of the existing chirality. This occurs during and after the construction of the carbon frame work. If the asymmetric induction is not complete in these steps, this is not fine, but also not fatal, because diastereoisomers are formed; these, in principle, can be separated.

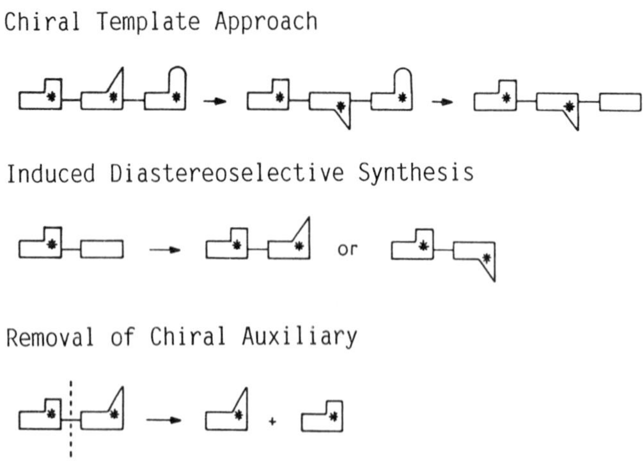

Scheme 1

In a general sence, the latter is one advantage, chemical synthesis has over enzymatic synthesis, which mostly deals with enantiomers. In many cases, the inducing chiral center does not fit for the structure of the target molecule; in these cases, it is removed after having done its job, it serves as a "chiral auxiliary".

Strategy 2, which is the subject of my lecture, depends on two conditions:

a. The availability of enantiomerically pure starting compounds, carrying one or few stereocenters. These are delivered by the "chiral pool", by enzymatic methods, or by resolution of racemates.
b. The availability of highly diastereoselective reactions for extension of the existing chirality.

Here the synthetic chemist has some advantages over enzymatic synthesis:

1. He is not limited to biogenetic pathways and reactions.
2. He is not limited to physiological reaction conditions, because he can freely select the temperature and the solvent; exclusion of water or air from the reaction flask is no problem.
3. If a good new reaction once is found, it usually is more general in respect to constitution and configuration of the target molecules than the enzymatic reactions are, where often the substrate specificity causes severe limitations.

2. REAGENTS FOR HOMOALDOL REACTION

2.1. Principle

Five years ago, we were searching for general methods for the stereocontrolled construction of γ-hydroxycarbonyl compounds 1 (scheme 2). We thought, it might be a good idea to construct the C-3 - C-4 bond by the combination of two prochiral centers, in particular, from a carbonyl compound 2 and an equivalent of a homoenolate ion 3. This would bring us in the position to control the relative - and, hopefully, also the absolute - configuration of the two newly formed asymmetric carbon atoms. To my knowledge, no biosynthetic transformation of this pattern is known; obviously nature does not utilize homoaldol reactions.

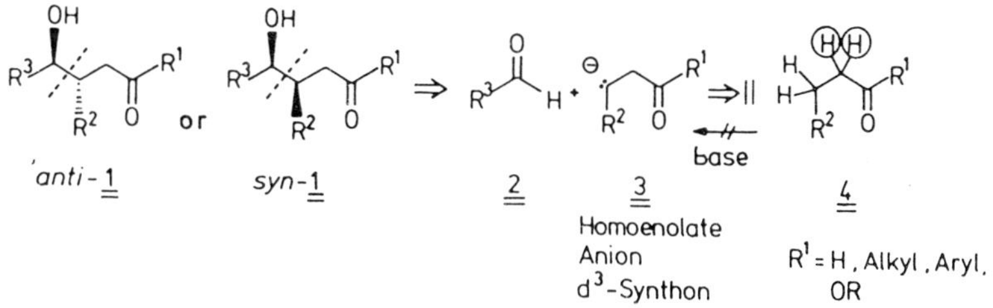

Scheme 2

Let's summarize the demands to a method for homoaldol reactions:
1. The method should provide diastereoselectively anti- or syn-configurated homoaldols 1, respectively. The option for asymmetric versions, which means the control of absolute configuration, should be given.
2. The method is desired to be general and highly flexible in respect to residues R^1 to R^3 in 1. It should be applicable to the synthesis of aldehydes (R^1 = H), to ketones (R^1 = alkyl or aryl), and to acid derivatives (R^1 = OH).
3. In addition, simplicity is desired. Particularly, the preparation of starting material must be simpler than the synthesis of the target compound itself.

When we started our work(1) in 1980, no method for diastereoselective homoaldol reaction was known and even the stereorandom versions, with few exceptions(1), were applicable only to very simple homoenolate equivalents like these of n-propanal or of propionic acid.

I am going to report on the development of a general method:
1. How the chemistry of the transformation was made working reliably.
2. How to achieve high non-induced diastereoselectivity(2).
3. Studies on synthesis of enantiomerically pure target molecules by additional diastereofacial selection (induced diastereoselectivity(2)). This chapter also includes results of several research groups.

Let us look more closely on the reagents needed: Useful concentrations of homoenolate ions(3) 3, of cause, cannot be prepared by deprotonation of carbonyl compounds 4, because the α-protons are at least 20 pk units more acidic than the β-protons. And even, if applying more sophisticated methods for their generation (like halogen metal exchange), several side reactions will destroy the reagent. At least, the electrophilic carbonyl group has to be protected, which also decreases the acidity of adjacent α-protons.

2.2 The Defensive Strategy

This defensive strategy was applied in 1969 by Büchi(4) who, to our knowledge, reported the first application of a homoenolate reagent (scheme 3): Here, the Grignard reagent 5 of a 3-bromopropanal acetal was prepared and added to the carbonyl group of a ketone. Acidic hydrolysis of the adduct 6 afforded the 4-hydroxy-aldehyde 7. Due to elimination reactions, which occur in higher substituted metallated acetals, the scope of the method is rather limited.

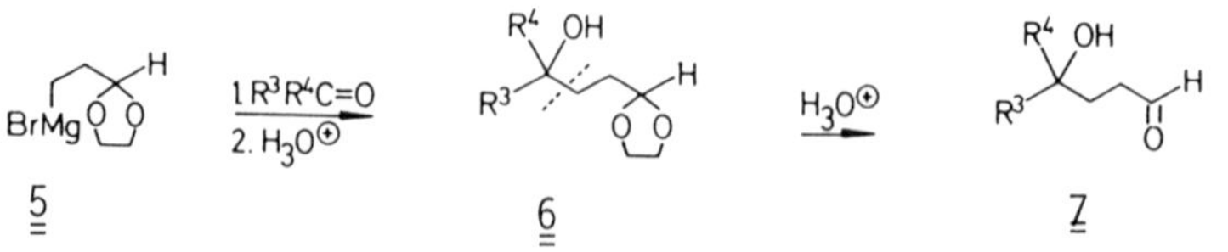

Scheme 3

This solution of the problem, also from another point of view, nowadays is unsatisfactory: It represents a defensive solution of the problem - but, modern synthetic chemistry should take the attack, not defense. Particularly, in the most elegant synthetic methods, an appearent handicap is turned to an advantage.

2.3. The Allylic Anion Strategy

The hetero-allylic anion strategy(1,5,6), first applied by Biellmann, Corey, Ahlbrecht, Seebach, Evans, Still, Schlosser and others in the early 70th, is an offensive strategy. Here the appearent draw back, the reactive carbonyl group at C-1, is used for creating allylic resonance, which contributes in stabilizing the anion 9; X is an oxygen, a nitrogen, or a sulfur group (scheme 4). The reagent 9 usually is prepared by deprotonation of allylic derivatives 8 by strong base. Its γ-addition to carbonyl compounds yields enol derivatives 10, which on acidic hydrolysis afford the homoaldols 12.

The elegant this solution is, the plentiful the problems are associated with. Time is to short for going into details, one may find these discussed in reviews(1,6). The problems concern:

1. Regioselectivity. From attack of the α-position in ambident allylic anions 9, a second isomer 11 arises as side product or even as the main product.
2. Structural limitations due to unsufficient stability of anions 9 and unsufficient kinetic acididy of precursors 8. Alkyl substituents decrease both; that is the reason why in most reagents of this type none or at most one alkyl group is tolerated.

3. Diastereoselectivity. No studies, concerning homoenolate reagents were known in 1980.

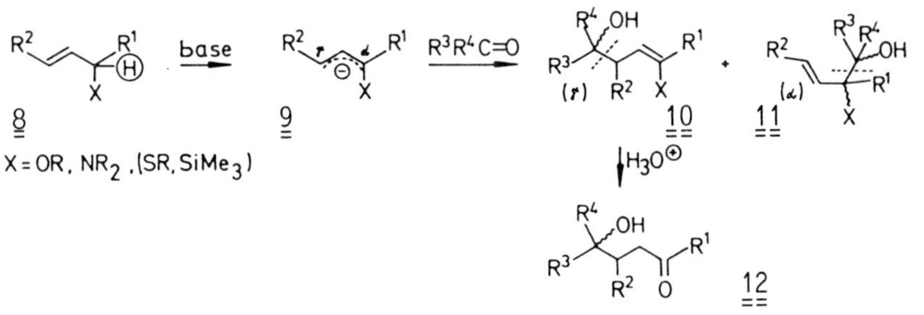

Scheme 4

The simple ideas, which brought us taking up the project was the following: Use the hetero substituent X offensively, too. Employ X for enhancement of the kinetic acidity in 8 and for stabilization of the allylic anion 9. Employ X also for holding the counter ion (which is not drawn) close to the α-position of the allylic anion 9. Since most carbonyl additions of allylmetals were recogneized to proceed in unpolar solvents via pericyclic processes(7), high γ-selectivity should result.

During our earlier work on carbanions, derived from nitrogen derivates of carbonic acid(8), we met several groups which seemed to be good candidates. After hard exploratory work at the lab desk, the one which proved to be the best was the N,N-diisopropylcarbamoyloxy group(9,10); abbreviated "OCb" in many of the following formulas.

2.4. Homoaldol Reactions of Lithiated 2-Alkenyl Carbamates

N,N-Diisopropylcarbamic acid 2-alkenylesters are easily obtained with high yields from allylic alcohols and N,N-diisopropyl carbamoylchloride(11). Deprotonation of the prenyl ester 13 by n-butyllithium and 1.0 - 1.1 equiv. of N,N,N',N'-tetramethylethylenediamine in ether at -78°C is complete within 30 min.(10,11), (scheme 5). Addition of 2-methyl-propanal to the solution of the intermediate lithium compound 14 furnishs the 4-hydroxy-1-alkenyl carbamate 15 with 87 % yield. The α-isomer is not detectable by ^{1}H-nmr. The double bond configuration is E, with 3 % contamination by the (Z)-enol derivative. In this respect, the example is untypical, because in most cases, exclusively (Z)-enol carbamates are formed(10,11).

Notice, that C-C bond formation had occurred at the most hindered position forming a quarternary carbon atom. From this, it is evident, that a close ion pair, bearing the cation at the α-carbon, if not a covalently bound, allyllithium compound 14 is the intermediate, which adds to the aldehyde with allylic inversion to form 15.

Deprotection of the surprisingly stable enol carbamates turned out to be difficult(11). Finally we found, that stirring adducts of type 15 in methanol in the presence of 1 mol-% of mercuric acetate and of 1 mol-equiv. of methane sulfonic acid at room temp. yields lactol methyl ethers of type 16 smoothly and quantitatively. Grieco oxidation(12,13) of 16 by m-chloro perbenzoic acid with boron trifluoride etherate as a catalyst gives the γ-lactone 17 with 93% yield. - Overall, the homoenolate ion of isovaleric acid was added to the aldehyde.

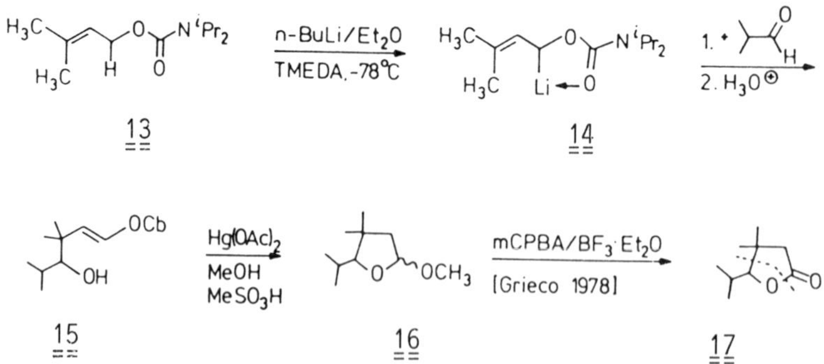

Scheme 5

The scope of the allyl carbamate lithiation is quite wide. The 2-propenyl derivative 18 and all possible monomethyl derivatives(10,11) 19 - 23 are lithiated in ether under standard conditions within 15 min. at -78°C (scheme 6). Dialkyl substituted compounds, like the 3-penten-2-yl(11) 25, the 2-cyclohexenyl(14) 26, the (1-cyclopentenyl)methyl(13) 27, the myrtenyl(10) 28, and the isopinocarveyl ester(14) 29 need 30 min. to 3 h for complete lithiation.

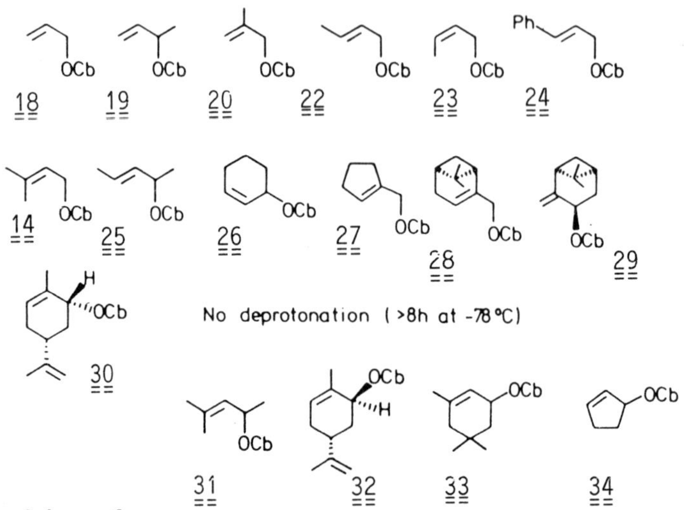

Scheme 6

The top is held by the N,N-diethyl cis-carveyl carbamate(10) 30, which represents a 1,2,3-trialkylated allyllithium. If electron-withdrawing groups are present, like phenyl in 24, the deprotonation takes only few seconds(11). All the anions add to ketones with high γ-regioselectivity.

Were are the limitations of our method for preparing homoenolate reagents by simple deprotonation? Three alkyl groups in the electron-rich α- and γ-positions are to much: The open-chain trimethyl derivative 31 and the isophoryl ester 33 are not lithiated(14). Also, if the stereoelectronic situation is unfavourable, that is, if the proton to be removed can not occupy an axial position easily, deprotonation is not observed(14); so for the trans-carveyl compound 32 or even for the cyclopentenyl ester 34.

3. HOMOALDOL REACTIONS WITH HIGH NON-INDUCED DIASTEREOSELECTIVITY

3.1. anti-Diastereoselective Additions

The reaction of a prochiral allylic reagent and a prochiral carbonyl compound gives rise to diastereoisomeric anti- and syn-adducts. For achieving high non-induced diastereoselectivities(2), usually two conditions are essential(1,7):

1. Uniform prochirality of the allylic double bond.
2. Introduction of a more or less covalently bound "cation" in order to assure a close six-membered transition state of the carbonyl addition.

We had our best results(13,15) if Li was exchanged by Ti-reagents, according to Reetz(16) and Seebach(17), or for diisobutyl aluminium(18): The (E)-butenyl carbamate 22 is lithiated by the standard procedure and a small excess of chloro-tris(isopropoxy)titanium is added (scheme 7).

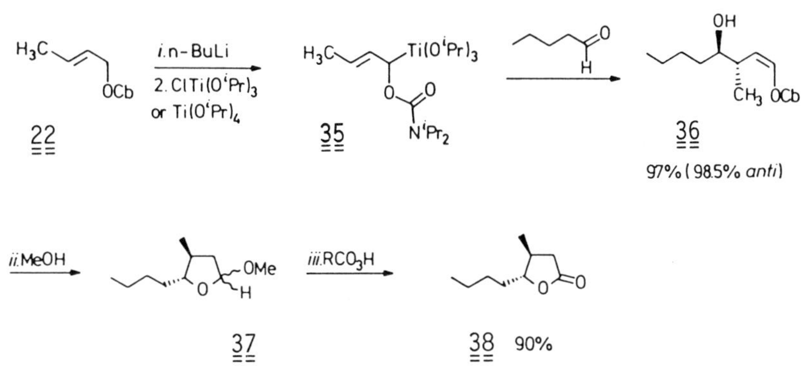

Scheme 7

To the intermediate titanium compound 35 formed, n-pentanal is added. After aqueous work up, with 97% yield, the γ-adduct 36 is obtained. In the reaction mixture, by capillary gas chromatography, beside 98.5 % of the (Z)-anti-36, 1.5 % of the (Z)-syn-diastereoisomer were detected(13). α-Adducts are not found. In most cases, the use of tetra(isopropoxy)-titanium gives equal results, but complete metal exchange may need longer reaction times. The mercuric salt assisted methanolysis of 36 and subsequent Grieco oxidation(12) yields racemic quercus lactone A (38), a constituent of oak wood, with high yield(13).

Scheme 8 gives an impression of the broad applicability of our anti-selective homoaldol reaction, just by choice of the homoenolate reagent and the carbonyl compound(1). Without going into detail, I like to point on few features: Yields, γ-regioselectivities, and both the non-induced diastereoselectivities (anti and Z) are very high in general.

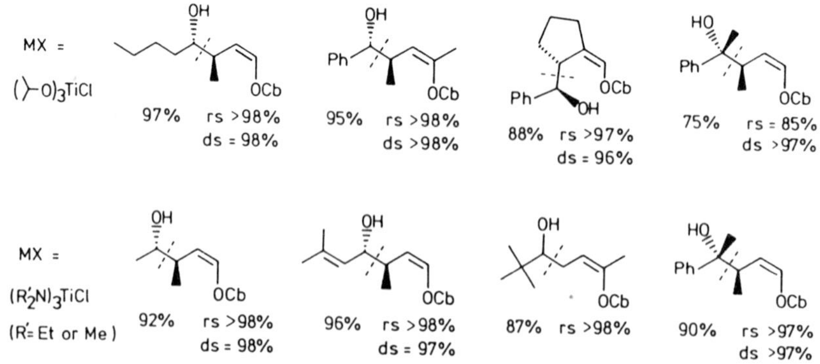

Scheme 8. % ds (% rs): percentage of the diastereoisomer (regio-isomer) shown in the mixture of diastereoisomers (regio-isomers)

Although still non-enantioselective, from this one advantage of a good "chemical" synthetic method is obvious: It is less sensitive to structural changes in the products than most of the enzymatic methods are. Therefore, once developped, it can be applied to the synthesis of a great variety of compounds without further exploratory work at the lab desk.

Including the subsequent methanolysis and Grieco oxidation, diastereomerically pure, racemic γ-lactones are obtained by a "brick box system" with few steps and high overall yields(13), (scheme 9):

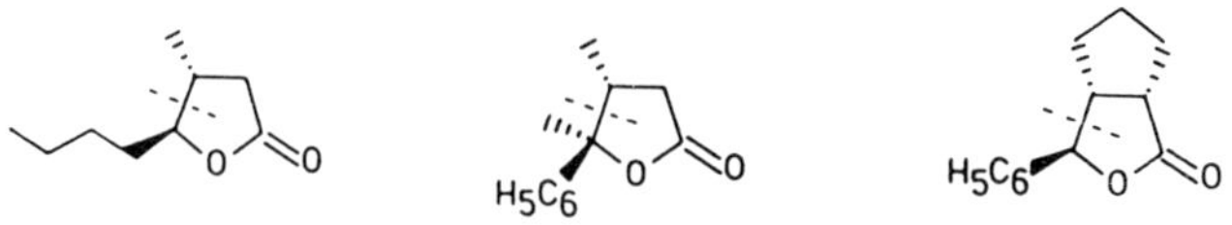

38 88% 39 80% 40 60%

Scheme 9

The "upper right" part of the molecules is delivered by the homoenolate reagent, the "lower left" one origins from the carbonyl compound. So lactone 39 is prepared from the (E)-2-butenyl carbamate 22 and acetophenone; lactone 40 is obtained from (1-cyclopentenyl)methyl carbamate 27 and benzaldehyde.

3.2. syn-Diastereoselective Additions

How to achieve syn-selective homoaldol reactions? From the Zimmerman-Traxler model, it is expected, that (Z)-allylmetals should afford syn-diastereoisomers(7).

(Z)-Butenyl carbamate 23 was prepared by cis-hydrogenation of the 2-butinyl carbamate(11). The lithiation of 23 proceeds with retention of the (2,3-Z)-double bond(11,18b), (scheme 10). The cation was exchanged by di(isobutyl)aluminium methanesulfonate 41, prepared from the hydride and methanesulfonic acid, and the aluminium reagent 42 is added to methyl pyruvate, affording the (E)-syn-adduct 43 diastereomerically pure after one recrystallization, yield 67 %. Methanolysis and oxidation yields racemic, so called "trans methyl crobarbatate"(18b) 44, yield 67 %.

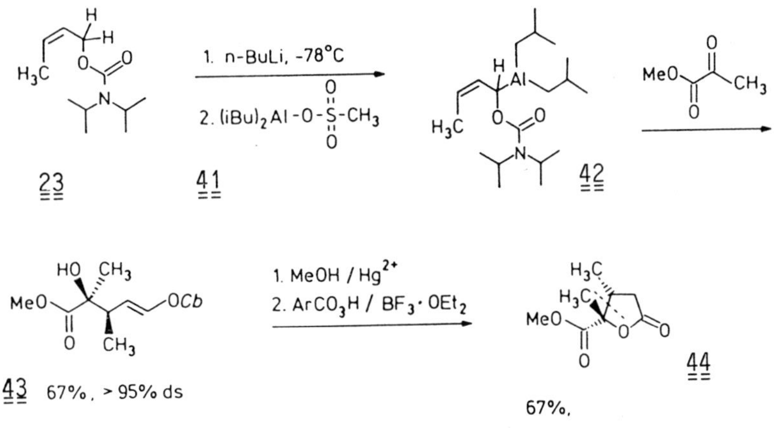

Scheme 10

4. ENANTIOSELECTIVE HOMOALDOL REACTIONS

4.1. Principle

Of cause, at that point, we are still dealing with racemic compounds. How to accomplish enantioselective homoaldol reactions? - The principle is simple (scheme 11):

One of the reaction partners, either the homoenolate reagent or the carbonyl compound has to be applied enantiomerically pure and the reaction conditions must be selected in a way, that high asymmetric induction occurs at prochiral reaction centers. That means, competing diastereoisomeric transition states should differ in energy by $\Delta G^{\ddagger}$ by 2 kcal/mol or more. Finding proper conditions is the more complicated task.

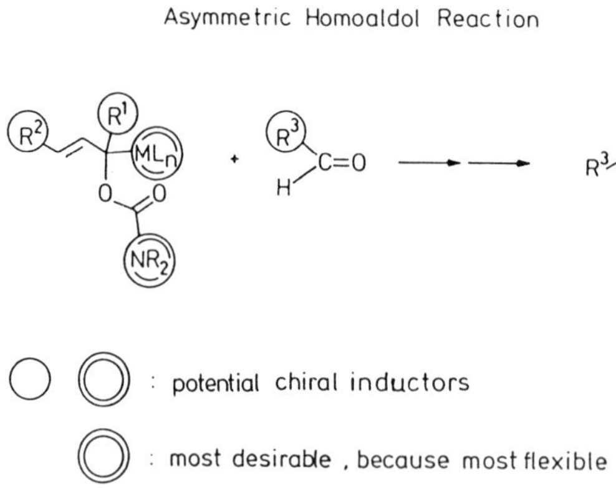

Scheme 11

The inducing center can be located in several positions of the allylic moiety of the homoenolate reagent, in the hetero substituent, or in the sphere of the "cation". If complete diastereo-facial selection in the γ-position of the allylic moiety or/and complete enantio-facial selection in the carbonyl compound occurs, enantiomerically pure homoaldols will be produced. The opposite strategy - utilization of an enantiomerically pure carbonyl compound and a prochiral homoenolate reagent also is envisaged.

4.2. Chiral Allylic Moiety

Let's see examples for the first strategy (scheme 12): The homoenolate reagent, derived from lithiation of the diisopropyl carbamate **28** of (-)-myrtenol is attacked by ketones (like cyclopentanone) exclusively from the _si_-face, affording a single diastereoisomeric γ-adduct(10) **45**. The diastereo-facial selection is due to the shielding of the bulky isopropylidene bridge and the rigid conformation of the bicyclus **28**. The optical purity of the product **45** equals that of the starting carbamate.

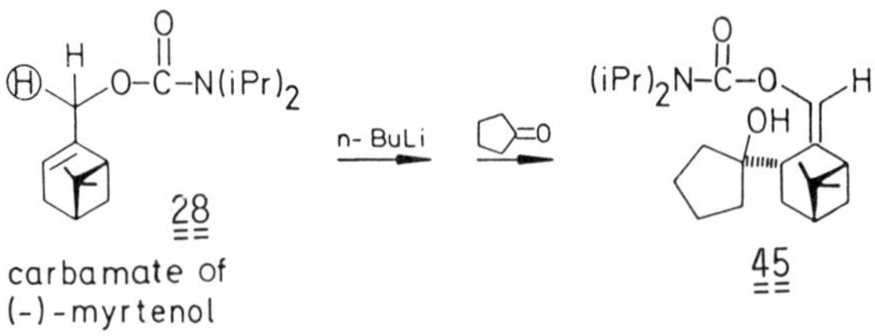

Scheme 12

Carbamates of similar terpenic allylic alcohols bring similar results(10,14). Although these reactions are very useful for the modifica-

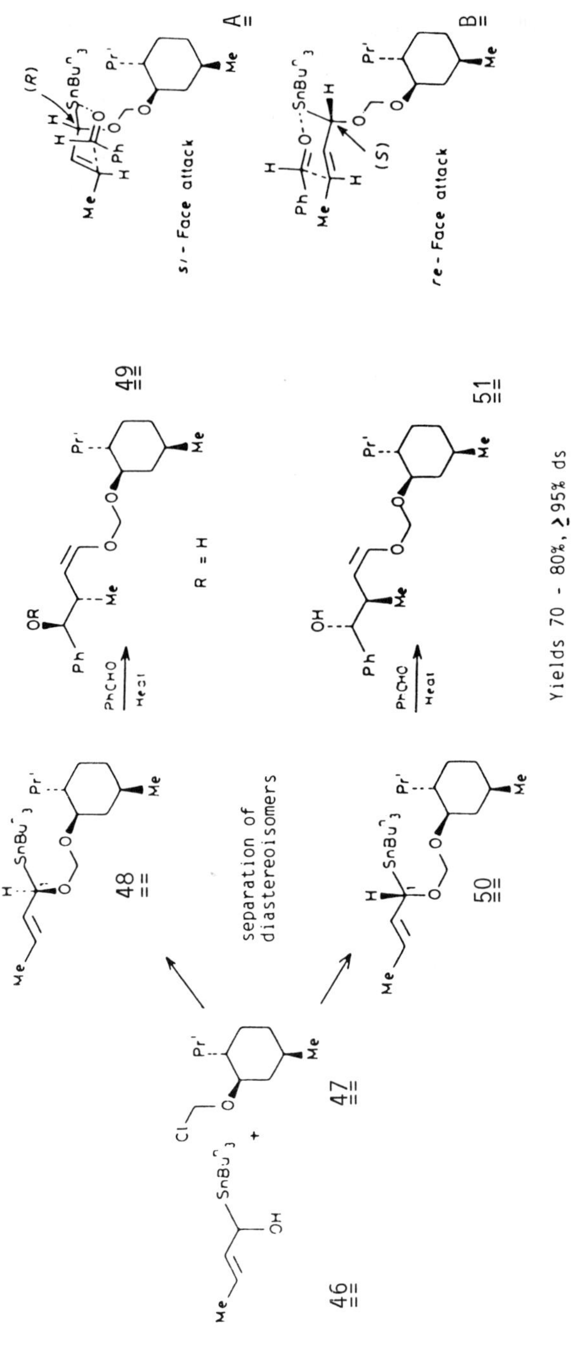
46
47
48
49
50
51
A
B
separation of diastereoisomers
R = H
Yields 70 - 80%, ≥95% ds
si - Face attack
re - Face attack

Scheme 13

tion of natural compounds, the structural limitation to rigid ring systems is a disadvantage.

An interesting approach was published very recently by Thomas et al.(19): Racemic (E)-1-tributylstannyl-2-butenol (46) is alkylated by the chloromethyl ether 47 of (-)-menthol; both optically active epimers 48 and 50 are separated by careful chromatography (scheme 13).

Both, 48 and 50, are heated separately with benzaldehyde, affording two optically active diastereoisomers 49 and 51 with greater than 95 % ds, each. The double bonds in 49 and 51 have (Z)- and the newly created stereogenic centers have anti-configuration in respect to each other; their absolute configurations are opposite in both diastereoisomers. So, after hydrolytic removal of the menthyl residue, both enantiomers of the homoaldol are obtained. This is a result of chirality transfer from the 1-carbon atom onto prochiral C-3 and of enantio-facial selection in the aldehyde, caused by the highly ordered, chair-like, six-membered transition states A and B. Similar observations are reported by R. W. Hoffmann et al.(20) by applying optically active allyl boronates.

This method exhibits fine chemistry, but some practical draw backs, from my personal point of view, should be considered:

1. The preparation of starting material requires several steps.
2. The separation of diastereoisomers seems to be very tough.
3. The uncatalyzed reactivity of allylic stannanes is low, which narrows the limits of application substantially.

4.3. Chiral Hetero Substituents

A method for asymmetric homoaldol reaction, in which the chiral information of the hetero substituent is utilized more directly and more efficiently comes from the group of Helmchen et al.(21), (scheme 21):

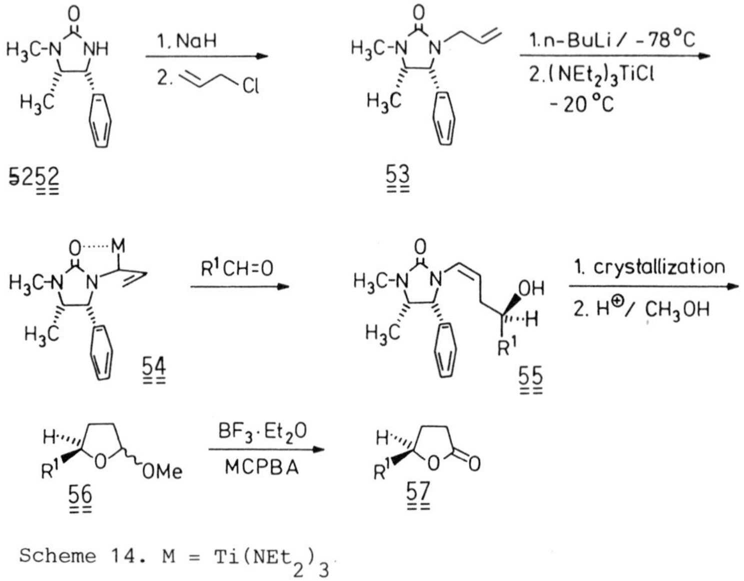

Scheme 14. M = $Ti(NEt_2)_3$

The cyclic urea 52 is prepared from (-)-norephedrin with one reaction step (scheme 14). It is N-allylated, and 53 is deprotonated by n-BuLi, then the cation is exchanged by chloro-tris(diethylamino)titanium. The intermediate 54 adds with high induced diastereoselectivity (88 - 96% ds) to aldehydes or to prochiral ketones to form (Z)-enamides 55. Usually, the pure major diastereoisomer 55 is obtained after one or two recrystallizations. Methanolysis of the enamide, followed by the Grieco oxidation affords enantiomerically pure γ-lactones 57.

Some of the derivatives prepared by this method are enantiomers of naturally occurring insect pheromones. - Prof. Helmchen has informed me, that work with 3-substituted homoenolate reagents of type 54 also is in progress.

4.4. Chiral "cation"

Besides similar, classical approaches(1), we try to induce asymmetry by a chiral ligand sphere at the "cation" (scheme 15). We attempted, for example, to prepare the titanium butenyl reagent 58 from lithio (E)-crotyl carbamate(22). Without isolation and characterization, it was reacted with 2-methyl-propanal. After hydrolysis, the mixture of enantiomeric homoaldol derivatives 59 and 60 was obtained in a 2 : 1 ratio, 34 % ee. (determined by ^{1}H-nmr in the presence of optically active shift reagents). There are some doubts, whether the titanium compound exists as a monomer 58.

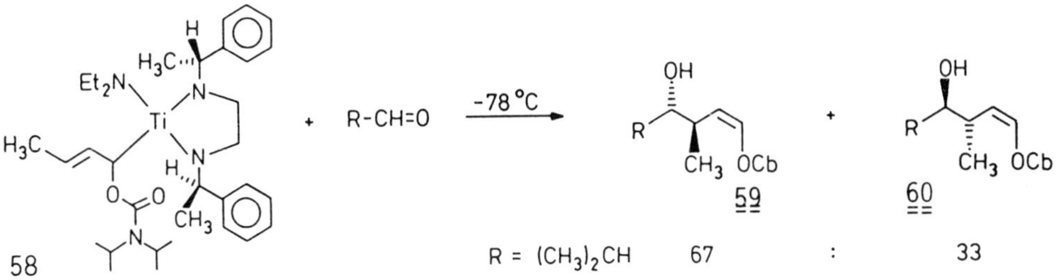

58

Scheme 15

Several further problems are associated with this approach. First, unlike to few examples concerning chiral titanium reagents, published by Seebach(23) and Reetz(24), the α-carbon in 58 is already chiral. If no rapid epimerization occurs under the reaction conditions, the degree of asymmetric induction is determined by the cation exchanging step.

Second, not separable diastereoisomers, but enantiomers are produced. This handicap is shared with many enzymatic reaction, if applied to unnatural substrates.

If the enantiomeric excess can not be improved by better ligands, the preparative value of the method will remain low. However, engagement in this most difficult strategy is rewarding, because any catalytic asymmetric homoaldol reaction has to overcome these problems.

4.5 Additions to Chiral Aldehydes

Let's follow now the easiest route via 1,2-diastereo-facial selection in chiral aldehydes:

As long as the simplest homoenolate synthon is concerned - the propanal homoenolate - the classical reagent of Büchi will work well. This was recently shown by Kelly et al.(25) for a de-novo synthesis of L-rhodinose (64), (scheme 16):

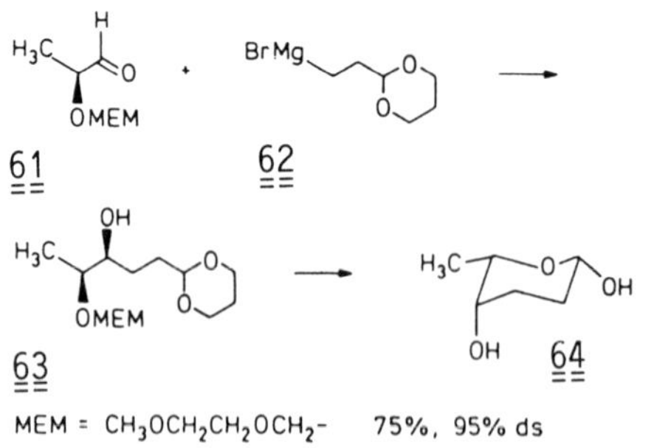

Scheme 16

By addition of the Grignard reagent 61 to MEM-protected (S)-lactaldehyde 63 at -100 °C, the chelation-controlled adduct 63 is formed with 95 % ds. A facile diastereomer separation and deprotection affords the antibiotic sugar 64.

Here plain 1,2-diastereo-facial selection is seen. No problem of non-induced diastereoselectivity is involved, because the reagent has no prochiral reaction site.

The stereochemical situation is more severe for 1- and 3-substituted homoenolate reagents, so for our (E)-butenyl derivative 66 (scheme 17). The reaction site (3-position) is prochiral and the 1-carbon atom, where the cation is held, is either prochiral or even chiral.

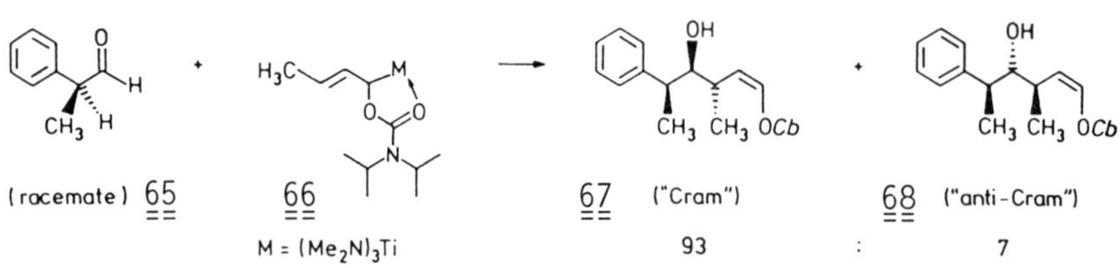

Scheme 17

By addition of the (E)-butenyl dimethylamino titanium compound 66 to racemic 2-phenyl-propanal (65), the racemic (Z)-3,4-anti-4,5-syn-adduct 67 is formed(26) by 93% ds. beside few 68. The relative configuration of 67 was elucidated by ozonolysis of the double bond, reduction of the resulting alkanal, followed by the comparison of the 1,3-diol with the known one(26).

Starting from the enantiomerically pure aldehyde 65, quite likely - but not certainly - the optical pure diastereoisomer 67 will be obtained with equal result.

Notice, that the value 93% ds. represents the product of three different diastereoselectivities: non-induced anti- and (Z)-selectivity, and (induced) 1,2-diastereo-facial selectivity, as well. So three centers of chirality and two centers of prochirality are set up efficiently in a single step.

We were not surprised when, by addition(26,27) of titanium reagent 70 to O-protected (S)-lactaldehydes 69, two out of eight possible optically active diastereoisomers, 71 and 72, were obtained (scheme 18); R' equals benzyl, tert.-butoxycarbonyl, or tert.-butyldimethylsilyl. The separation of the (Z)-L-ribo- and the (Z)-L-lyxo-derivatives 71 and 72, respectively, is very facile by silica gel chromatography. In every case, the lyxo-isomers are the less polar due to their greater tendency for intramolecular hydrogen bridging.

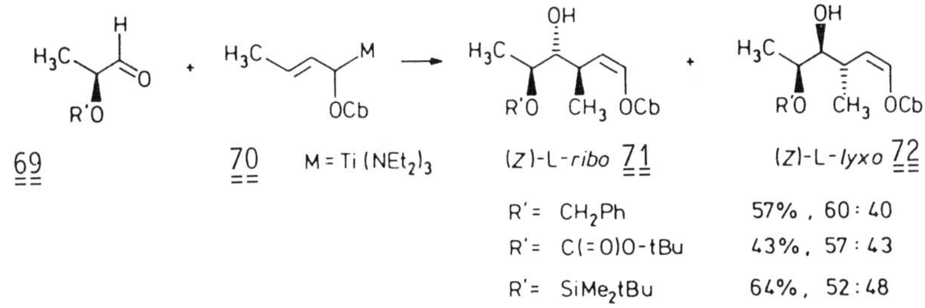

Scheme 18

The products are enol derivatives of branched desoxy aldoses. By variation of the homoenolate reagent and the oxy-aldehyde, a wide number of unnatural sugars now is accessible within few steps. Since D-lactic esters (which are the starting materials for aldehydes (R)-69) also are commercially available, an entry to enantiomeric D series also is given.

We think, that the flexibility in constitution, relative and absolute configuration of the target molecules, coupled with the predictability of the reaction course, do outweigh one easy diastereoisomer separation.

In addition, the (Z)-double bonds are best situated for further diastereo-face selecting functionalizations(28).

On treatment of the O-benzyl protected (Z)-L-ribo adduct 71a with tert.-butyl hydroperoxide in presence of catalytic amounts of vanadin salts(28), according to Sharpless(29), a single epoxide 73 is formed (scheme 19). Its L-allo-configuration was secured by a single crystal X-ray analysis, which was performed by Dr. Jones et al.(30).

In the same way, and equally efficiently, the (Z)-lyxo isomer 72a affords the L-talo epoxide 74. It is not crystalline, but from the investigation of several related epoxides(27) by X-ray analyses(30), the relative configuration *syn* for the 4-OH group and the epoxide oxygen is well established.

Studies of stereospecific substitution reactions of this novel class of epoxides are in progress(31). A number of highly flexible methods for the synthesis of enantiomerically pure, polystereogenic synthetic building blocks will result.

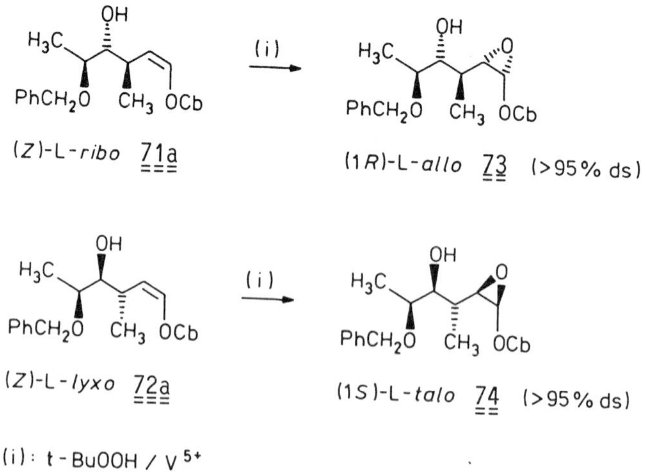

Scheme 19

Small, enantiomerically pure, poly-functionalized synthetic building blocks often are not easily available. So degradation of a carbon chain and destruction of excess chirality may be a constructive synthetic operation - although it hurts the heart of a synthetic chemist.

For example (scheme 20), the L-*lyxo*-diastereoisomer 72b, obtained from Boc-protected (*S*)-lactaldehyde 69b and the (*E*)-crotyl reagent 70, was cyclized by catalytic methanolate to form the 1,3-dioxolan-2-one (27) 75. Now, with 2,2-dimethyl-1,3-propanediol in presence of 1 mol-% of mercuric acetate, the cyclic acetal is formed and, subsequently, hydrolyzed to give the diol 76. Periodate cleavage affords the enantiomerically pure, partially protected isoprenoic dialdehyde (*S*)-77. Sodium borohydride reduction affords the alcohol 78. The yields of each of the very simple reaction steps are greater than 90 %.

From the L-*ribo* isomer 71b, by the same sequence, the enantiomeric aldehyde (*R*)-77 will be obtained. Overall, the reaction sequence performs a formylation of the homoenolate reagent, combined with optical resolution.

Again, high structural flexibility is predicted, because the limitations are given by the accessibility of the appropriate (*E*)-configurated homoenolate reagents; any alkyl group is supposed to be tolerated in the 1- and in the 3-position.

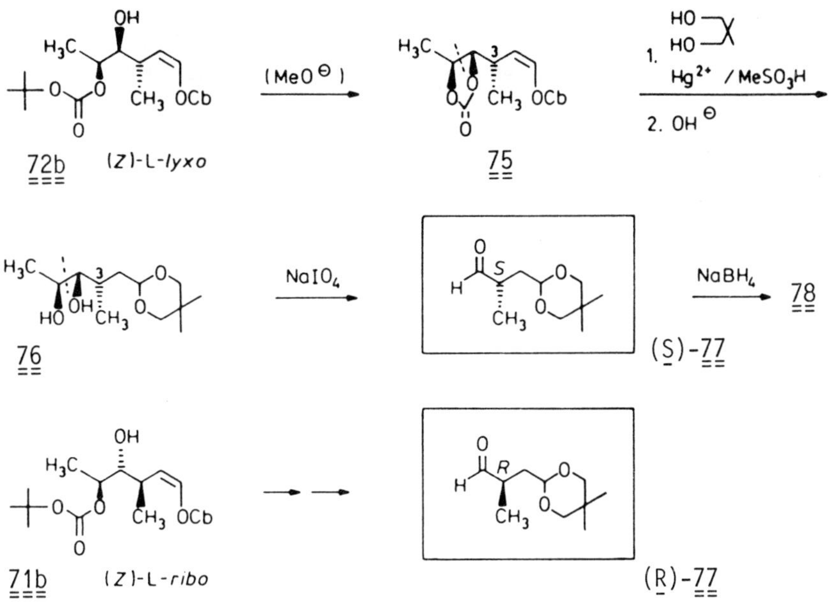

Scheme 20

5. HIGHLY STEREOSELECTIVE AND FLEXIBLE SYNTHESIS OF 1-OXY-1,3-ALKADIENES

Sometimes, in a multi-step synthesis, it is not wise, to introduce further chiral centers in the beginning, but to leave this for a later step. Reasons for that may be disturbing functionallity or a better diastereo-facial selection in advanced intermediates. Here, the introduction of a C-C double bond having uniform prochirality, is one of the best investments. This also can achieved by our methods for homoaldol reactions (scheme 21):

No matter, whether the (E)-, the (Z)-3-(trimethylsilyl)-2-propenyl carbamate 79 or mixtures thereof are lithiated by n-butyllithium, after 3 min. stirring at -78°C, only the 3-endo-anion 80 is present(32) (as it was proven by trapping the intermediate 80 by chlorotrimethylsilane). The Ti-mediated homoaldol reaction affords the (Z)-anti hydroxysilanes 82, but Al-mediated carbonyl addition yields the (E)-anti adducts 81. Subsequent Peterson elimination estabilishs the configuration of the 3,4-double bond, either (Z) in 84 or 856 by syn-stereospecific elimination with sodium hydride, or (E) in 83 or 85 by the anti-stereospecific elimination, mediated by boron trifluoride etherate at -78°C.

According to the terms of Schlosser(33), the synthesis is stereoconvergent in the first step, but doubly stereo-divergent in the two following reaction steps.

So starting from a single stereorandom precursor, just by two selections of the reagents, all of the four possible geometric isomers of 1,3-alkadienyl carbamates 83 - 86 can be obtained with high purities(32). For example, with acetaldehyde (R = CH_3), in all cases, overall yields

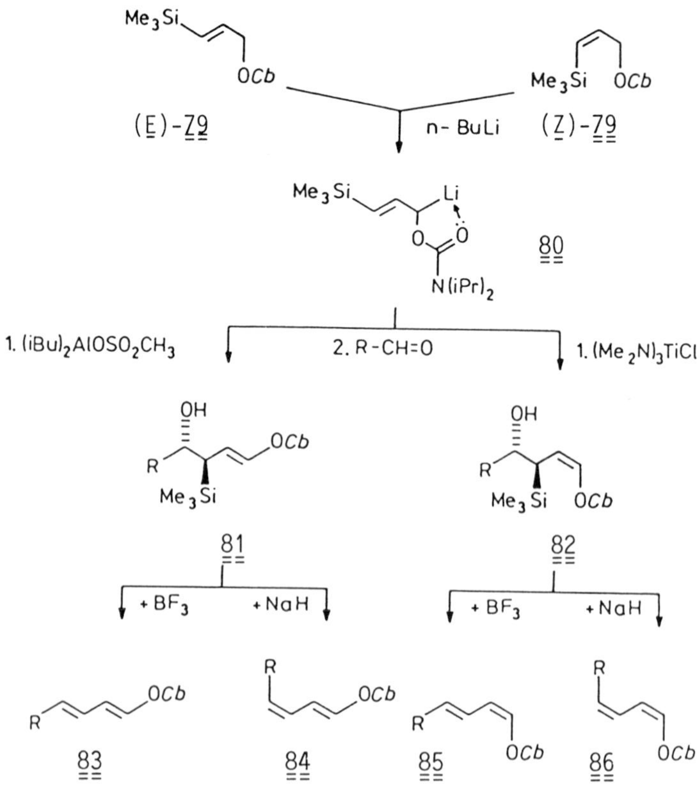

Scheme 21

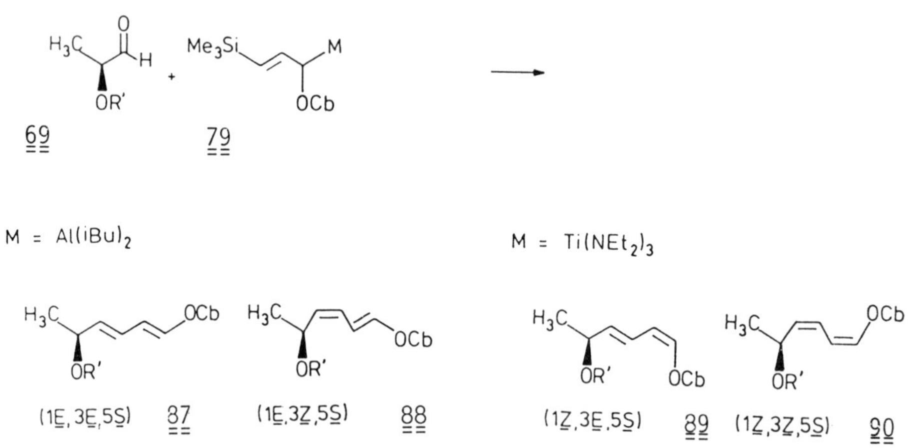

Scheme 22

are around 80% and diastereomeric purities range between 99.5 and 99.9% (determined by gaschromatography).

If the residue R contains a chiral center, it is introduced to the diene and its configuration is expected not to be touched (scheme 22).

Starting from O-benzyl (S)-lactaldehyde **69**, we obtained by these routes(34) the (1E,3Z,5S)- and the (1Z,3Z,5S)-hexadiene **88** and **90** (R' = benzyl). The (3E)-dienes **87** and **89**, at present, cause some troubles, because under condition of boron trifluoride mediated elimination, the products are prone to decomposition (presumably via dienyl carbenium ions). I am sure that this problem will be solved soon, too.

At present, we are exploring the potential of this novel class of dienes for cycloaddition reactions and diastereo-face selecting hydroxylation of the double bonds.

6. CONCLUSION

In the last slide, I like to summarize some advantages of metallated 2-alkenyl-carbamates in their application for homoenolate reagents for the nucleophilic introduction of substituted three-carbon-units to aldehydes and ketones (scheme 23):

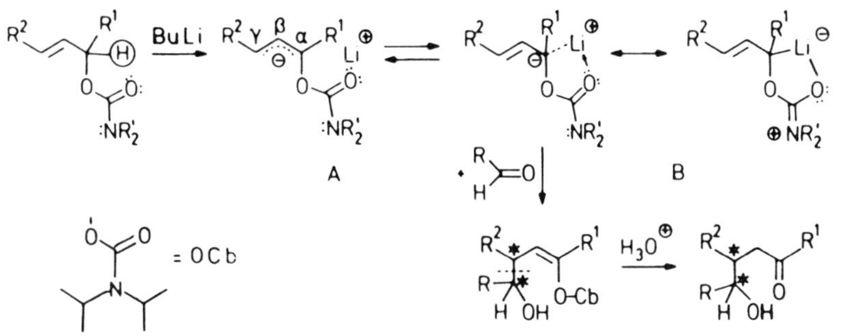

Scheme 23

1. The non-electrophilic carbamoyloxy group enhances the kinetic acidity of the precursors and also contributes to the stability of these reagents, presumably by "induced dipole stabilization"(1), formula B. This secures a wide applicability of our methods for homoaldol reactions.
2. The group holds the counter ion close to the α-position by formation of five-membered chelate complexes. This is the origin for high γ-regioselectivity of carbonyl additions and for high torsional stability of the 2,3-double bond in 3-alkylated reagents, which is one reason the for high non-induced diastereoselectivities observed.
3. The enolic carbamates formed are stable to many acid and basic reaction conditions, therefore, these represent ideally protected 4-hydroxy-alkanals and -alkanones. Nevertheless, deprotection or further

highly stereoselective substitutions of the enolic double bonds proceed smoothly under mild conditions.

4. Due to the flexibility in respect to constitution and configuration, several advantageous applications are given or are anticipated in the synthesis of enantiomerically pure, natural and unnatural target molecules of pharmaceutical interest.

It should be added, that O-carbamoyl groups also proved a success in the generation of dianions(35), oxybenzyl(36), and oxy-2-alkinyl(37) anions. The latter are the basis for a highly diastereoselective syntheses of allenes and 1,3-alkadienes(38) by Claisen rearrangement.

Acknowledgements: It is a great pleasure to me, to express my thanks to engaged and able coworkers, whose names appear in the references. The answer on stereochemical questions was greatly facilitated by the stimulating cooperation with one of the leading X-ray groups, namely, Prof. George M. Sheldrick, Dr. Ernst Egert, Dr. Peter Jones and cand. phil. Dieter Schmidt, all members of the Department of Inorganic Chemistry at University of Göttingen. Financial support by the Deutsche Forschungsgemeinschaft and the Fonds der Chemischen Industrie is gratefully acknowledged.

REFERENCES AND FOOTNOTES

(*) New address: Department of Organic Chemistry, University of Kiel, Olshausenstr. 40, 2300 KIEL 1.

(1) Review: D. Hoppe, Angew. Chem. **96** (1984) 930; Angew. Chem. Int. Ed. Engl. **23** (1984) 926.

(2) The term "non-induced diastereoselectivity" refers to diastereoselectivities being observed in reactions of only prochiral centers. Unless coupled with "induced diastereoselectivity" (under the influence of existing chiral centers), on principle, racemic diastereomers are formed; unpublished proposal by Prof. L. F. Tietze (University of Göttingen).

(3) Review: N. H. Werstiuk, Tetrahedron **39** (1983) 205.

(4) G. Büchi, H. Wüest, J. Org. Chem. **34** (1969) 1122.

(5) Reviews: a) H. Ahlbrecht, Chimia **31** (1977) 391; b) J. C. Stowell, Chem. Rev. **84**, 409 (1984).

(6) Reviews, dealing with regioselectivity in the application of d^3-reagents: a) D. Seebach, Angew. Chem. **91** (1979) 259; Angew. Chem. Int. Ed. Engl. **18** (1979) 239. b) D. Seebach und K.-H. Geiss, J. Organomet. Chem., Libr. **1** (1976) 1; c) J. F. Biellmann, J.-P. Ducep, Org. Reactions **27** (1982).

(7) Reviews: a) R. W. Hoffmann, Angew. Chem. **94** (1982) 569; Angew. Chem.

Int. Ed. Engl. 22 (1982) 555; b) Y. Yamamoto, K. Maruyama, Heterocycles 18 (1982) 357.

(8) D. Hoppe, L. Beckmann, R. Follmann, Angew. Chem. 92 (1980) 305; Angew. Chem. Int. Ed. Engl. 19 (1980) 303.

(9) D. Hoppe, R. Hanko, A. Brönneke, Angew. Chem. 92 (1980) 637; Angew. Chem. Int. Ed. Engl. 19 (1980) 625.

(10) D. Hoppe, R. Hanko, A. Brönneke, F. Lichtenberg, Angew. Chem. 93 (1981) 1106; Angew. Chem. Int. Ed. Engl. 20 (1981) 1024.

(11) D. Hoppe, R. Hanko, A. Brönneke, F. Lichtenberg, E. van Hülsen, Chem. Ber. 118 (1985), in press.

(12) P. A. Grieco, T. Oguri, Y. Yokoyama, Tetrahedron Lett. 1978, 419.

(13) D. Hoppe, A. Brönneke, Tetrahedron Lett. 24 (1983) 1687.

(14) A. Brönneke, Dissertation, University of Göttingen 1983.

(15) R. Hanko, D. Hoppe, Angew. Chem. 94 (1982) 378; Angew. Chem. Int. Ed. Engl. 21 (1982) 372.

(16) Reviews: M. T. Reetz, Top. Curr. Chem. 106 (1982) 1; M. T. Reetz, R. Steinbach, B. Wenderoth, J. Westermann, Chem. and Ind. (London) 1981, 541.

(17) Reviews: B. Weidmann, D. Seebach, Angew. Chem. 95 (1983) 12; Angew. Chem. Int. Ed. Engl. 22 (1983) 31; D. Seebach, B. Weidmann, L. Widler in (R. Scheffold) Modern Synthetic Methods 1983, p. 217.

(18) a) D. Hoppe, F. Lichtenberg, Angew. Chem. 94 (1982) 378; Angew. Chem. Int. Ed. Engl. 21 (1982) 372; Angew. Chem. Suppl. 1982, 961 - 972. b) D. Hoppe, F. Lichtenberg, Angew. Chem. 96 (1984) 241; Angew. Chem. Int. Ed. Engl. 23 (1984) 239.

(19) V. J. Jephcote, A. J. Pratt, E. J. Thomas, J. Chem. Soc. Chem. Commun. 1984, 800.

(20) R. W. Hoffmann, B. Landmann, Angew. Chem. 96 (1984) 427; Angew. Chem. Int. Ed. Engl. 23 (1984) 437.

(21) H. Roder, G. Helmchen, E.-M. Peters, K. Peters, H.-G. von Schnering, Angew. Chem. 96 (1984) 895; Angew. Chem. Int. Ed. Engl. 23 (1984) 898.

(22) M. Reggelin, Diploma Thesis, in preparation, University of Göttingen 1985.

(23) A. G. Olivero, B. Weidmann, D. Seebach, Helv. Chim. Acta 64 (1981) 2485; and ref.(17).

(24) M. T. Reetz, S.-H. Kyung, J. Westermann, Organometallics **3** (1984) 1716, and ref.(16).

(25) T. R. Kelly, P. N. Kaul, J. Org. Chem. **48** (1983) 2775.

(26) M. Wilckens, Diploma Thesis, University of Göttingen 1984.

(27) G. Tarara, Diploma Thesis, University of Göttingen 1985.

(28) J. Lüßmann, Diploma Thesis, University of Göttingen 1984.

(29) Review: K. B. Sharpless, T. Verhoeven, Aldrichim. Acta **12** (1979) 63.

(30) P. G. Jones, D. Schmidt, J. J. Stezowski, unpublished results.

(31) D. Hoppe, J. Lüßmann, unpublished results.

(32) E. van Hülsen, D. Hoppe, Tetrahedron Lett. **26** (1985), 411.

(33) H. Bosshardt, M. Schlosser, Helv. Chim. Acta **63** (1980) 2393.

(34) D. Hoppe, E. van Hülsen, unpublished results.

(35) R. Hanko, D. Hoppe, Angew. Chem. **93** (1981) 115; Angew. Chem. Int. Ed. Engl. **20** (1981) 127.

(36) D. Hoppe, A. Brönneke, Synthesis **1982**, 1045.

(37) D. Hoppe, C. Riemenschneider, Angew. Chem. **95** (1983) 64; Angew. Chem. Int. Ed. Engl. **22** (1983) 54.

(38) D. Hoppe, C. Gonschorrek, E. Egert, D. Schmidt, Angew. Chem. **97** (1985), issue 8, in press; Angew. Chem. Int. Ed. Engl. **25** (1985), issue 8, in press.

ALDOLASES AS CATALYSTS IN ORGANIC SYNTHESIS

Chi-Huey Wong
Department of Chemistry
Texas A&M University
College Station
Texas 77843, USA

ABSTRACT. This manuscript describes studies of the synthetic utility of more than 20 enzymes which catalyze aldol addition reactions. The enantio- and diastereo-specific carbon-carbon bond formation catalyzed by aldolases is one of the most efficient and practical route to common and uncommon sugar derivatives. Detailed studies on the synthetic application of FDP aldolase-catalyzed condensation indicate that enzymatic aldol reactions seem to have genuine value as a method of synthesis for polyhydroxy compounds. The enzymatic aldol reactions provide an alternative route to carbohydrates which are difficult to prepare based on organic asymmetric aldol addition reactions and complement methods based on asymmetric epoxidations and asymmetric Diels-Alder reactions.

1. INTRODUCTION

Asymmetric carbon-carbon bond formation based on aldol addition reactions remains one of the most interesting and challenging subjects in synthetic organic chemistry.[1-4] Many complex molecules in principle can be built up efficiently by this type of reaction. Most of the asymmetric aldol reactions developed recently require the use of a metal or metal-like complex of an enolate to achieve stereoselectivity. Because the enolate complex is unstable and tends to dissociate in aqueous solution, the asymmetric aldol condensations must be carried out in organic solvents at low temperature. This requirement limits the application of organic aldol reactions to synthesis of molecules soluble in organic media. The synthesis of polyfunctional compounds such as carbohydrates still relies upon other methods. The aldol addition reactions catalyzed by the aldolases, however, are performed in aqueous solution using polyfunctional substrates. Both chemical and enzymatic aldol reactions thus could complement each other and extend the synthetic utility of aldol reactions.

There have been about 25 different aldolases isolated so far. Each of them catalyzes a different type of aldol reaction using different substrates. Most of the reactions are involved in the biosynthesis of carbohydrates and amino acids. The work reported in this manuscript is

M. P. Schneider (ed.), Enzymes as Catalysts in Organic Synthesis, 199–216.

part of a program to evaluate the potential of aldolase-catalyzed aldol reactions as a route to isotopically labelled sugars and uncommon sugars which may be difficult to prepare based on organic chemical aldol reactions.

2. TYPES OF ENZYMATIC ALDOL REACTIONS

There are two types of aldolases. The enzymes from plants and animals (class I aldolases) use an enzyme amino group to form a Schiff base intermediate for the activation of the aldol donors; those from lower organisms (class II aldolases) use a metal ion, usually Zn^{++}, at the active site to form an enolate intermediate for the similar reaction.

2.1. Fructose-1,6-Diphosphate Aldolases

These enzymes catalyze the reversible condensation of dihydroxyacetone phosphate **1** and D-glyceraldehyde-3-phosphate **2** to form fructose-1,6-diphosphate **3**, an essential reaction in the pathways for synthesis and fermentation of hexoses. Equations (1) and (2) indicate the enantio-

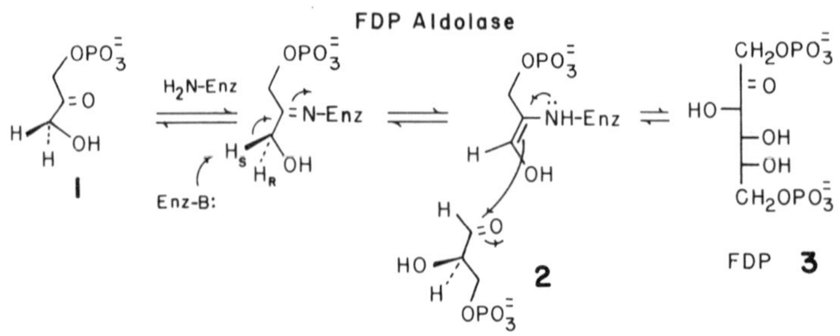

(1)

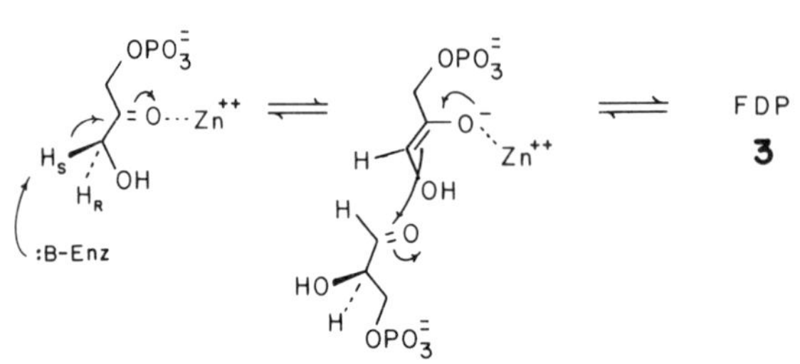

(2)

and diastereo-specific carbon-carbon bond formation in aldol reactions catalyzed by FDP aldolase from rabbit muscle[5] and from yeast[6], respectively. Two interesting features of these enzymes deserve a brief comment. First, these enzymes require **1** as one substrate, but will accept a wide variety of aldehydes as the second.[7] Second, the yeast aldolases have the zinc ion replaced with other metal ions by equilibrium dialysis and still retain significant activity. Table 1 indicates the relative activities of some modified FDP aldolases which contain metal ions other than Zn^{++}. Since the zinc ion of FDP aldolases plays a very important role in the catalysis, these modified enzymes may have some synthetic utility. Whether they would accept different substrates remains to be investigated.

TABLE I. Fructose Diphosphate Cleavage Catalyzed by Yeast Aldolase

Enzyme	Metal Ion	Rel Activity (a)	Rel Activity (b)
Apoenzyme	-	2	< 0.5
	+ Zn(II)	100	100
	+ Co(II)	85	51
	+ Fe(II)	67	38
	+ Mn(II)	15	14
	+ Ni(II)	11	< 0.5
	+ Cu(II)	< 2	< 0.5
	+ Mg(II)	< 2	< 0.5
	+ Fe(III) or Hg(II) or Cd(II)	< 2	-

(a). R.D. Kobes, R.T. Simpson, B.L. Vallee. W.J. Rutter Biochemistry **8** (1969) 585.
(b). J.T. Kadonaya, J.R. Knowles Biochemistry **22** (1983) 130.

2.2. Fuculose-1-Phosphate and Rhamnulose-1-Phosphate Aldolases.

These two types of aldolases use the same substrates (**1** and L-lactaldehyde **4**) in natural reactions with different stereospecificity in C-C bond formation (Equations 3 and 4). Both enzymes can be isolated from *E. coli* mutants.[8,9] The former isolated from *E. coli* 0111-B_4 has a specific activity of 4 U/mg (1 U = 1 umol product formed per min) and the latter isolated from *E. coli* K_{40} has a specific activity of 18 U/mg. These enzymes are responsible for the syntheses of L-fucose and L-rhamnose, important components of a number of cell-surface glycoprotein determinants currently difficult to prepare. In the same species the aldol products (fuculuoe-1-phosphate and rhamnulose-1-phosphate) are subsequently converted to the phosphate free ketose species catalyzed by phosphatases and isomerized to the aldose species **5** and **6** catalyzed by fucose isomerase and rhamnose isomerase respectively. These two aldolases seem to accept other aldehydes as a second substrate, but no detailed study on the substrate specificity has been carried out.

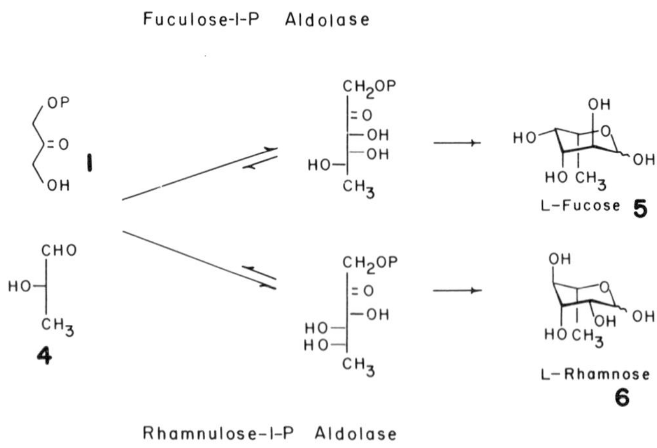

(3)

(4)

2.3. 3-Deoxy-D-Manno-Octulosonate (KDO) Aldolases and Sialic Acid Aldolases

These two classes of enzymes are present in E. coli mutants and other species and use pyruvate as a common natural aldol donor. The KDO aldolases, which can be isolated from E. coli 0111-B_4 use D-arabinose as a natural aldol acceptor to form the eight-carbon sugar KDO[10] (Equation 5), an important lipopolysaccharide determinant of Gram-negative bacterial cell walls. The specific activity is about 6 U/mg. The sialic acid aldolases which can be isolated from E. coli K-235 use N-acetylmannosamine as a natural acceptor to form the nine-carbon sugar sialic acid[11] (Equation 6), a terminal messenger for a number of polysaccharides. The specific activity is about 10 U/mg. The sialic acid aldolases have been used in a small scale synthesis of sialic acid and analogs[12]. One of the problems in using these two types of aldol reactions is the reversibility of the reactions; the equilbrium constant for each reactions is about 0.1 M. Fortunately, two other enzymes (KDO synthetase[13] and sialic acid synthetase[14]) which use phosphoenolpyruvate **12** as an aldol donor catalyze the irreversible aldol reactions for the synthesis of these two important sugars (Equations 7 and 8). The irreversibility of the reactions would simplify the isolation procedures and make the processes feasible for large-scale operation. Given the fact that compound **12** can be prepared on large scales economically[15], these two enzymes would be useful as catalysts for large-scale preparation of these two sugars.

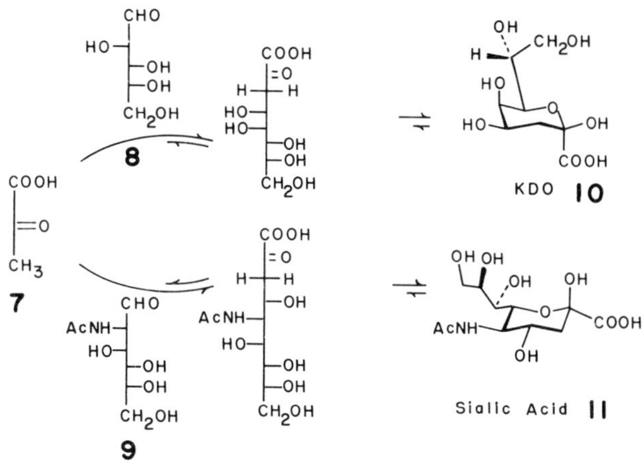

(5)

(6)

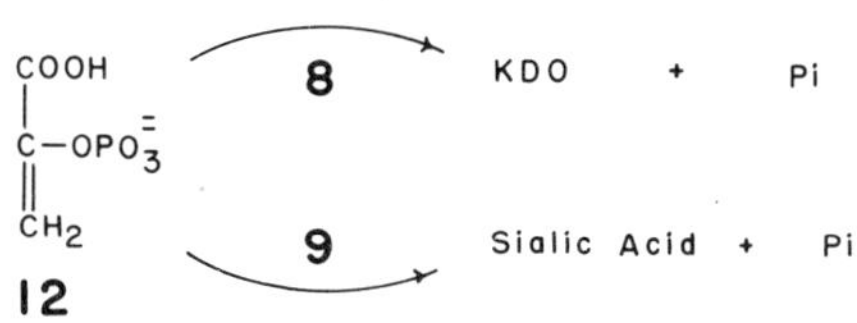

(7)

(8)

2.4. 2-Deoxyribose-5-Phosphate (**14**) Aldolases.

These enzymes catalyze an aldol reaction using acetaldehyde as an aldol donor and D-glyceraldehyde-3-phosphate **13** as an acceptor for the synthesis of the deoxysugar **14** (Equation 9), an important component of DNA molecules. The enzymes can be easily isolated from Lactobacillus plantarum (ATCC 8041)[16]. The enzyme from this species has a reasonably high catalytic activity (640 U/mg). The problems are that it requires compound **13** which is not readily available and that the equilibrium of the reaction is not quite in favor of condensation ($K_{eq} = 2.4 \times 10^{-4}$ M). One possible solution is to start with an excess of acetaldehyde and compound **1** which is easy to prepare on large scale (see below) and which can be converted into **13** in situ catalyzed by triose phosphate isomerase.

The aldolases discussed so far are those involved in the biosynthesis of polyhydroxy compounds (or carbohydrates). I should mention that many other aldolases isolated before could also be useful

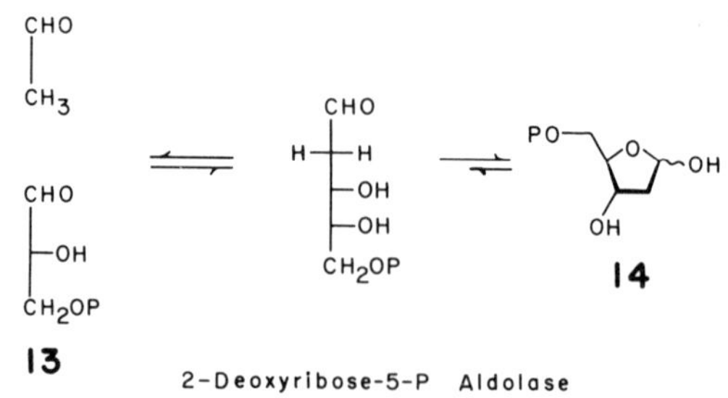

(9)

as synthetic catalysts in carbohydrate synthesis. Some of these enzymes are summarized in Table 2[17] based on the stereochemistry of the carbon-carbon bond formation and substrate specificity.

2.5. Threonine Aldolases and *allo*-Threonine Aldolases.

The aldolases involved in the synthesis of threonine and *allo*-threonine using acetaldehyde and glycine as substrates are D- and L-threonine aldolases and D- and L-*allo*-threonine aldolases. Equation (10) indicates the synthesis of L-threonine **16** catalyzed by L-threonine aldolase from *Candida humicola* (ATCC 14438)[18] and D-threonine **17** catalyzed by D-threonine aldolase from an *Arthrobacter* species[19]. Previous studies indicate that both enzymes also accept analogous

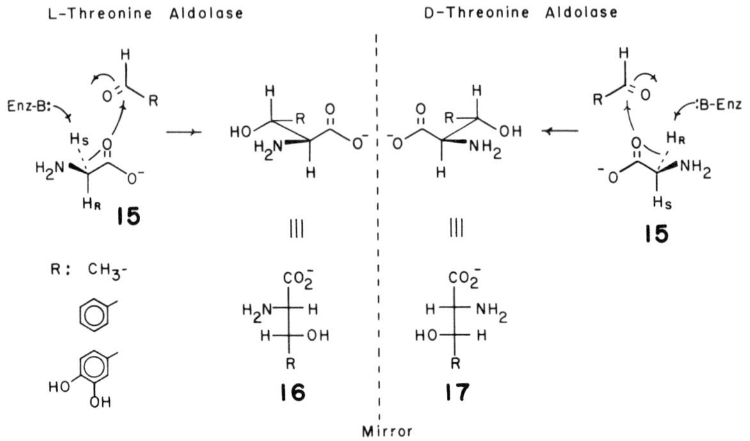

(10)

TABLE II. Enzymes Which Catalyze Aldol Reactions.

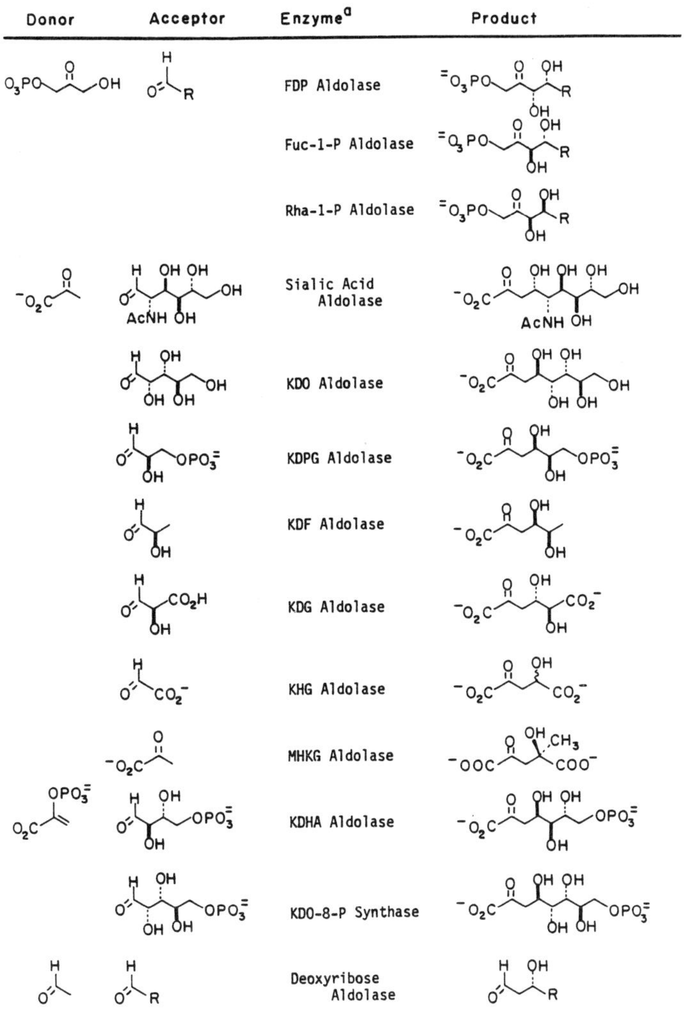

[a]Abbreviations: FDP, D-fructose-1,6-diphosphate; Fuc-1-P, L-fuculose-1-phosphate; Rha-1-P, L-rhamnulose-1-phosphate; KDPG, 2-keto-3-deoxy-6-phosphogluconate; KDF, 2-keto-3-deoxy-D-fuconate; KDG, 2-keto-3-deoxy-D-glucarate; KHG, 2-keto-4-hydroxyglutarate; MHKG, 4-methyl-4-hydroxy-2-ketoglutarate; KDHA, 2-keto-3-deoxyheptonate.

aldehydes as acceptors which would allow development of new methods for the synthesis of other amino acids. For example, in the L-threonine aldolase-catalyzed reaction using benzaldehyde as an acceptor (Equation 11), the aldol product could be acetylated and converted to **18** which, upon hydrogenolysis, would give N-acetyl-L-phenylalanine **19** in high yield[20]. This process could be an alternative approach for the manufacture of L-phenylalanine, a precursor of the artificial sweetener Aspartame.

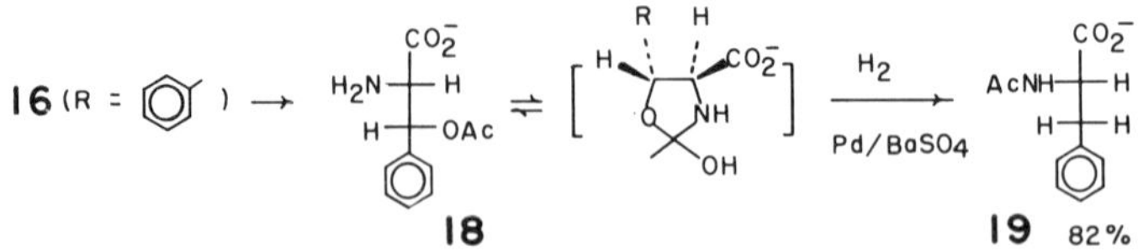

(11)

3. SYNTHESIS BY FDP ALDOLASE-CATALYZED REACTIONS

The FDP aldolase from rabbit muscle is commercially available. It is inexpensive, and it has a high specific activity and good stability. As mentioned above, the enzyme is quite specific for **1**; any change in the structure of **1** greatly reduces reactivity in the aldol reactions. The analogs of **1** reported to be accepted as weak substrates are acetol phosphate (about 0.01% as active as **1**)[21] and 4-hydroxy-3-oxo-1-phosphonate (about 10% as active as **1**)[22]. The specificity for the aldehydes, however, is quite broad and which provides a practical route to a number of simple sugars and sugar derivatives, and especially to ^{13}C-labelled sugars important in the study of carbohydrate metabolism. The procedure for the synthesis has the advantage that it is applicable to unprotected sugars in aqueous solution at pH 7. Using immobilized enzymes and <u>in situ</u> cofactor regeneration, these methods should be applicable to preparations carried out on mole scales.

3.1. Preparation of Isotopically Labelled Sugars

In preparing isotopically labelled sugars, it is useful to incorporate the labels to the two C_3 pieces which are assembled to make the C_6 sugar. Scheme 1 outlines a synthesis of glucose (and intermediates) labelled with ^{13}C at the C-2 and C-2,5 positions. Starting with (2-$^{13}C_1$)glycerol the target molecules were prepared with high yield in a combined chemical and enzymatic approach[7]. Although the process for the reactions shown in the scheme looks very complicated, it is in fact quite straightforward. Several enzymes are coimmobilized in the same polymer matrix and work together to make the target molocules efficiently -- an unique feature of enzyme catalysis which is difficult to excute based on transition metal catalysis. Figure 1 indicates the ^{13}C-NMR spectra of glucose prepared by using these procedures. Similar sequences of reactions are also applicable to the synthesis of glucose and fructose having different labelling patterns and to L-sorbose derivatives. The aldol acceptors used in these reactions need not be optically pure. They can be used as a racemic mixture and the epimeric mixtures produced in the aldolase-catalyzed reactions can be separated without chromatography by taking advantage of substrate-selective enzymatic phosphorylation. The reactions shown in Scheme 2 illustrates a representative example. As shown in the Scheme, after acid-catalyzed hydrolysis to remove the phosphate moiety from the mixture of fructose and sorbose phosphates, the fructose in the mixture can be converted to D-fructose-6-phosphate **24** with hexokinase and thus separated from the

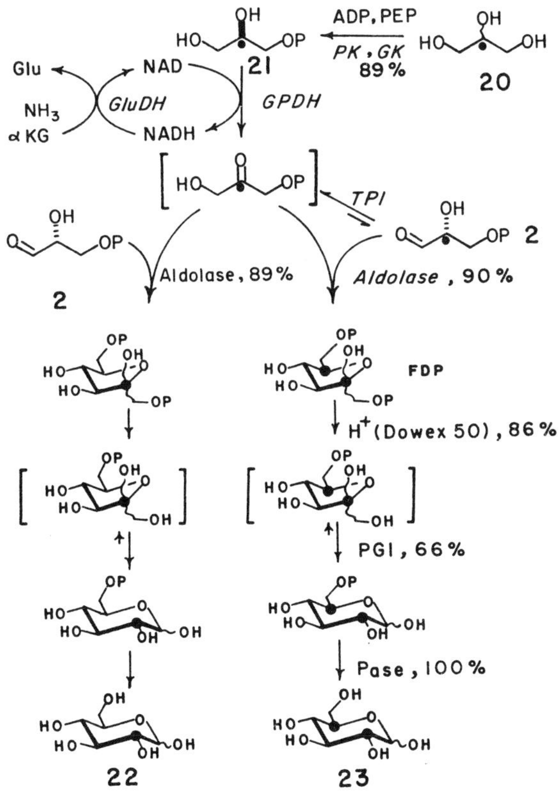

Scheme 1. Synthesis of (2-$^{13}C_1$)Glucose **22** and (2,5-$^{13}C_2$)Glucose **23** and Intermediates (Heavy Circles indicate Positions of ^{13}C Label). Abbreviations: PEP, phosphoenolpyruvate; PK, pyruvate kinase; GK, glycerokinase; KG, 2-ketoglutarate; Glu, L-glutamate; GPDH, glycerophosphate dehydrogenase; TPI, triosephosphate isomerase; PGI, phosphoglucoisomerase; Pase, inorganic phosphatase; FDP, D-fructose-1,6-diphosphate.

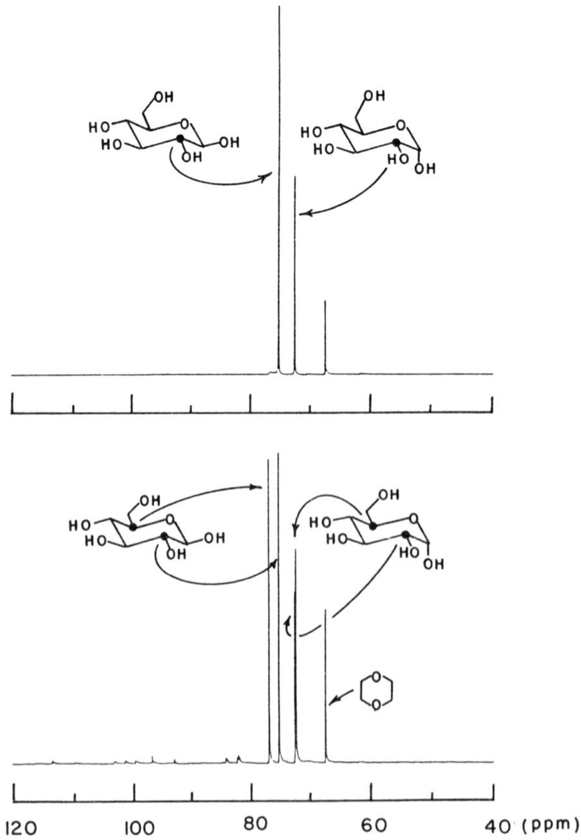

Figure 1. ^{13}C-NMR Spectra of Compounds **22** and **23**.

unreacted L-sorbose **25**. The sugar phosphates prepared are 80-90% pure. The impurities are mainly inorganic phosphates, and are of no consequence in most further enzymatic transformations.

3.2. Synthesis of Deoxysugars

The specific activity of FDP aldolase as a catalyst for the aldol condensation of **1** with lactaldehyde (**4** or **26**) or 2-hydroxybutyraldehyde **30** (5U/mg) is about 10% of that observed with use of its natural substrates **1** and **2**. Scheme 3 outlines the synthesis of 6-deoxy-L-

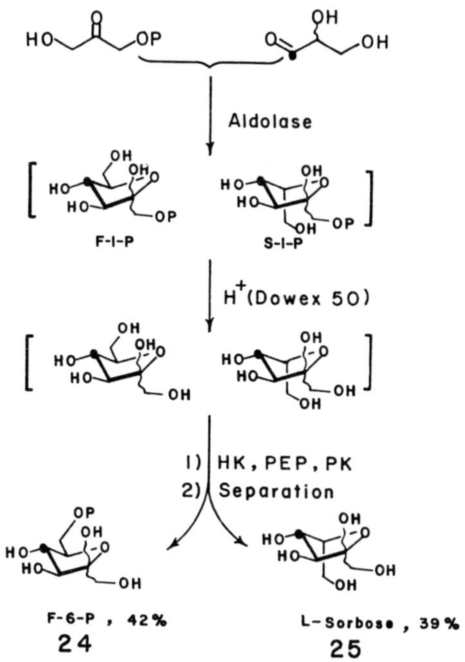

Scheme 2. Preparation of (4-^{13}C)-D-Fructose **24** and (4-^{13}C)-L-Sorbose **25**. HK, hexokinase. The aldolase is from rabbit muscle.

sorbose **27** and 6-deoxy-D-fructose **28**[23]. These two deoxysugars can be converted into 2,5-dimethyl-4-hydroxy-2,3-dihydrofuran-3-one (Furaneol **29**, a caramel flavor component) at 80 °C in 80% yield, using a mixture of piperidine and acetic acid (6:10 by weight) as catalyst and absolute ethanol as solvent. The mechanism for the conversion of 6-deoxysugars to Furaneol is not clear. It may involve acid- and base-catalyzed dehydration and rearrangement as shown in Scheme 4. This proposed mechanism can be used to rationalize the conversion of 6,7-dideoxyheptoses prepared by FDP-aldolase-catalyzed condensation of **1** and DL-2-hydroxybutyraldehyde to the Furaneol analogs **31** and **32** as shown in Scheme 5[23].

3.3. Other Syntheses[25]

The synthesis of sugars based on FDP aldolase catalysis is not limited to isotopically labelled and 6-deoxy sugars as described above. A

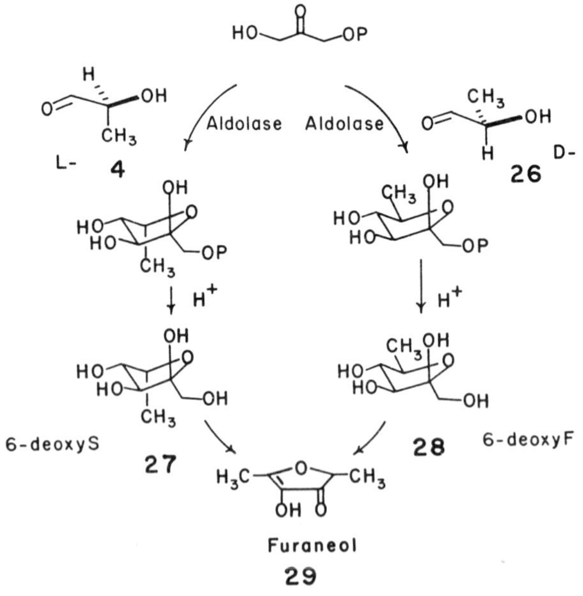

Scheme 3. Preparation of 6-Deoxyhexoses and Furaneol **29**

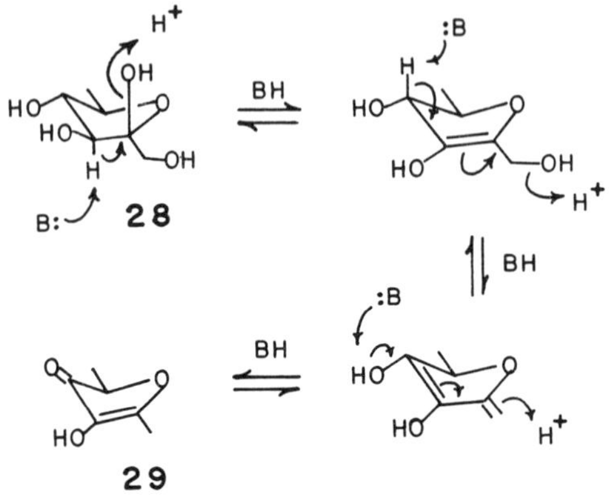

Scheme 4. Proposed Mechanism for the Conversion of **28** to **29**.

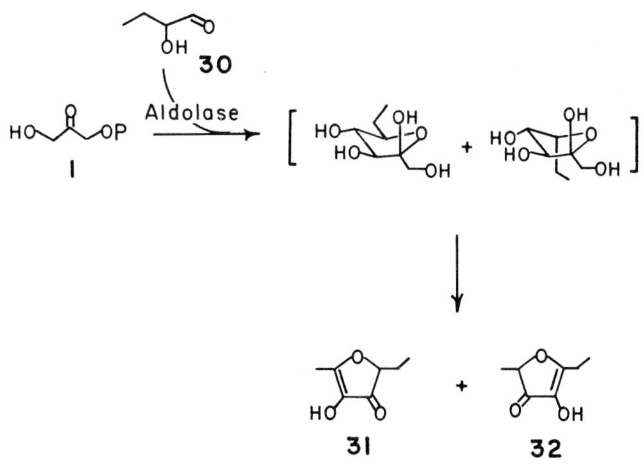

Scheme 5. Synthesis of Furaneol Analogs.

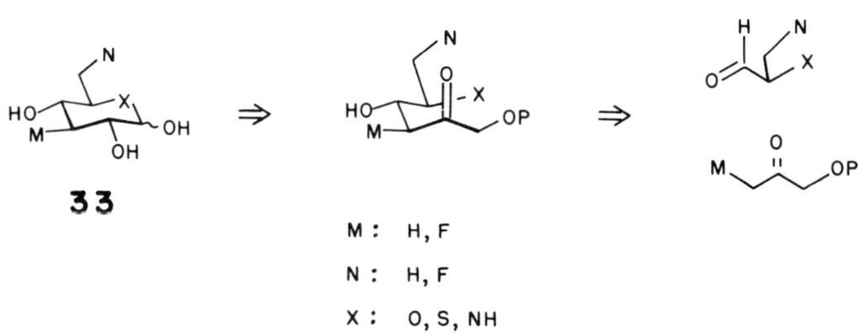

(12)

FDP Aldolase Catalyzed Synthesis

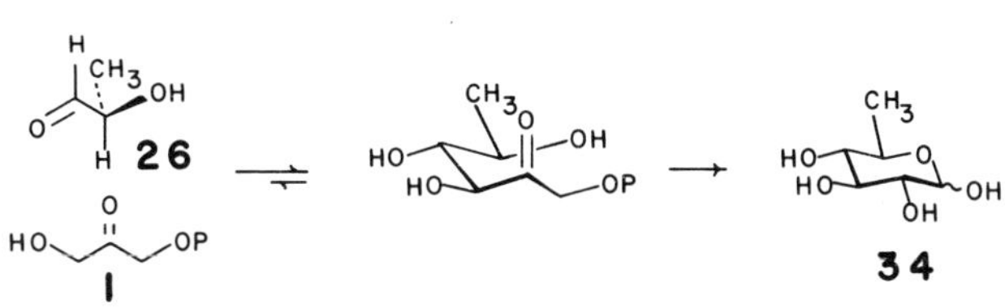

(13)

retrosynthetic analysis (Equation 12) indicates that several other interesting uncommon sugars could be prepared by this type of aldol reactions. Fluorinated sugars in which one of the hydroxy groups is replaced with the F group, 3- or 5-deoxyhexoses, and hexoses with a different heteroatom in the ring (**33**) could be prepared using FDP aldolases as catalysts. These uncommon sugar derivatives are of potential interest in medicinal chemistry; they are currently prepared by chemical methods which require multiple protection and deprotection steps. The reactions illustrated in Equation 13 provide a representative procedure for the synthesis of 6-deoxy-D-glucose **34**. The FDP aldolase catalyzed condensation of **1** with D-lactaldehyde **26** gave 6-deoxy-D-fructose-1-phosphate which, after acid hydrolysis to remove the phosphate group, was converted to **34** catalyzed by glucose isomerase. The isomerization reaction catalyzed by glucose isomerase gives an approximately equal mixture of 6-deoxy-D-fructose and **34** which can be easily separated by passing through a Ba^{++}-Dowex 50 column with water as the eluant. If the aldol acceptor is changed to 3-hydroxypropanal, the reaction product is 5-deoxy-D-fructose-1-phosphate which could be converted to 5-deoxy-D-fructose **36** after hydrolysis. The ketose can be further isomerized to 5-deoxy-D-glucose **37** (Equation 14). Since compound **36** locks the hexose in a 6-member ring conformation and compound **37** in a 5-member ring conformation, they may have some interesting applications.

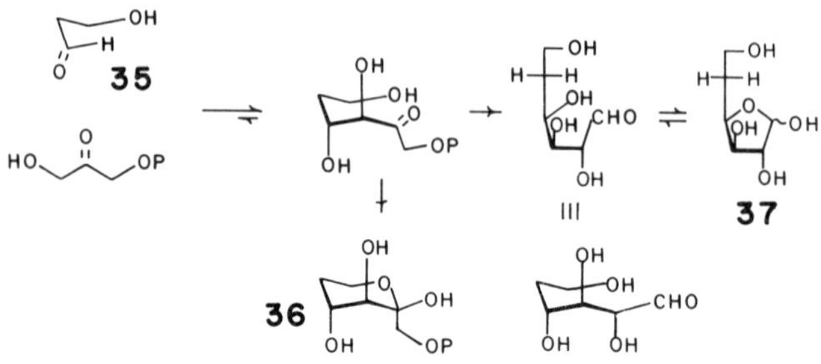

(14)

4. CONCLUSION AND OUTLOOK

The FDP aldolase-catalyzed condensations as illustrated provide a practical route to a wide variety of simple sugars and sugar derivatives. The ketose phosphate product produced can be subsequently converted to aldose phosphate derivatives catalyzed by phosphoglucoisomerase and to aldose derivatives after hydrolysis to remove the phosphate groups. The phosphate free ketose derivatives also can be directly isomerized to aldose derivatives catalyzed by glucose isomerase. The enzyme glucose isomerase from Flavobacterium arborescens was originally named xylose isomerase which catalyzes the isomerization of xylulose to xylose _in vivo_. The unnatural catalytic property of this enzyme in catalyzing the isomerization of glucose to fructose has been used for large-scale production of high fructose corn syrup from glucose or starch. Millions of metric tons of high fructose corn syrup have been produced in the United States every year. This reaction of course is not of major interest to synthetic organic chemists. The enzyme is, however, readily available in large quantities, and the range of substrates which the enzyme will accept has not been explored in detail except some 5- and 6-carbon sugar analogs. If the glucose isomerase would accept a broad spectrum of substrates, the combination of FDP aldolases and glucose isomerase would provide a practical synthetic route to a number of sugar derivatives. A limitation of the enzymatic method is that dihydroxyacetone phosphate **1** is required as one component and which must be prepared separately from dihydroxyacetone, either by chemical or enzymatic phosphorylation[7]. A second disadvantage is that there is no way of changing the stereochemistry of the centers formed in the FDP aldolase-catalyzed condensation. Although different types of aldol condensations with respect to the stereochemistry of C-C bond formation exist in nature, they use different substrates and are available only by isolation. Whether these other aldolases are stable enough for use as synthetic catalysts and whether they have a broad spectrum of substrate specificity remain to be explored.

Synthesis based on enzymatic aldol reactions has so far been restricted to research application. Current research in this area is to establish the real practicality of enzymatic aldol reactions and isomerizations. How broadly applicable are they? Can the substrates **1** and **12** and the aldol products be prepared on large scales economically? In what circumstances do they provide methods superior to existing or easily developed organic techniques for the synthesis of same molecules? If the demand for these enzymatic reactions can be justified, the immediate future will see the application of the techniques of recombinant DNA technology to the large-scale production of aldolases and modified aldolases with better properties.

In summary, the enzymatic aldol reactions seem to have genuine value as a method of synthesis for polyhydroxy compounds and to complement methods based on organic chemical aldol reactions[1-4], asymmetric epoxidation reactions[26], and asymmetric Diels-Alder reactions[27] developed by a community of synthetic chemists.

5. ACKNOWLEDGMENT

The author is indebted to Professor George M. Whitesides for his previous guidance. Part of this work was carried out by Dale Drueckhammer and Rob Durrwachter, both are graduate students at Texas A&M. Supply of the enzyme glucose isomerase from Dr. R. Weiss of Miles Laboratory and support of this research by the National Science Foundation are gratefully acknowledged.

6. REFERENCES

1. C.H. Heathcock "The Aldol Addition Reactions" in J.D. Morrison Asymmetric Synthesis **3** (1984) 111, Academic Press.

2. D.A. Evans, J.V. Nelson and T.R. Tabor Top. Stereochem. **13** (1982) 1.

3. S. Masamune, T. Kaiho, D.S. Garvey "Aldol Methodology: Synthesis of Versatile Intermediates 3-Hydroxy-2-vinylcarbonyl Compounds" J. Am. Chem. Soc. **104** (1982) 5521.

4. P.A. Bartlett "Stereocontrol in the Synthesis of Acylic Systems: Applications to Natural Product Synthesis" Tetrahedron **36** (1980) 3; B.M. Trost "Sculpting Horizons in Organic Chemistry" Science **227** (1985) 908.

5. I.A. Rose "The Absolute Configuration of Dihydroxyacetone Phosphate Tritiated by Aldolase Reaction" J. Am. Chem. Soc. **80** (1958) 5835.

6. J.T. Kadonaga, J.R. Knowles "Role of Mono- and Divalent Metal Cations in the Catalysis by Yeast Aldolase" Biochemistry **22** (1983) 130.

7. C.-H. Wong, G.M. Whitesides "Synthesis of Sugars by Aldolase-Catalyzed Condensation Reactions" J. Org. Chem. **48** (1983) 3199.

8. M.A. Ghalambor, E.C. Heath "The Metabolism of L-Fucose II. The Enzymatic Cleavage of L-Fuculose-1-Phosphate" J. Biol. Chem. **237** (1962) 2427.

9. T.H. Chiu and D.S. Feingold "L-Rmamnulose-1-Phosphate Aldolase from Escherichia coli. Crystallization and Properties" Biochemistry **8** (1969) 98.

10. A. Weissbach and J. Hurwitz "The Formation of 2-Keto-3-Deoxyheptonic Acid in Extracts of Escherichia coli B" J. Biol. Chem. **234** (1959) 705; M.A. Ghalambor, E.C. Heath "The Biosynthesis of Cell Wall Lipopolysaccharide in Escherichia coli V. Purification and Properties of 3-Deoxy-D-Manno-Octulosonate Aldolase" J. Biol. Chem. **241** (1966) 3222.

11. D.G. Comb, S. Roseman "The Sialic Acids I. The Structure and Enzymatic Synthesis of N-Acetylneuraminic Acid" J. Biol. Chem. **235** (1960) 2529.

12. C. Auge, S. David, C. Gautheron "Synthesis with Immobilized Enzyme of the Most Important Sialic Acid" Tetrahedron Lett. **25** (1984) 4663; C. Auge, S. David, C. Gautheron and A. Veyrieres "Synthesis with an Immobilized Enzyme of N-Acetyl-9-O-Acetyl-Neuraminic Acid, A Sugar Reported as a Component of Embryonic and Tumor Antigens" ibid **26** (1985) 2439.

13. P.H. Ray "Purification and Characterization of 3-Deoxy-D-Manno-Octulosonate-8-Phosphate Synthetase from Echerichia coli" J. Bacteriol. **141** (1980) 635.

14. R.S. Blacklow and L. Warren "Biosynthesis of Sialic Acid Acids by Neisseria meningitidis" J. Biol. Chem. **237** (1962) 3520.

15. B.L. Hirschbein, F.P. Mazenod, G.M. Whitesides "Synthesis of Phosphoenolpyruvate and Its Use in Adenosine Triphosphate Cofactor Regeneration" J. Org. Chem. **47** (1982) 3766.

16. W.E. Pricer, B.L. Horecker "Deoxyribose Aldolase from Lactobacillus plantarum" J. Biol. Chem. **235** (1960) 1292; P. Hoffee, O.M. Rosen, B.L. Horecker "The Mechanism of Action of Aldolases VI. Crystallization of Deoxyribose-5-Phosphate Aldolase and The Unmber of Active Sites" ibid **240** (1965) 1512.

17. G.M. Whitesides and C.-H. Wong "Enzymes as Catalysts in Synthetic Organic Chemistry" Angew. Chem. Int. Eng. Ed. In Press.

18. H. Nakazawa, H. Enei, K. Kubota, O. Kanagawa "Serine and Serinol Derivatives" Ger Offen 2310471 (1973); Chem. Abst. **79**:135298e (1973).

19. Denki Kagaku Kogyo K.K. "D-Threonine Aldolase" Jpn Kokai Tokkyo Koho JP 5811680 (1983); Chem. Abstract **99**:174232 (1983).

20. J.S. Tou, B.D. Vineyard "Novel Synthesis of L-Phenylalanine" J. Org. Chem. **49** (1984) 1135.

21. O.C. Richards, W.J. Rutter "Comparative Properties of Yeast and Muscle Aldolase" J. Biol. Chem. **236** (1961) 3185.

22. D. Stribling "Properties of the Phosphonomethyl Isosteres of Two Phosphate Ester Glycolytic Intermediates" Biochem. J. **141** (1974) 725.

23. C.-H. Wong, F.P. Mazenod and G.M. Whitesides "Chemical and Enzymatic Synthesis of 6-Deoxyhexoses. Conversion to 2,5-Dimethyl-4-Hydroxy-2,3-Dihydrofuran-3-one (Furaneol) and Analogs" J. Org. Chem. **48** (1983) 3493.

24. S.A. Firmenich, Geneva, Switzerland. Furneol is a registered trademark of Firmenich, Inc.

25. C.-H. Wong, D.G. Drueckhammer, J.R. Durrwachter, Unpublished.

26. T. Katsuki, K.B. Sharpless "The First Practical Method for Asymmetric Epoxidation" J. Am. Chem. Soc. **102** (1980) 5974; S.Y. Ko, A.W.M. Lee, S. Masamune, L.A. Reed, III, K.B. Sharpless, F.J. Walker "Total Synthesis of the L-Hexoses" Science **949** (1983) 220.

27. S.J. Danishefsky, C.J. Maring "A New Approach to the Synthesis of Hexoses: An Entry to (±)-Fucose and (±)-Daunosamine *J. Am. Chem. Soc.* **107** (1985) 1269; S.J. Danishefsky, W.H. Pearson, B.E. Segmuller "Total Synthesis of (±)-3-Deoxy-D-Manno-2-Octulopyranosate (KDO) *ibid* **107** (1985) 1280.

IMMOBILISED REDOX ENZYMES AND THEIR USE AS CATALYSTS FOR FINE CHEMICAL SYNTHESIS

Shirley Y.R. Pugh
Biotechnology Centre
Cranfield Institute of Technology, Cranfield
Bedfordshire. MK43 OAL
U.K.

ABSTRACT. The use of redox enzymes as catalysts in bioelectrochemical reactors is reviewed. Particular emphasis is placed on cofactor recycling, especially in terms of electrochemical regeneration of $NAD(P)^+$ and NAD(P)H and on the choice of, and immobilisation of, redox enzymes. Model reactors constructed for the production of 6-phosphogluconate and cinnamyl alcohol are discussed in detail. The future of enzymes in catalysing hydroxylation reactions in terms of functionalising carbon skeletons is explored.

1. INTRODUCTION

There is now considerable interest in the use of purified enzymes in the manufacture of fine chemicals and pharmaceuticals. In the future these proteinaceous biological catalysts will be of particular importance in the catalysis of oxidation-reduction reactions and especially in terms of functionalising carbon skeletons with hydroxyl groups, which presently is difficult to achieve by simple chemical routes.

Enzymes have three special properties to offer (1):

- Speed - Enzymes will catalyse reactions at rates of 10^6-10^8 elementary acts per second.
- Specificity - Enzymes promote only specific transformations within a fixed range of compounds. These transformations are both stereoselective and regioselective.
- Reaction conditions - Enzymes work at ambient temperatures and at approximate neutral pH.

In this discussion I will consider only one particular class of biological catalyst, the redox enzyme which promotes the addition or removal of electrons from the compound of interest; this is normally achieved by the interchange of molecular oxygen or hydrogen atoms. The majority of redox enzymes require an external source of redox power, normally termed a cofactor, for the catalysis of the reaction. The most common cofactors are nicotinamide adenine dinucleotide (NAD^+) and nicotinamide adenine dinucleotide phosphate ($NADP^+$) and their reduced forms NADH and NADPH (Figure 1). These cofactors are expensive,

M. P. Schneider (ed.), Enzymes as Catalysts in Organic Synthesis, 217–232.

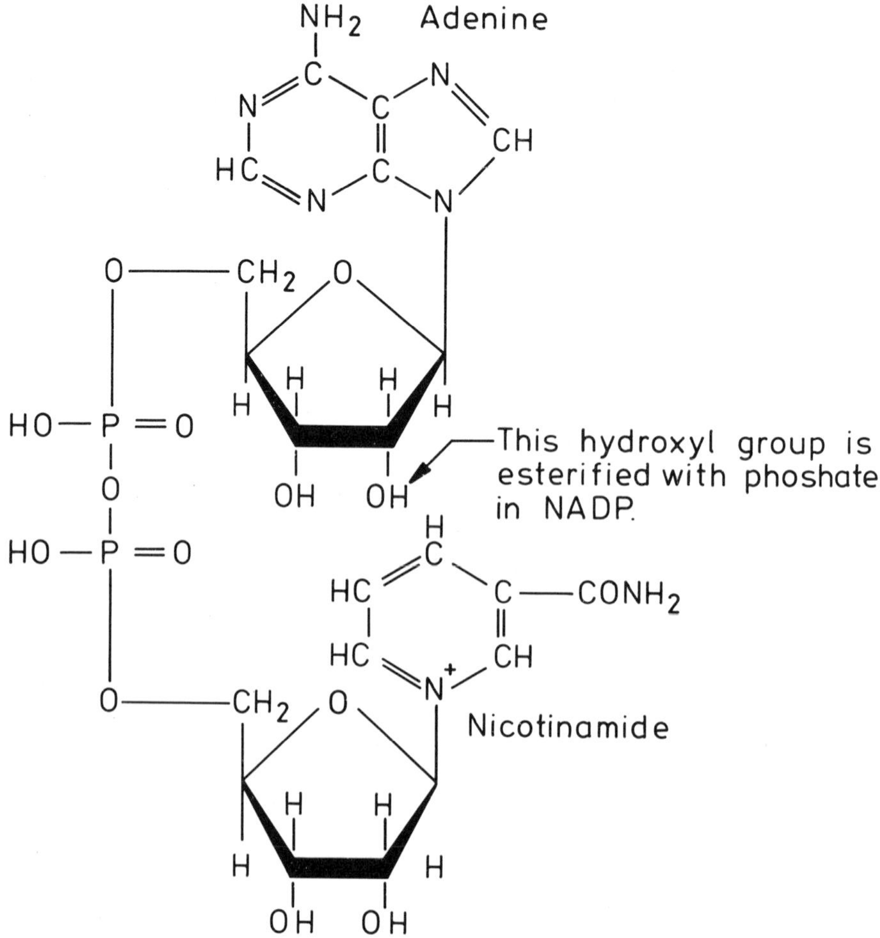

$$NAD(P) + 2e + H^+ \rightleftharpoons NAD(P)H$$

Figure 1. The structures of nicotinamide adenine dinucleotide and nicotinamide adenine dinucleotide phosphate.

£1000-2000 a mole (2) and this prevents their stochiometric use in the manufacture of fine chemicals. Thus, a major goal in this area of research is the efficient *in situ* recycling of these cofactors.

2. COFACTOR RECYLING

There are currently two methods available for cofactor regeneration. The first involves the use of a cheap sacrifical co-substrate, which in the presence of a second enzyme, continually regnerates the desired cofactor form which is usually NADH or NADPH. Examples of recycle couples are the enzyme formate dehydrogenase (3), EC 1.2.1.2, (FDH), which in the presence of formate and NAD^+ produces NADH and the co-product gaseous carbon dioxide, and the enzyme glucose-6- phosphate dehydrogenase, EC 1.1.1.49, (G6PDH), which produces NADH and 6-phophogluconate (6PG) from glucose-6-phosphate (G6P) and NAD^+ (Figure 2) (4). Of the two enzymes mentioned FDH is preferable in that the co-product, CO_2, is easily removed from the reactor.

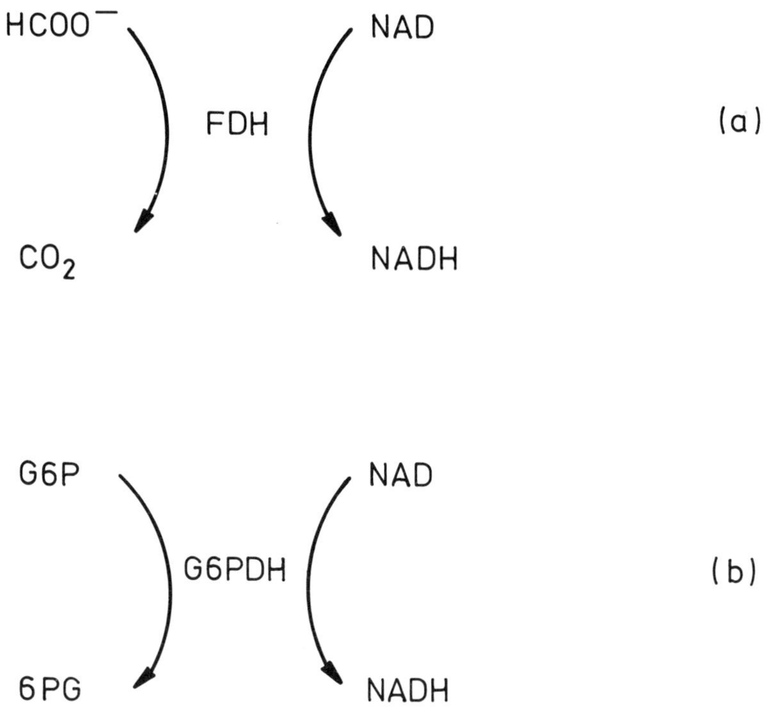

Figure 2. Use of sacrificial co-substrates for cofactor recycling.
(a) Formate dehydrogenase.
(b) Glucose-6-phosphate dehydrogenase.

The second method of cofactor recycling and the one our group has chosen to concentrate its efforts on is electrochemical recycling. This is intrinsically a much more difficult problem to solve and this is reflected by the lack of electrochemistry in this field of fine chemical synthesis.

Electrochemistry, in theory, allows the use of an electrode potentiostated at the correct redox potential for the regeneration of the correct cofactor form. In practice direct electrochemical production of NAD^+ is relatively easy to achieve (Figure 3a) (5), but the reverse reaction is somewhat more difficult and regeneration has only been achieved in the presence of a mobile electro-active mediator and a diaphorase enzyme (Figure 3b). The mediator shuttles electrons from the reducing electrode surface to the enzyme, which uses this 'reducing power' for the production of NAD(P)H. Two common enzymes possess this diaphorase ability; lipoamide dehydrogenase (6), EC 1.6.4.3, (LipDH), which is specific for NADH production and ferredoxin reductase (6), EC 1.18.1.2, (FNR), which catalyses the formation of NADPH. Electrochemical production of $NAD(P)^+$ can be achieved, in the absence of both mediators and diaphorase enzyme, at a strongly oxidising electrode surface.

I will now consider three aspects of biological reactors; reactor design, enzyme immobilisation and the choice of enzymes for fine chemical synthesis.

3. REACTOR DESIGN

Enzyme reactors exist in three main forms (7):

Free solution - As the name suggests the enzyme is mobile and free in solution together with the substrates and cofactors.

Membrane retained - In this reactor the enzyme is retained in free solution within an ultrafiltration module into which substrates are fed and from which products are removed.

Immobilised - In this case the enzyme is chemically attached to, or entrapped within, a particular inert support. The catalyst normally takes the form of a column of immobilised enzyme through which substrates pass and from which products are removed.

Immobilised systems have a decided advantage over the use of enzymes in free solution and a couple of areas are pertinent.

Product yield - Yields tend to be better as higher substrate concentrations can be used due to the protection afforded on immobilisation against substrate or product inhibition (8).

Process control - Enzymes only catalyse the conversion of interest and no side products are produced. Thus efficiency is high and columns can be re-used, greatly extending the working lifetime of the catalyst.

Product separation - Immobilisation of the enzymes prevents contamination of the product with the catalyst, making product removal easy.

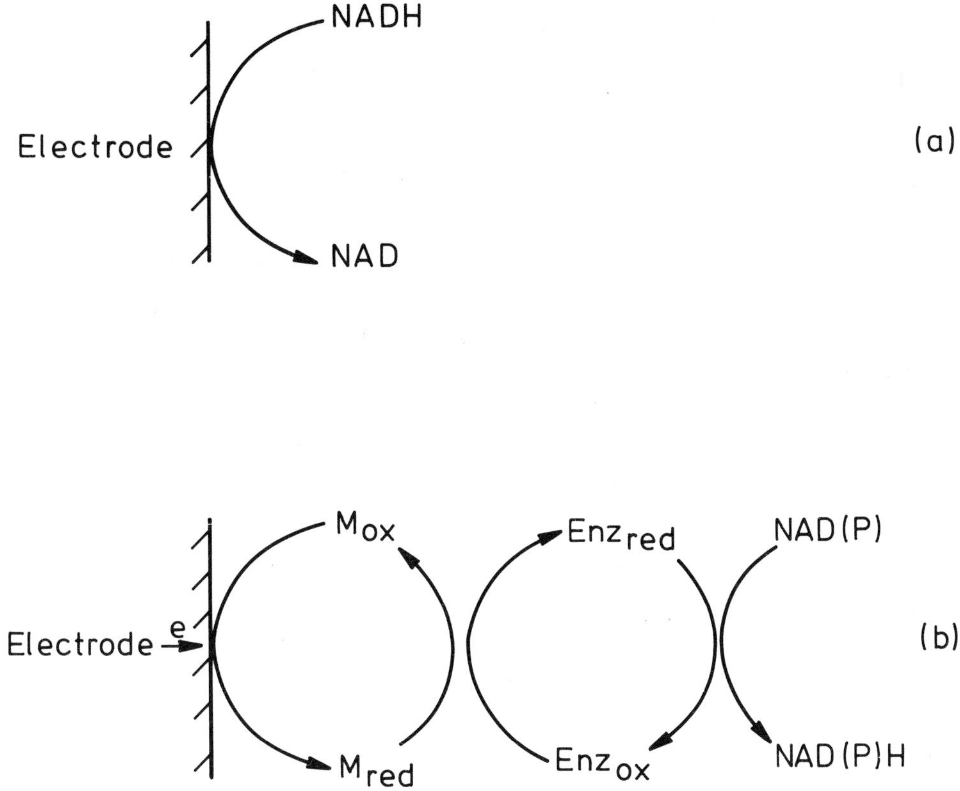

Figure 3. Electrochemical recycling of cofactors.
(a) Direct oxidation of NADH.
(b) Mediated reduction of NAD.

4. ENZYME IMMOBILISATION

Four methods of enzyme immobilisation have been pursued in our laboratories and I will briefly discuss these procedures.

<u>CNBr-activated Sepharose 4B</u> - The agarose beads are bought in the pre-activated form. CNBr activates the hydroxyl groups on the polysaccharide producing the reactive imidocarbonate group, >C=NH, which covalently binds the amino group, $-NH_2$, of proteins at alkaline pHs giving N-substituted imidocarbonate, >C=N-protein, N-substituted carbamate, -C(=O)-NH-protein and N-substituted isourea, -C(=NH)-NH-proteins. N-substituted ureas are the main product (9) (Figure 4).

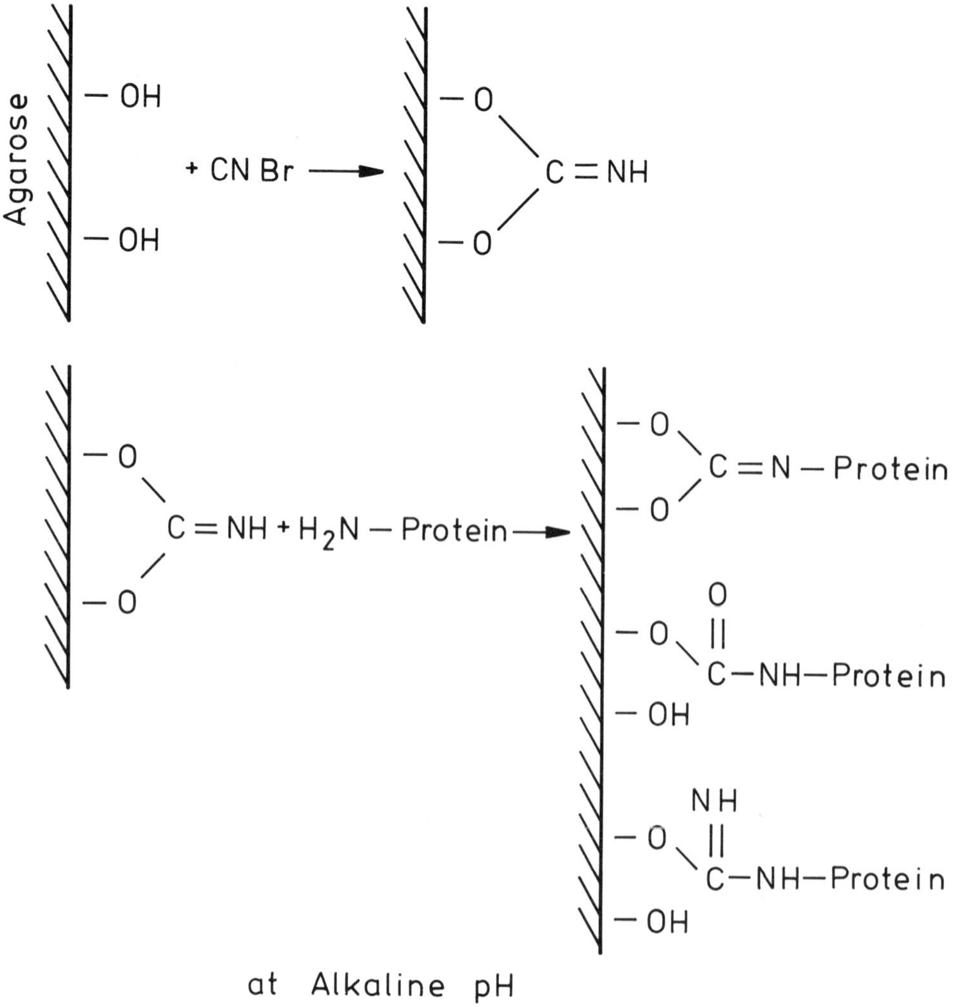

Figure 4. Immobilisation of enzymes on CNBr-activated Sepharose 4B.

<u>Oxirane Acrylic beads</u> - The tradename for these beads is 'Eupergit C' (10). The beads are bought in the pre-activated form and function by binding the amino group of protein. An advantage of this support is that immobilisation is pH independent (10) (Figure 5).

<u>Carbodiimide</u> - This has been used for the immobilisation, in our case, of enzymes onto working carbon electrodes. Initially the carbon felt is oxidised by a mixture of concentrated H_2SO_4 and HNO_3 producing reactive hydroxyl and

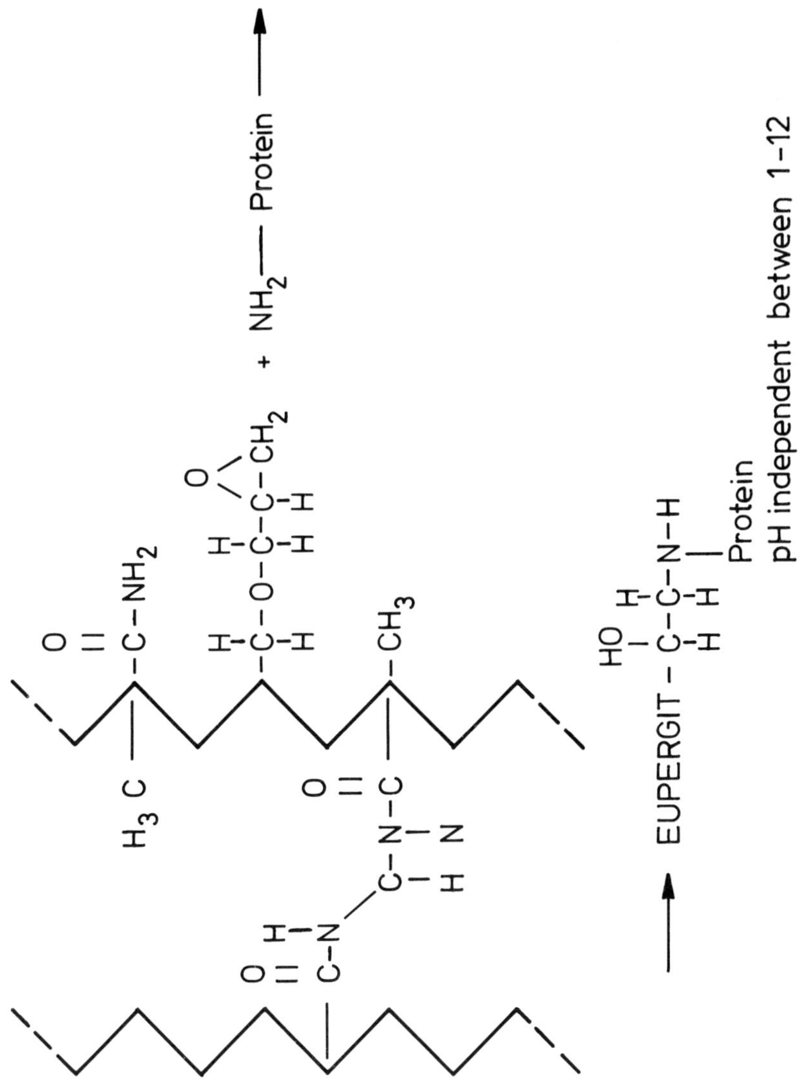

Figure 5. Immobilisation of enzymes on oxirane acrylic beads.

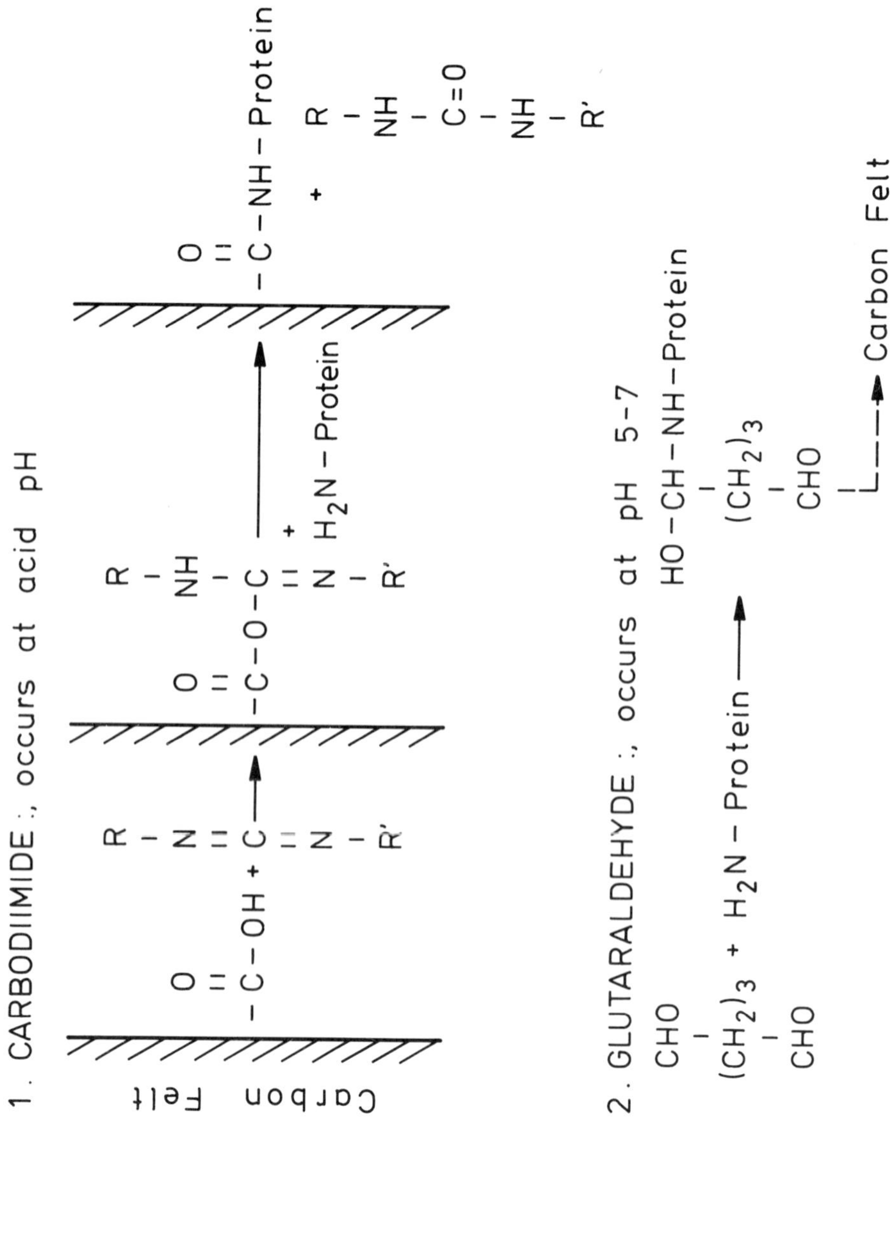

Figure 6. Immobilisation of enzymes on carbon felts.

carboxyl groups. At acid pHs carbodiimide reacts with the carboxyl group to give O-acyl isourea derivatives which covalently bind the amino groups of proteins (9) (Figure 6.1).

Glutaraldehyde - This immobilisation technique has also been used in the context of carbon felt electrodes. Although the reaction is not entirely understood, immobilisation is presumed to occur via hydroxyl groups on the electrode surface at neutral pH (9) (Figure 6.2).

It is important to choose the correct immobilisation technique, especially in terms of the pH to which the enzyme is exposed during immobilisation as many enzymes are inactivated by low or high pH.

5. PROTOTYPE REACTORS

I would now like to consider three prototype bioelectrochemical reactors that we have constructed for the production of fine chemicals and give examples of other enzymes which are being considered as catalysts of viable conversions.

Three different methods of cofactor recyle have been explored and I will consider the most elegant system first. This system satisfies all the criteria laid down previously for a bioelectrochemical reactor. It involves direct electrochemical regeneration of cofactor and the formation of a single pure product, which is easily removed from the reactor by precipitation as the barium salt. The principal involved in our first reactor is illustrated in Figure 7. NAD^+ produced directly at the electrode surface is used as the source of oxidising power for the enzyme G6PDH, which catalyses the conversion of G6P to 6PG, the product of interest. The NADH formed in this conversion is re-oxidised to NAD^+ at the electrode surface (RVG 1000 carbon felt) and so the reaction perpetuates.

The physical arrangement of the reactor is summarised in Figure 8. The electrochemical cell is responsible for the regeneration of NAD^+ and the enzyme G6PDH is immobilised into a column reactor on Sepharose 4B beads. Cofactor and G6P-6PG are continually pumped through the system until completion of the reaction is achieved. The data from a typical experiment shows the time course of 6PG production, at a starting concentration of 50 mM G6P, and the disappearance of NADH (3 mM) (Figure 9) (11). The NADH within the reactor is continually recycled back to NAD^+ at the electrode surface. As the G6P concentration falls proportionally less of the NAD^+ is converted back to NADH and thus eventually all cofactor is converted to the oxidised form (Figure 9).

The second reactor, or strictly speaking half reactor, involves electrochemical reduction of NAD^+. In this case, reduction is mediated via the electro-active redox carrier methyl viologen (MV), which shuttles electrons from the electrode surface to the enzyme LipDH which reduces NAD^+ to NADH (Figure 10). A novelty of this reactor is that the LipDH is immobilised directly on the carbon felt electrode surface and thus electrode, mediator and catalyst are in very close proximity. It must be stressed that this reactor is extremely O_2-sensitive, reduced MV will tend to donate its electrons to O_2 forming the superoxide radical

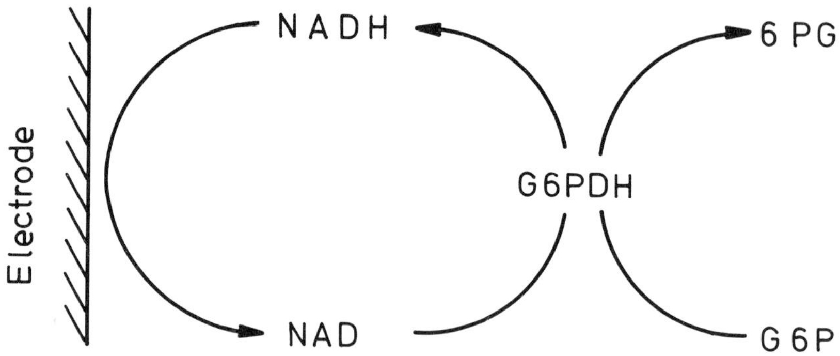

Figure 7. Direct electrochemical oxidation of NADH linked to the production of 6-phosphogluconate (6PG) using the enzyme glucose-6-phosphate dehydrogenase (G6PDH).

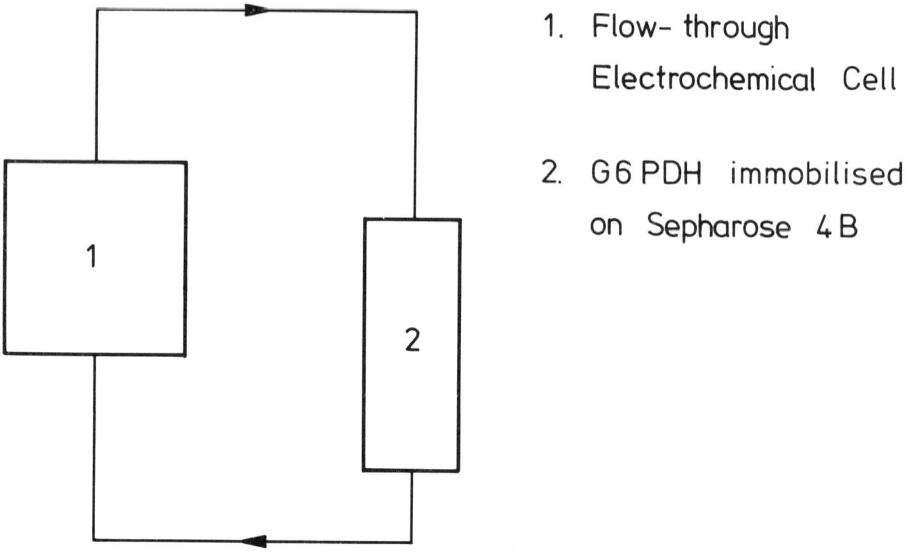

Figure 8. Schematic diagram of the reactor used for the production of 6-phosphogluconate.

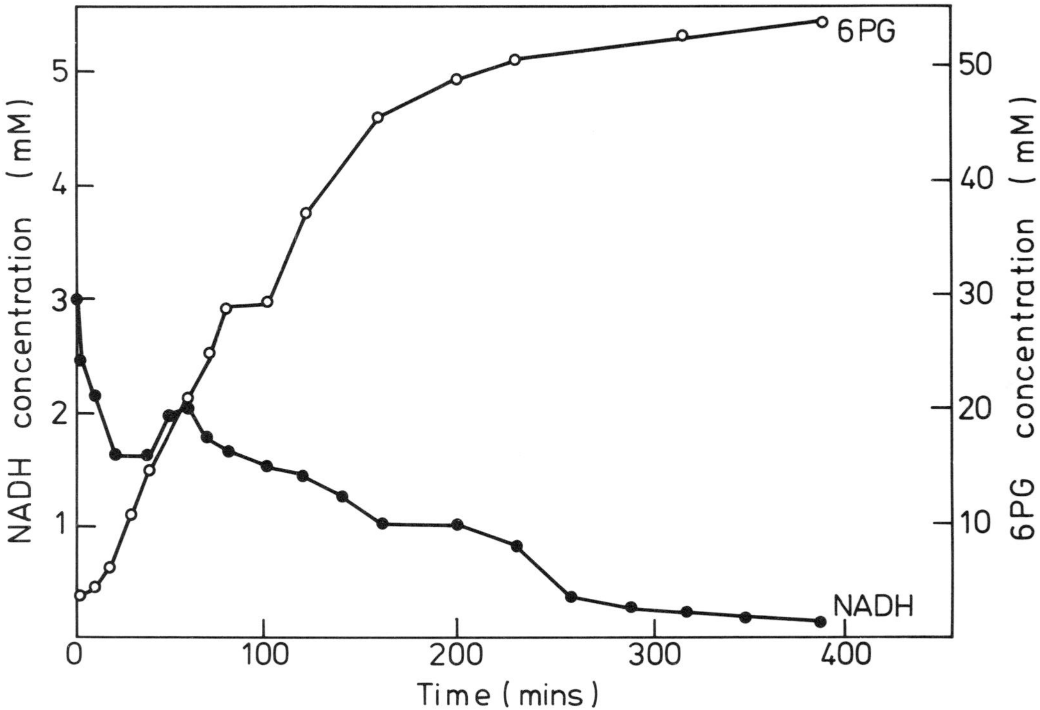

Figure 9. Time course for the production of 6PG in the G6PDH reactor.

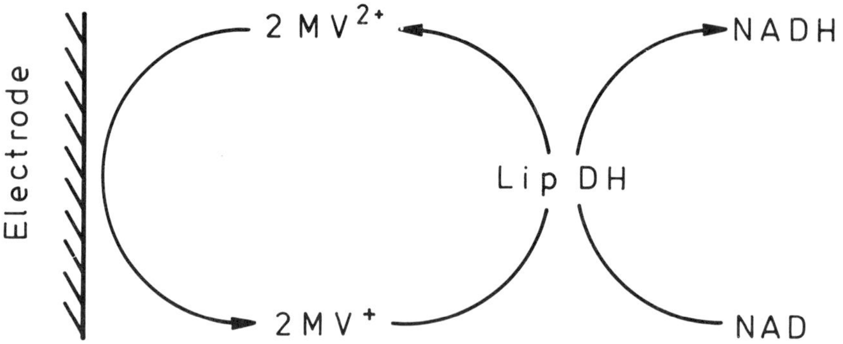

Figure 10. Mediated electrochemical production of NADH using the enzyme lipoamide dehydrogenase (LipDH) and methyl viologen (MV) as the mediator.

($O_2^{\bullet}$). This will short circuit the reactor, taking electrons away from the production of NADH and would also lead to enzyme inactivation by the reactive $O_2^{\bullet}$ species and other radicals derived from it.

Production of NADH within this half reactor was monitored over a 60 min time period with an initial NAD^+ concentration of 2 mM (Figure 11) (12), 1-3 are successive runs with the same electrode, but using different NAD^+ and MV solutions. In each successive run the rate of production and yield of NAD^+ is progressively decreased; a likely explanation for this is the possible inhibition of LipDH by high concentrations of NADH (12). This however is not a problem when the reactor module is linked to a second enzyme, which keeps the NADH concentration at a low steady state level, as the cofactor is used for the production of fine chemicals.

The last reactor I will discuss uses biochemical regeneration employing a second enzyme and sacrificial co-substrates for the production of NADH. The enzymatic conversion of interest was that of cinnamaldehyde to cinnamyl alcohol (13), the alcohol being an important intermediate in the perfumery industry (14). The principal of this reactor is summarised in Figure 12, the enzyme yeast alcohol dehydrogenase EC 1.1.1.1 (ADH) was chosen for its stability and cheapness. The cofactor is recycled using the enzyme G6PDH and the sacrificial co-substate G6P (4). In this case both enzymes are co-immobilised in a column reactor on either Sepharose 4B or Eupergit C beads and G6P-6PG, cofactor and aldehyde are mobile throughout the system (Figure 13).

A characteristic of cinnamaldehyde and cinnamyl alcohol are their low solubilities, circa 10 mM, in aqueous environments, but high solubility, circa 2 M, in organic solvents. This property proved to be useful in the extraction of the product cinnamyl alcohol from the reactor. The reactor was set up so that one pass thorugh the column gave 100% conversion to the alcohol which was directly extracted into

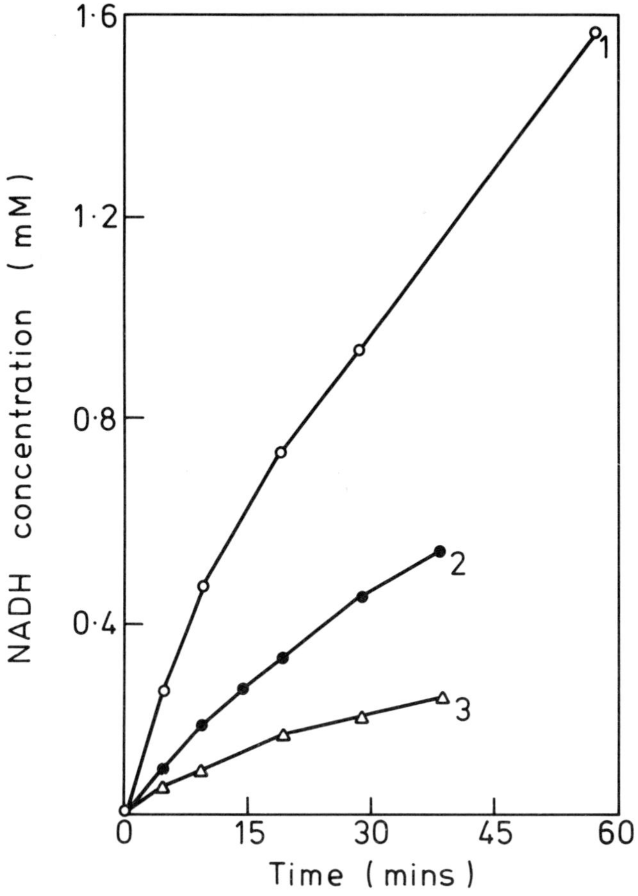

Figure 11. Electrochemical reduction of NAD^+; 1-3 are successive runs using the same electrode.

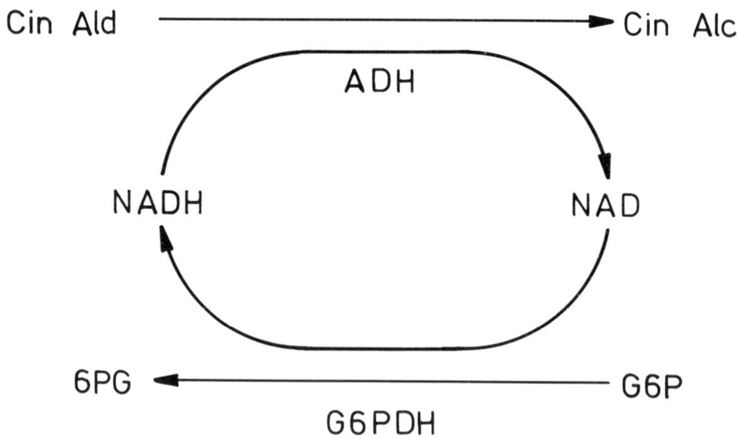

Figure 12. Coupled enzyme system for the production of cinnamyl alcohol (Cin Alc) from cinnamaldehyde (Cin Ald), using the enzyme alcohol dehydrogenase. Cofactor is recycled using glucose-6-phosphate dehydrogenase (G6PDH).

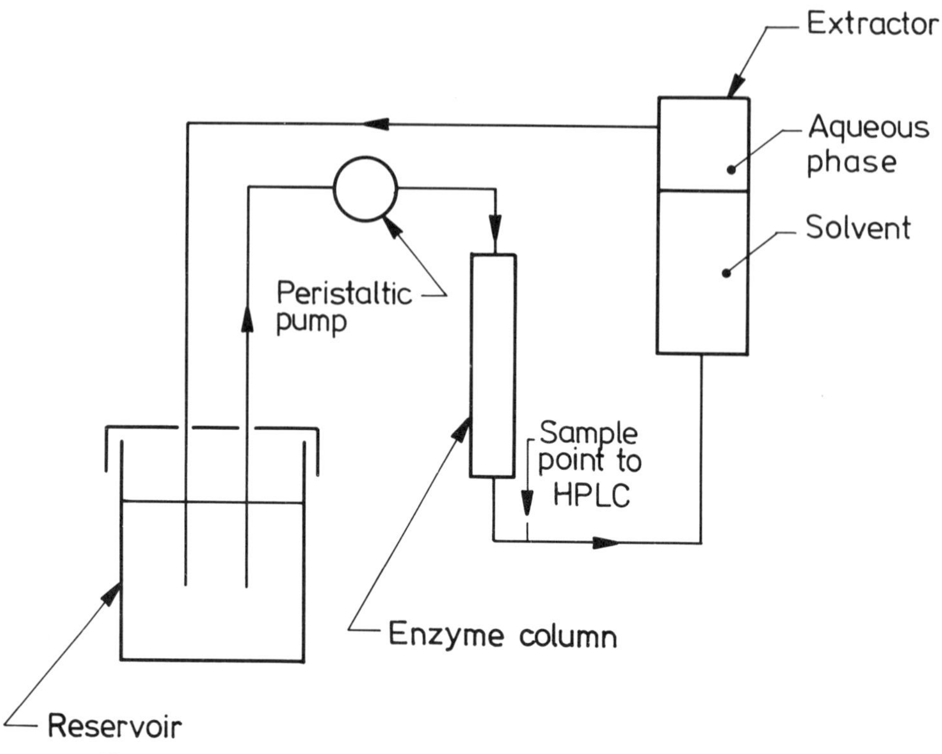

Figure 13. Schematic diagram of the ADH reactor used for the production of cinnamyl alcohol.

dichloromethane, the solvent of choice. The cofactors and co-substrate were soluble in the aqueous phase and were continually recycled through the system (Figure 13). This proved to be a very convenient method of product retrieval and the solvent did not affect enzyme activity or efficiency over a 5 day period of continual use. The 6PG, formed as a product of cofactor recycling, was recovered by barium precipitation and also proved to be a useful co-product itself with a considerable market value.

6. FUTURE REACTORS

An important extenson of this work is the use of redox enzymes for the hydroxylation of aliphatic and aromatic compounds. At present very few hydroxylating enzymes are available commercially and of these horse radish peroxidase is perhaps the most stable and inexpensive. Many other groups of oxidising enzymes show great potential including yeast and mammalian cytochrome P_{450} mono-oxygenases (16), methane mono-oxygenases (17) and flavoprotein hydroxylases (18). The conversions these enzymes catalyse are very interesting, but their use at the moment is restricted by both availability and the instrinsic instability which many of these multicomponent oxygenases possess. The way forward in the use of enzymes as catalysts of fine chemical synthesis thus depends on much more research on improving enzyme stability and operational efficiency (19).

7. ACKNOWLEDGEMENTS

I would like to acknowledge the excellent work of Dr. D.J. Best, Mr. B.H. Schneider and Mr. F. Taylor from Cranfield Institute of Technology; Dr. P.M. Allen and Dr. N. Lambert from Oxford University and Dr. W.R. Bowen, Miss J. Homewood and Miss N.J.D. Schomburgk from AERE Harwell, all of whom have been involved with this project.

This work was funded by the Department of Trade and Industry through the Materials, Chemicals and Vehicles Requirement Board.

8. REFERENCES

1. Jones, B.J., in Enzymic and Non Enzymic Catalysis, (Dunhill, P., Ed.), 1980, p. 54-83.
2. Baricos, W.H., Chambers, R.P. and Cohen, W., Anal. Letts., 1976, **9**, 257-276.
3. Wichmann, R., Wandrey, C., Bückmann, A.F. and Kula, M-R., Biotechnol. Bioeng., 1981, **23**, 2789.
4. Wong, C-H., Gordon, J., Cooney, C.L. and Whitesides, G.M., J. Org. Chem., 1981, **46**, 4676-4679.
5. Samec, Z. and Elving, P.J., J. Electroanal. Chem., 1983, **144**, 217-234.
6. DiCosimo, R., Wong, C-H., Daniels, L. and Whitesides, G.M., J. Org. Chem., 1981, **46**, 4622-4623.
7. Bowen, W.R. and Pugh, S.Y.R., Chemistry and Industry, 1985, **10**, 323-326.

8. Bowen, W.R., Pugh, S.Y.R. and Schomburgk, N.J.D., Submitted to J. Chem. Technol. Biotechnol., 1985.
9. Goldstein, L. and Manecke, G., in Immobilised Enzyme Principles, Applied Biochem. Bioeng., 1, (Wingard, L.B., Katchalski-Katzir, E. and Goldstein, L., Eds.), Academic Press, Lodnon, 1976, p. 23-126.
10. Krämer, D.M., 6th Enz. Eng. Conference, 1981, from Röhm Pharma, Weiterstadt, West Germany.
11. Bowen, W.R. and Pugh, S.Y.R., paper in preparation.
12. Schneider, B.H. and Lambert, N., personal communication.
13. Bowen, W.R., Lambert, N., Pugh, S.Y.R. and Taylor, F., submitted to J. Chem. Technol. Biotechnol., 1985.
14. Poucher, W.A., Perfumes, cosmestics and soaps, **1**, Chapman and Hall, London, 1975, p. 104.
15. Klibanov, A.M., Berman, Z. and Alberti, B.N., J. Am. Chem. Soc., 1981, **103**, 6263-6264.
16. Wiseman, A., in Topics in Enzyme and Fermentation Biotechnology, Ellis Horwood, Chichester, 1978, **2**, 172.
17. Higgins, I.J., Best, D.J., Hammond, R.C. and Scott, D., Microbiol. Rev., 1981, **45**, 556.
18. Walsh, C., Enzymatic Reaction Mechanisms, 1980, p. 411-431.
19. Klibanov, A.M., Biochem. Soc. Trans., 1983, **11**, 19-20.

APPLICATIONS OF MICROBES AND MICROBIAL ENZYMES IN ENVIRONMENTAL CONTROL AND ORGANIC SYNTHESIS

Oreste Ghisalba, Hans-Peter Schär and Gerardo M. Ramos Tombo
Central Research Laboratories of Ciba-Geigy Ltd.
CH-4002 Basel
Switzerland

ABSTRACT. Specialized methylotrophs, originally isolated for the specific biodegradation of chemical wastes, were found to be an excellent source of novel microbial enzymes, such as methylsulfatase, N,N-dimethylformamidase (DMFase), esterases, methanol dehydrogenase etc. The properties and possible applications of these microbes and enzymes are discussed. Mutant strains of Nocardia mediterranei (rifamycin producer) were successfully applied for the microbial production of D(-)-ribulose and shikimic acid.

1. INTRODUCTION

As the selection of enzymes commercially available in significant amounts is still rather limited, the set up of enzymic or chemico-enzymic routes to defined organic compounds, as well as the design of useful biosensors, must necessarily include screening activities. Therefore, a highly integrated approach was chosen, combining the fields of defined waste biodegradation (production of biomass containing useful enzymes), bioanalysis (application of highly specific enzymes for biosensors), enzymic synthesis (application of enzymes with good regio-, stereo- and/or enantio-selectivity) and biotransformations with whole cells of microorganisms (including mutants and waste biomass).

Screening procedures are involved on three different levels in the development of a biotechnical process:

- screening for suitable microorganisms (plant cells etc.) and/or enzymes with the required properties;
- screening for the optimum conditions for culture growth and production of the desired enzyme(s);
- screening for the optimum reaction conditions with whole cells, crude or purified enzymes (including technological aspects such as cofactor regeneration, immobilization, type of reactor etc.).

The most tricky step in process development is the optimization of enzyme production in cases where the desired enzyme is an inducible one and the inducer (natural substrate) is unknown, or in cases where the amount of enzyme within the cells shows strong variations during the growth cycle.

M. P. Schneider (ed.), Enzymes as Catalysts in Organic Synthesis, 233–250.

The best results are obtained under operating conditions where the desired enzyme is essential for cell growth (substrate utilization).

2. ENZYMES FROM WASTE TREATMENT PROCESSES

For a number of years we have investigated methylotrophs for their biodegradative activities in order to replace physical methods of waste disposal with high specific energy consumption by waste specific microbial treatments (1-3). Microbial processes were established for three model waste streams containing either monomethyl sulfate, trimethylethylammonium chloride or N,N-dimethylformamide as the main organic compound (1-3). Monomethyl sulfate is formed as a by-product in stoichiometric amounts in methylation reactions with dimethyl sulfate as the methyl donor. Trimethylethylammonium chloride is a stoichiometric by-product formed in deethylation reactions with trimethylamine as the ethyl acceptor. N,N-Dimethylformamide is widely used in the chemical industry as an aprotic solvent, solubilizing agent or reaction medium.

As monomethyl sulfate (MS), trimethylethylammonium chloride (TMEA) and N,N-dimethylformamide (DMF) are C_1-compounds (TMEA at least partially), we decided to investigate and isolate methylotrophs (C_1-utilizing microorganisms). From the literature it was known that the following C_1-compounds are utilized by specialized methylotrophs as the sole source of carbon and energy (4): methane, methanol, dimethyl ether, methylamine, dimethylamine, trimethylamine, trimethylamine-N-oxide, tetramethylammonium salts, trimethylsulfonium salts, formaldehyde, formate, methyl sulfides, dimethyl sulfoxide, dichloromethane etc.

More than 30 methanol and/or methylamine degrading methylotrophs from culture collections were tested for MS, TMEA and DMF utilization. None of the investigated strains was able to utilize MS, TMEA or DMF at significant rates ! Thus it seems that the utilization of these three substrates is not a common property of methylotrophs (1-3).

For the biodegradation of MS, TMEA and DMF three novel types of methylotrophs were enriched and isolated from sewage sludge, effluents from sanitary landfills or mud ponds: *Hyphomicrobium* sp. utilizing monomethyl sulfate (MS), *Pseudomonas* sp. utilizing trimethylethylammonium chloride (TMEA) and related compounds, and *Pseudomonas* sp. utilizing N,N-dimethylformamide and related compounds. The degradation pathways for MS, TMEA and DMF were investigated and pilot fermentations were carried out (1-3).

The considerable amount of defined biomass produced in such waste treatment processes is a cheap source for interesting microbial enzymes. However, if no such waste stream is at hand, the biomass can also be produced in fermentations with the pure substrates.

At least three novel enzymes were discovered so far: methylsulfatase from *Hyphomicrobium* sp., N,N-dimethylformamidase (DMFase) from *Pseudomonas* and esterase from *Pseudomonas* sp. The PQQ-dependent methanol dehydrogenase from *Hyphomicrobium* sp. has a very broad substrate specificity for primary and secondary alcohols and aliphatic aldehydes. Applications of these four particular enzymes will be discussed below. The substrate spectra of our specialized methylotrophs indicate the presence of a wide range of potentially interesting enzymes suitable for analytical and synthetic applications.

3. ENZYMES FROM MONOMETHYL SULFATE UTILIZING *HYPHOMICROBIUM* STRAINS

3.1. Substrate Spectrum of the *Hyphomicrobium* MS Strains

Pure cultures of sodium monomethyl sulfate utilizing *Hyphomicrobium* MS strains (MS 72, MS 75, MS 219, MS 223 and MS 246) can utilize the following substrates as the sole source of carbon and energy (1,3): sodium monomethyl sulfate, methanol, ethanol, formate, acetate, methylamine, dimethylamine, methylethylamine, N-formylglycine ethylester; and after prolonged incubation : dimethyl phosphite, trimethyl phosphite and trimethylamine (only strains MS 219 and MS 223). Dimethyl sulfate is also utilized as a carbon source, but in this case the chemical hydrolysis of DMS to MS and methanol seems to be the rate-limiting step. Therefore, not DMS itself but only its hydrolysis product monomethyl sulfate acts as a true substrate. Methyl-sulfur compounds with an oxidation number different from that of monomethyl sulfate (MS) are not utilized.

3.2. Degradation Pathway for Monomethyl Sulfate

As the spontaneous hydrolysis of MS to methanol and sulfate is extremely slow at low temperature (30°C) and under neutral conditions, an enzyme (methylsulfatase) is needed to achieve an efficient degradation of MS. After enzymatic hydrolysis of MS the methanol formed then acts as a carbon source for *Hyphomicrobium* MS strains. Most probably the assimilation of methanol (via formaldehyde) follows the serine pathway (similar to other species of *Hyphomicrobium*). The biodegradation pathway for MS is depicted in Fig. 1. A comparison with data reported in the literature seems to indicate that our strains are different from other species of *Hyphomicrobium* described so far (1,3).

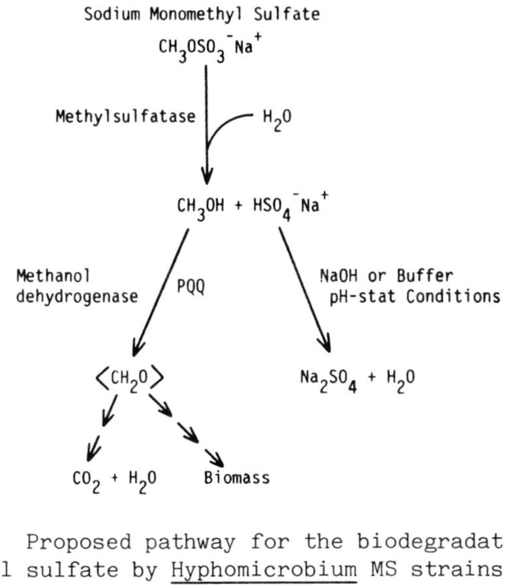

Figure 1. Proposed pathway for the biodegradation and utilization of monomethyl sulfate by *Hyphomicrobium* MS strains.

3.3. Methylsulfatase

Monomethyl sulfate is utilized as the sole source of carbon and energy by Hyphomicrobium MS but not monoethyl sulfate, sec-propyl sulfate, tert-butyl sulfate or dodecyl sulfate. As Hyphomicrobia show a distinct preference for C_1- and C_2-compounds and not only methanol but also ethanol is utilized, one would expect at least monoethyl sulfate to be utilized. From the negative results with monoethyl sulfate and longer chain alkyl sulfates, however, it can be concluded that the sulfatase present in Hyphomicrobium MS strains is methyl specific (1,3). To our knowledge a methyl specific sulfatase has not been reported before. The sulfatases described in the literature are: glycosulfatase, chondrosulfatase, myrosulfatase, steroid sulfatase, arylsulfatases, phenolsulfatases and alkylsulfatases. Typical substrates for alkylsulfatases are sodium 2-(2,4-dichlorophenoxy) ethyl sulfate, sodium dodecyltriethoxy sulfate and n-pentyl sulfate up to n-dodecyl sulfate (5-8).

The novel enzyme methylsulfatase is membrane-bound. After cell disruption and centrifugation the methylsulfatase activity is found in the pellet only (cell fragments), whereas the protein containing supernatant was found to be completely inactive. Attempts to separate the active enzyme from the membrane fragments by treatment with detergents were not successful. The isolation of the free enzyme thus needs further work.

3.4. Application of Methylsulfatase for Process Control (Electrode)

For process control a biosensor consisting of physically entrapped Hyphomicrobium MS 219 and a combined glass electrode has been developed (9). The construction of this bacterial electrode and the complete sensor system are depicted in Figs. 2 and 3.

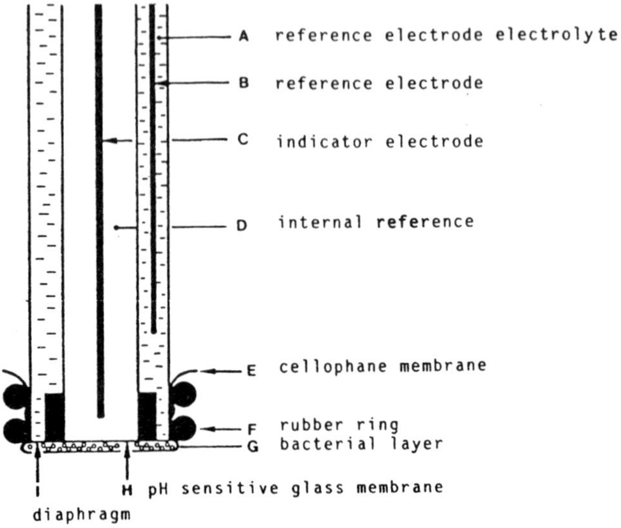

Figure 2. Schematic diagram of the bacterial electrode.

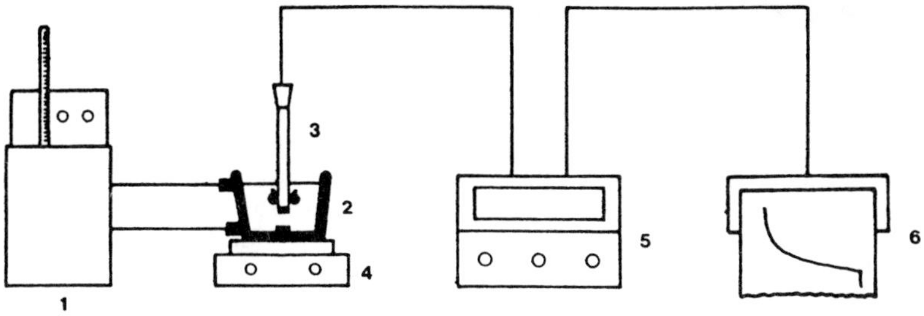

1 water bath, 2 reaction vessel, 3 microbial electrode, 4 magnetic stirrer
5 pH-meter, 6 recorder

Figure 3. Schematic diagram of the sensor system.

The response of the bacterial electrode is linear between $2.5 \cdot 10^{-2}$ M and $6.3 \cdot 10^{-1}$ M monomethyl sulfate with an effective response to concentrations as low as 10^{-3} M and as high as 1 M MS. The probe has an average slope of 8 mV per concentration decade over the linear range. Response times vary from five minutes in the linear range to thirty minutes at the detection limit. The sensor has a lifetime of at least one week and shows high selectivity to methyl sulfate in the absence of other growth substrates. The determination of monomethyl sulfate is important not only for industrial waste water process control but also in environmental pollution control. Fly ash and airborne particulate matter from coal combustion processes contain high concentrations of mono and dimethyl sulfate. MS is usually determined by ion chromatography, isotachophoresis or by gravimetric methods. These methods are rather time consuming and not ideal for continuous process control. The bacterial electrode would allow a continuous operation. Using bulb shaped glass membrane electrodes instead of a flat membrane the sensitivity of the bacterial electrode could even be increased to concentrations as low as $2. 10^{-5}$ M MS.

The electrode shows no response to alkyl sulfates other than MS. Dimethyl sulfate produces a response due to the non-enzymic hydrolysis to MS and methanol followed by the enzymatic hydrolysis of MS. Interference was observed with methylamine, dimethylamine, dimethyl phosphite, trimethyl phosphite, glucose, formate and acetate. These interferences could be excluded by using cell fragments (methylsulfatase is membrane-bound !) after separation from the soluble protein fraction instead of whole cells. Preliminary experiments with cell fragments indicate that the MS-electrode becomes even more sensitive, due to the facilitated diffusion of the substrate MS. The MS-biosensor was also investigated in combination with a flow injection analysis (FIA) device. This combination resulted in a significantly shorter reaction time (measurements in 60-90 sec intervals) and in a prolonged lifetime of the bacterial electrode (operation for several weeks continuously)(10).

The Biomass of Hyphomicrobium MS (grown on MS) can be stored either as lyophilized preparations or deep-frozen (-20°C). For application in the bacterial electrode the cells must be grown on MS. Methylsulfatase is an inducible enzyme and therefore absent in Hyphomicrobium MS cells grown on methanol or other carbon sources different from monomethyl sulfate.

3.5. PQQ-Dependent Methanol Dehydrogenase from Hyphomicrobium MS

Hyphomicrobium MS (strain MS 223) contains a dye-linked methanol dehydrogenase (EC 1.1.99.8) which was isolated and characterized (11). The enzyme requires ammonium ions for activation (in vitro). It has a molecular weight of 118,000 and is composed of two subunits of identical molecular weight. The enzyme shows a broad substrate specificity for primary alcohols and is able to oxidize secondary alcohols and several alipahtic aldehydes. The enzyme was purified by ammonium sulfate fractionation, ion-exchange chromatography and Sephacryl S 200 sf gel filtration and a 7.6-fold purification was achieved (37% yield, specific activity 3.8 U/mg). A single band was obtained when the enzyme preparation was subjected to SDS gel electrophoresis. The methanol dehydrogenase (MDH) fraction accounts for 10-20% of the total protein content of the crude cell extract (of methanol-grown Hyphomicrobium MS 223) as estimated on polyacrylamide gels specifically stained for MDH activity. The isoelectric point as determined by isoelectric focussing was 9.7 and the catalytic activity was found to be maximal at pH 9.5 (with methanol as a substrate). Methylamine and ethylamine can substitute for ammonium, whereas dimethylamine and trimethylamine do not act as activators.

The following compounds are substrates for the purified MDH from Hyphomicrobium MS 223 (relative activity in brackets):

- primary alcohols: methanol (100%), ethanol (110%), n-propanol (99%), n-butanol (76%), n-pentanol (39%), n-hexanol (31%),
- aldehydes: formaldehyde (97%), acetaldehyde (111%), propionaldehyde (73%), n-butyraldehyde (50%),
- secondary alcohols: propane-2-ol (3.1%), butane-2-ol (2.4%), pentane-2-ol (8.0%), hexane-2-ol (12.4%).

Primary alcohols are first oxidized to the corresponding aldehydes and then, in a second step, to the carboxylic acids. The secondary alcohols yield the corresponding ketones.

The methanol dehydrogenase of Hyphomicrobium MS 223 is of particular interest as only very few dye-linked dehydrogenases are reported to oxidize secondary alcohols (enzymes from Acinetobacter calcoaceticus, Rhodopseudomonas acidophila, Bacterium PAR and Pseudomonas C). A common characteristic of most MDH's described so far is that formaldehyde and in rare cases acetaldehyde are the only aldehydes which are oxidized. The MDH from Hyphomicrobium MS 223 is similar (with respect to the substrate spectrum) to the MDH from the photosynthetic Rhodopseudomonas acidophila and to the alcohol dehydrogenase from the non-methylotrophic Acinetobacter calcoaceticus in its ability to oxidize aliphatic aldehydes (for details see 11). The MDH isolated from Hyphomicrobium MS 223 has different properties than the enzyme from Hyphomicrobium X (12,13). For the latter aldehydes and secondary alcohols are poor substrates or no substrates at all. In addition, it seems to exist in different forms (12).

Acetaldehyde oxime was identified as a new reversible inhibitor for alcohol and aldehyde substrates (11). This compound inhibits the aldehyde oxidation (with MDH from Hyphomicrobium MS 223) much more strongly (I_{50} = 4 μM) than alcohol oxidation (I_{50} = 34 μM), an effect that may be explained by different affinities for ethanol and acetaldehyde ($K_{m,eth}$ = 1.1 mM, $K_{m,acet}$ = 13 mM). Similar results were obtained with hydroxylamine. Since aldehyde oximes are formed easily from NH_2OH and aldehydes, it is possible that even in the case of apparent inhibition of methanol oxidation by NH_2OH the true inhibitor is formaldehyde oxime formed from NH_2OH and enzymatically produced formaldehyde. Cyclopropanol is a suicide inhibitor for the MDH from Hyphomicrobium MS 223.

Methanol dehydrogenase from Hyphomicrobium MS can be used for the construction of fuel cells (with methanol as fuel), for analytical applications (biosensors for environmental control or process control) or for the synthesis of ketones. PQQ (methoxatin), the cofactor of MDH has been synthesized and derivatized in our laboratories (Fig. 4).

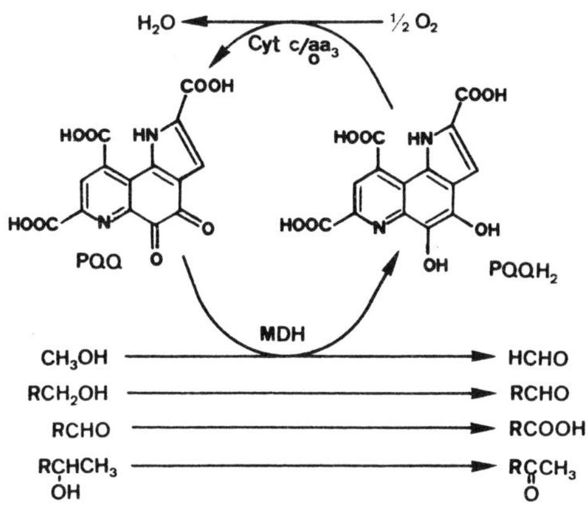

Figure 4. The PQQ-dependent methanol dehydrogenase system (in vivo).

In collaboration with Duine and Frank (Technical University, Delft) inhibition and reconstitution experiments were carried out with PQQ-dependent glucose dehydrogenase apoenzyme and PQQ-derivatives. Not only an intact o-quinone structure but also free carboxy groups are important for the binding and the catalytic activity of the coenzyme. $PQQH_2$ is rapidly converted into PQQ under aerobic conditions.

Until now, only oxidation reactions involving PQQ-dependent enzymes are described. As PQQ (in its oxidized form) is an extremely stable compound, PQQ-dependent enzymes could possibly be a good alternative to NAD-dependent enzymes such as Horse Liver ADH or Yeast-ADH. The hydrolytic instability and the relatively short lifetime of the still rather expensive coenzymes NAD and NADH is often prohibiting the application of classical oxidoreductases in large-scale chemico-enzymic synthesis. PQQ can be synthesized in large quantities although the total synthesis is rather time-consuming. Several attempts have been made to produce PQQ by fermentation but the reported titers are still very low (14). The best results so far were reported for the obligate methylotroph M5 and some isolated methylotrophs, which have not been characterized taxonomically. The PQQ content in the medium reached more than 10 mg/l after 2 days of incubation (14).

4. ENZYMES FROM TRIMETHYLETHYLAMMONIUM CHLORIDE UTILIZING PSEUDOMONAS STRAINS

4.1. Substrate Spectra of the *Pseudomonas* TMEA Strains

Pure cultures of trimethylethylammonium chloride (TMEA) utilizing *Pseudomonas* TMEA strains (TMEA 14, TMEA 81, TMEA 83, TMEA 84, TMEA 86, TMEA 87, TMEA 89, TMEA 199 and TMEA 211) can utilize the following substrates as the sole source of carbon and energy (2,3): trimethylethylammonium chloride, tetramethylammonium chloride, methylamine, dimethylamine, trimethylamine, trimethylamine-N-oxide, methylethylamine, dimethylethylamine, ethanol, acetamide, creatinine, creatine, acetate, glucose, glyoxylate, acetylcholine, betaine, dimethyl glycine, sarcosine, glycine, N-formylglycine ethylester, N-formyl- and N-acetyl amino acids, methanol (only TMEA 14, TMEA 199 and TMEA 211); and after prolonged incubation: formate, isobutyronitrile, hydantoin and barbituric acid.

4.2. Degradation Pathway for Trimethylethylammonium Chloride

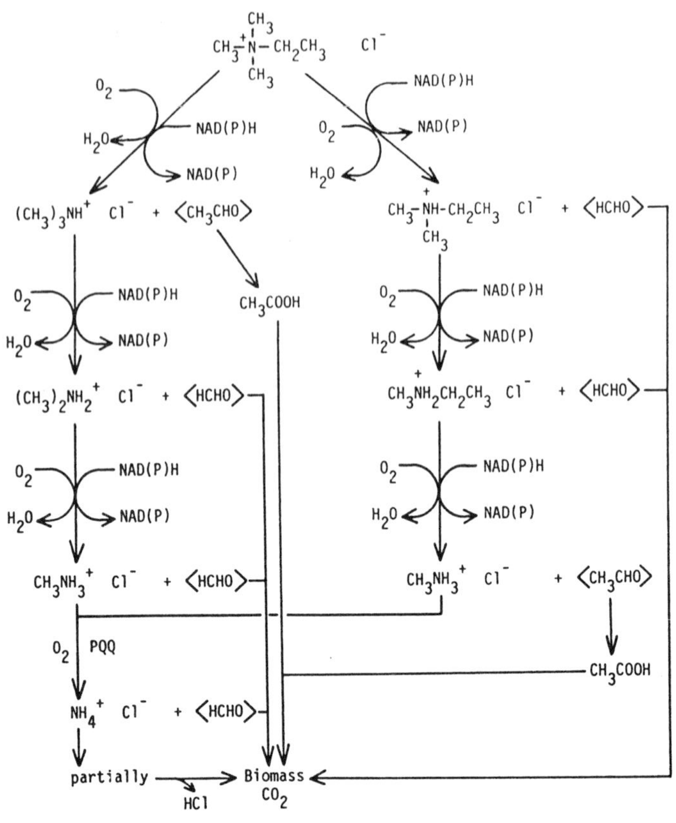

Figure 5. Possible pathways for the biodegradation and utilization of trimethylethylammonium chloride by *Pseudomonas* TMEA strains.

TMEA (or tetramethylammonium chloride) is degraded by subsequent N-demethylations with NAD(P)H-dependent monooxygenases, or by deethylation respectively, and PQQ-dependent methylamine dehydrogenase (Fig. 5).

Growth on methylamine, dimethylamine, trimethylamine, methylethylamine or dimethylethylamine is faster than growth on TMEA or tetramethylammonium chloride. This seems to indicate that the degradation of the quaternary alkylammonium salt to the tertiary alkylamine might be the rate-limiting step in the biodegradation of TMEA or tetramethylammonium chloride. Tetraethylammonium chloride, methyltriethylammonium chloride and dimethyldiethylammonium chloride are practically not utilized. Thus, only one methyl group in tetramethylammonium chloride can be replaced by an ethyl group to allow significant growth of our TMEA strains with these salts as the sole carbon source.

4.3. *Pseudomonas* TMEA Strains as an Enzyme Source

The substrate spectrum indicates the presence of the following enzymes: monooxygenases for N-demethylation/dealkylation , alcohol dehydrogenase, methanol dehydrogenase (only strains TMEA 14, TMEA 199 and TMEA 211), acetamidase, acylases, esterases, nitrilase (or nitrile hydratase), hydantoinase, creatinine/creatine degrading enzymes etc. Some of them are being evaluated. Enzymes such as hydantoinase, nitrilase and nitrile hydratase are of special interest in organic synthesis (enantioselective synthesis of amino acids, production of acrylamide etc.).

5. ENZYMES FROM N,N-DIMETHYLFORMAMIDE UTILIZING *PSEUDOMONAS* STRAINS

5.1. Substrate Spectra of the *Pseudomonas* DMF Strains

Two types of N,N-dimethylformamide (DMF) utilizing strains have been isolated: a) *Pseudomonas* strains DMF 3/3, DMF 3/4, DMF 3/5, DMF 3/6, DMF 3/11, DMF 3/12 and DMF/HW1-5; b) *Pseudomonas* strains DMF 4/4, DMF 5/3, DMF 5/5, DMF 5/7, DMF 5/8, DMF 5/9 and DMF 5/10. All of these strains can utilize the following substrates as the sole source of carbon and energy (3): N,N-dimethylformamide, N-methylformamide, N-ethylformamide, acetamide, methylamine, dimethylamine, trimethylamine, ethanol, ethylamine, isobutyronitrile, creatinine, creatine, barbituric acid, methylformate, formate, acetate, glucose, glyoxylate, acetylcholine, choline, betaine, dimethyl glycine, sarcosine, glycine, N-formylglycine ethylester, N-formyl- and N-acetyl amino acids, formamide; and after prolonged incubation: hydantoin, acetonitrile and acrylonitrile. The following substrates are utilized only by the strains of group a): methanol, DMF as carbon *and* nitrogen source, N,N-diethylformamide, N,N-dimethylacetamide, dimethyl phosphite and trimethyl phosphite.

5.2. Degradation Pathways for N,N-Dimethylformamide

The strains of group a) contain the novel enzyme DMFase and split DMF into dimethylamine and formate which are then further degraded to NH_4^+ and CO_2. The strains of group b) do not contain this new enzyme but degrade DMF by subsequent N-demethylations to formamide which is then split by formamidase to formate and NH_4^+. The two pathways are depicted in Fig. 6.

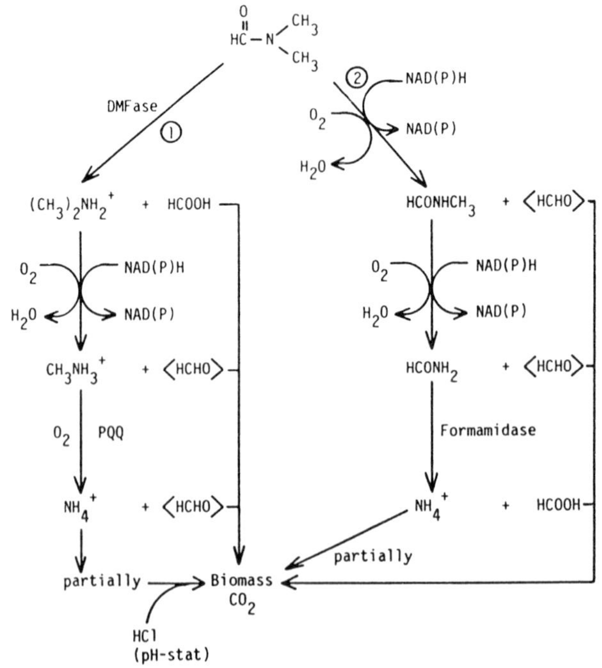

Figure 6. Proposed pathways for the biodegradation and utilization of N,N-dimethylformamide by Pseudomonas DMF strains.
Pathway 1: strains DMF 3/3, DMF 3/4, DMF 3/5, DMF 3/6, DMF 3/11, DMF 3/12 and DMF/HW1-5;
Pathway 2: strains DMF 4/4, DMF 5/3, DMF 5/5, DMF 5/7, DMF 5/8, DMF 5/8 and DMF 5/10.

Surprisingly, the DMF utilizers accept choline and betaine as growth substrates but not the structurally related trimethylethylammonium chloride (TMEA). This indicates that the N-demethylating enzymes for choline/betaine etc. are not identical with the N-demethylating enzymes for TMEA (see Fig. 5). The strains of group b) must possess a very efficient formamidase as formamide is a very good carbon source. In contrast to that, the strains of group a) show only moderate growth with formamide as the sole carbon source.

5.3. N,N-Dimethylformamidase (DMFase)

DMFase from Pseudomonas DMF 3/3 was isolated and partially characterized. It is an inducible metalloprotein; the enzyme is inhibited by EDTA and present in significant amounts only when the cells were grown with DMF as the sole carbon (and nitrogen) source. The molecular weight is approximately 230,000. The purified enzyme shows a specific activity of 50 U/mg (colorimetric determination of dimethylamine formed with sodium nitroprusside and acetaldehyde as reagents). DMF, N,N-dimethylacetamide and N,N-diethylformamide are substrates for DMFase.

Possible applications for DMFase are:

- quantitative analysis of DMF (on-line process control with biosensor);
- synthetic application: removal of dimethylamino or methylamino protecting groups.

Besides DMFase and formamidase the enzymes present in Pseudomonas DMF strains are those already listed for TMEA utilizers (see 4.3).
Of special interest is the enzyme 2,4-dimethylglutaric acid dimethyl ester esterase. This type of enzyme is found not only in Pseudomonas DMF strains but also in many other species of microorganisms. The evaluation of such microbes and enzymes is described in the following section.

6. MICROBES AND MICROBIAL ESTERASES FOR THE PREPARATION OF CHIRAL 2,4-DIMETHYLGLUTARIC ACID MONOESTERS

6.1. (2R,4R)-2,4-Dimethylglutaric Acid Monoethyl Ester

(2R,4R)-2,4-Dimethylglutaric acid monoethylester (or (2R,4R)-4-(ethoxycarbonyl)-2,4-dimethylbutyric acid) is an important chiral building block for the synthesis of the angiotensin converting enzyme (ACE) inhibitor 1[(2R,4R)-4-ethoxycarbonyl-2,4-dimethylbutyryl]-indoline-2(S)-carboxylic acid (15). (2R,4R)-2,4-Dimethylglutaric acid monoethylester can either be obtained by chemical resolution of 2,4-dimethylglutaric acid (meso, dl mixture) via its (+)-2-methylbenzylamine salt followed by esterification of (2R,4R)-2,4-dimethylglutaric anhydride (15) or by enantioselective enzymatic (or microbial) hydrolysis of racemic 2,4-dimethylglutaric acid diethyl ester (16).

6.2. Microbial or Enzymatic Hydrolysis of Racemic 2,4-Dimethylglutaric Acid Dimethyl and Diethyl Esters

The enantioselective microbial hydrolysis of racemic 2,4-dimethylglutaric acid dimethyl ester to (2R,4R)-2,4-dimethylglutaric acid monomethylester by whole cells of Gliocladium roseum has been reported by Chen et al.(17). When the reaction was stopped at 50% turnover, the isolated monoester had an enantiomeric excess of 68% (2R,4R)(see Table 2). After one recycling step an optical purity >98% (2R,4R) was obtained. No reports on the enantioselective microbial hydrolysis of racemic 2,4-dimethylglutaric acid diethyl ester were found in literature. Therefore we have screened microorganisms (in total approximately 100 strains) from strain collections (methylotrophs) and isolates from sewage treatment plants (methylotrophs and others), waste biomass (Actinomycetes) and soil (Streptomyces) for the presence of 2,4-dimethylglutaric acid diethyl ester esterase (DEE-esterase) and 2,4-dimethylglutaric acid dimethyl ester esterase (DME-esterase).

The strains were grown with racemic 2,4-dimethylglutaric acid diethyl ester (DMGADEE) or 2,4-dimethylglutaric acid dimethyl ester (DMGADME) as the sole source of carbon and energy. Thus the first selection principle was the utilization of methanol or ethanol cleaved off from the substrates by action of esterases. In some cases the resulting monoesters or even free 2,4-dimethylglutaric acid were further degraded. This approach yielded a selection of 6 strains for further evaluation: Pseudomonas M27 (DSM 1339, methylotroph), Isolate K1 (methylotroph), Isolate EE-9 (ethanol utilizer), Isolate EE-210 (methylotroph), Pseudomonas DMF 5/8 (methylotroph, DMF utilizer) and Streptomyces R 1186 (isolated by W. Zimmermann, Ciba-Geigy Biotechnology Department). In addition, two different strains of Gliocladium roseum (DSM 1165 = ATCC 10521 and DSM 62726) were evaluated. The reactions were carried out with whole cells in phosphate buffer without addition of organic solvents and the diester concentration was 10-20 mM. The turnover was determined by CGC and ee-values were calculated from the $[\alpha]^{25}_{365nm}$ -values of the isolated reaction products. E (enantiomeric ratio) was computed according to Chen et al.(J. Am. Chem. Soc. 104, 7249 (1982)). The results are summarized in Tables 1 and 2.

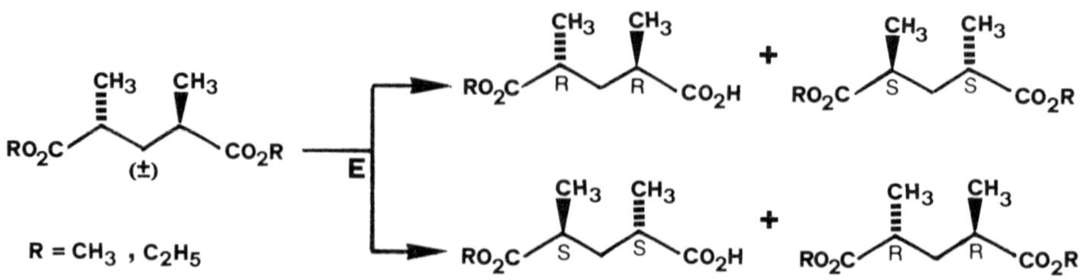

Figure 7. Enantioselective microbial (enzymatic) hydrolysis of racemic 2,4-dimethylglutaric acid dimethyl and diethyl esters. Depending on the nature of the organism or enzyme(s) either the R,R- or the S,S monoester is formed preferentially. E = enzyme or microbe.

TABLE 1. Results of microbial hydrolysis of racemic 2,4-dimethylglutaric acid diethyl ester by DEE-esterase.

Substrate racemic	Strain	Conditions pH, °C	% Monoester (CGC)	ee %	E
DMGADEE	Pseudomonas M27	7, 28	40	50 (-,RR)	4.1
DMGADEE	Isolate K1	7, 28	50	50 (+,SS)	4.8
DMGADEE	Isolate EE-9	8, 25	40	27 (+,SS)	2.0
DMGADEE	Isolate EE-210	8, 25	75	21 (+,SS)	2.7
DMGADEE	Streptomyces R1186	8, 25	47.5	66 (-,RR)	8.8
DMGADEE	Gl. roseum DSM 1165	7, 28	43	31 (-,RR)	2.4
DMGADEE	Gl. roseum DSM 62726	7, 28	56	11 (-,RR)	1.4

DMGADEE = racemic 2,4-dimethylglutaric acid diethyl ester
Gliocladium roseum DSM 1165 is identical with strain ATCC 10521 used by Chen et al.(17). Gl. roseum was grown in a medium containing peptone (1.0 g/l), yeast extract (0.2), glucose (2.0) and $CoCl_2$ (0.001) at pH 7, 28°C.

TABLE 2. Results of microbial hydrolysis of racemic 2,4-dimethylglutaric acid dimethyl ester by DME-esterase.

Substrate racemic	Strain	Conditions pH, °C	% Monoester (CGC)	ee %	E
DMGADME	Pseudomonas M27	7, 28	35	47 (+,SS)	3.5
DMGADME	Isolate K1	7, 28	54	10 (+,SS)	1.3
DMGADME	Pseudomonas DMF 5/8	7, 37	30	85*(+,SS)	17.5*
DMGADME	Isolate EE-210	8, 28	36	∿20 (-,RR)	∿1
DMGADME	Isolate EE-9	8, 28	-	-	-
DMGADME	Streptomyces R1186	8, 28	-	-	-**
DMGADME	Gl. roseum DSM 1165	7, 28	54	43 (-,RR)	4.0
DMGADME	Gl. roseum DSM 62726	7, 28	47	59 (-,RR)	6.4
DMGADME	Gl. roseum ATCC 10521 (results from Chen et al. (17))	7, 25	50	68 (-,RR)	10.5

DMGADME = racemic 2,4-dimethylglutaric acid dimethyl ester
*) degradation of monoester (partially)
**) total degradation of substrate and product

The following conclusions can be drawn from the data presented in tables 1 and 2:

- Gliocladium roseum is more selective with respect to DMGADME rather than DMGADEE.
- Streptomyces R 1186 is the strain of choice for the preparation of (2R, 4R)-2,4-dimethylglutaric acid monoethyl ester. With one recycling step an ee >90% can be obtained (calculated from E-value). Whole cells of strain R1186 are not suitable for the production of the monomethyl ester.
- Pseudomonas M27 and Isolate EE-210 show inverted enantioselectivities with respect to the hydrolysis of dimethyl and diethyl esters.

With Streptomyces R 1186 not only the whole cells but also the culture filtrate shows esterase activity, indicating that significant amounts of the DEE-esterase are excreted into the medium or liberated by lysis of cells. The enzyme is thus easy to produce. Enzyme fractions from Isolates EE-9 and EE-210 show a high hydrolytic activity and the same enantioselectivity as the whole cells.

7. MICROBIAL PRODUCTION OF D(-)-RIBULOSE AND SHIKIMIC ACID BY MUTANT STRAINS OF NOCARDIA MEDITERRANEI

7.1. Biosynthetic Studies with Rifamycins (Ansamycins)

In order to elucidate the biogenetic pathway for the seven-carbon amino unit (later identified as 3-amino-5-hydroxybenzoic acid) of the rifamycin chromophore, auxotrophic mutants deficient in the biosynthesis of aromatic amino acids (shikimate pathway) were selected and characterized (see reviews 18 and 19).

It was demonstrated that the C_7N-unit of the rifamycins and other ansamycins is derived from an intermediate of the shikimate pathway between 3-deoxy-D-arabinoheptulosonic acid-7-phosphate and shikimate. 3-Amino-5-hydroxybenzoic acid (AHBA) is thus a novel product of the shikimate pathway (18,19). The two characterized mutants of Nocardia mediterranei are a transketolase$^-$-mutant A8 excreting D(-)-ribulose and a mutant A10 (probably lacking shikimate kinase) excreting shikimate (Fig. 8).

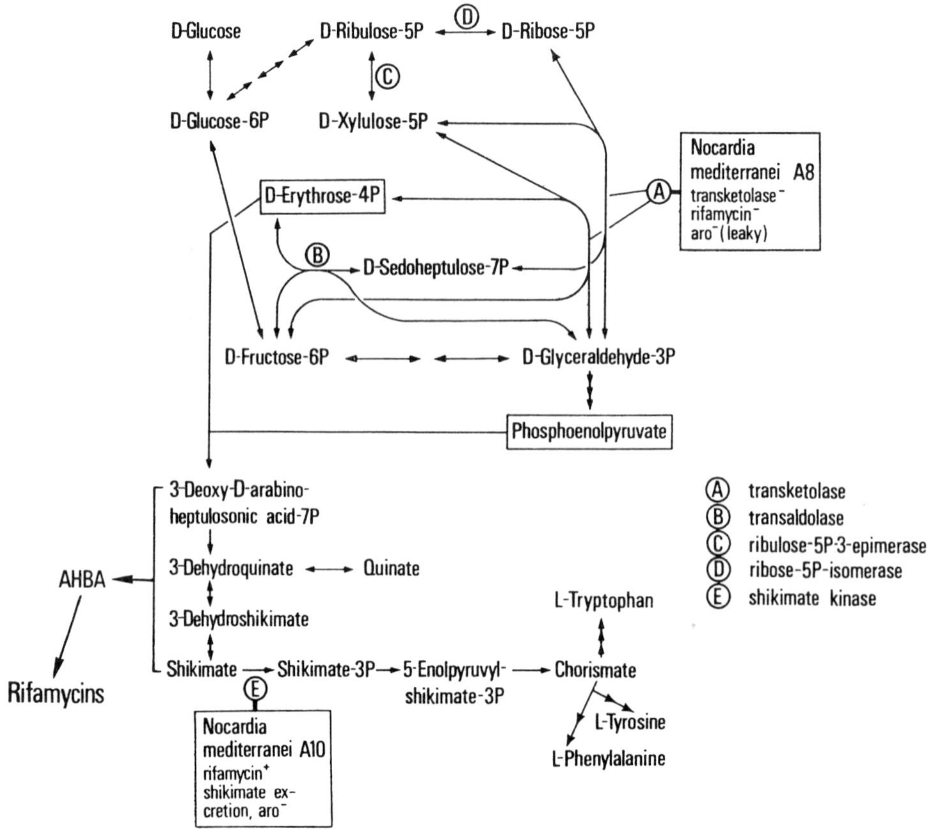

Figure 8. Biosynthetic pathways: pentose shunt, glycolysis and shikimate pathway.

7.2. Microbial Production of D(-)-Ribulose

D(-)-Ribulose is commercially available at extremely high prices and relatively low purity. The transketolase$^-$-mutant Nocardia mediterranei A8 accumulates a mixture of pentoses with D-ribulose as the major product and arabinose, xylulose and xylose as the identified minor products. In shake flask cultures titers up to 6-8 g/l D-ribulose can be obtained. Batch fermentations in 30-l fermenters using an industrial production medium yielded titers up to 10-15 g/l of D-ribulose after 8 days with a complete depletion of the glucose (carbon source) present in the medium. By feeding an additional amount of glucose (fed-batch fermentation) the titers can easily be increased up to 20 and more g/l of D-ribulose (19). So far the best results were obtained with a double mutant derived from strain A8 by spontaneous mutation. This double mutant A8/6 has a second block directly in the biosynthesis of rifamycin. D-ribulose can also be produced in a biotransformation system with glucose, buffer, and mycelium of _Nocardia mediterranei_ A8 or A8/6. D-ribulose of >95% purity (determined with ribitol dehydrogenase) is obtained from fermentation broths by ion-exchange chromatography.

Although D-ribulose (Fig. 9) is an interesting chiral compound it is rarely used for chemical synthesis, probably due to its extremely high price. On condition that D-ribulose becomes available at high purities and reasonable costs (e.g. when produced by fermentation on a large scale) it would be an excellent starting material for chemical synthesis and enzymatic transformations using enzymes from microorganisms such as _N. mediterranei_, _E. coli_ etc., plants (spinach) or liver (Fig. 9).

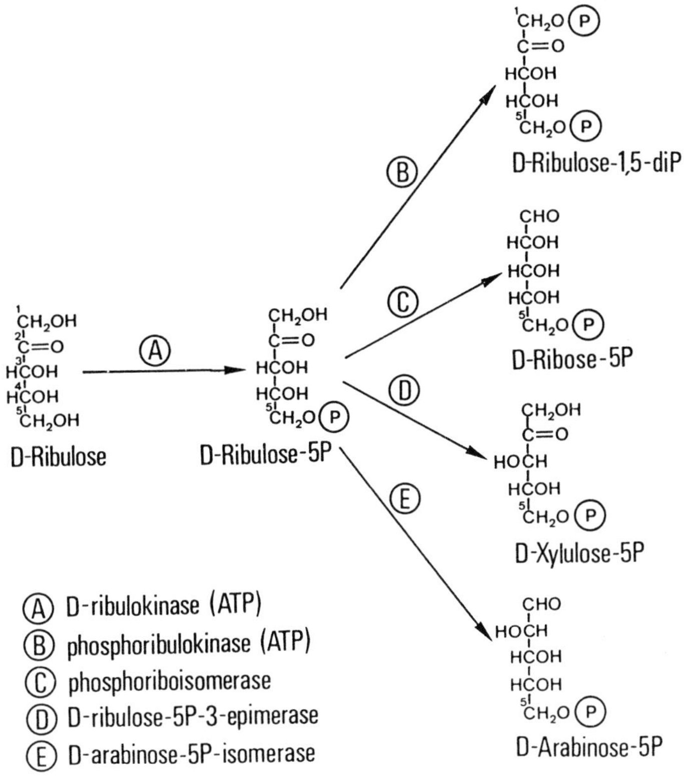

Figure 9. Enzymic synthesis of important enzyme substrates starting from D-ribulose.

The products available in simple one-step or two-step transformations are: D-ribulose-5-phosphate, D-ribulose-1,5-diphosphate, D-ribose-5-phoshate, D-xylulose-5-phosphate, D-arabinose-5-phosphate etc. All of these compounds are very important enzyme substrates and intermediates for biochemical research and all of them are very expensive. Specifically labelled ^{13}C or ^{14}C substrates could be prepared starting from labelled D-ribulose produced by fermentation with N. mediterranei A8 or A8/6 and labelled precursors (glucose, acetate etc.).

7.3. Microbial Production of Shikimic Acid

(-)-Shikimic acid (3,4,5-trihydroxy-1-cyclohexene-1-carboxylic acid) is commercially available at high prices. The aro$^-$-mutant Nocardia mediterranei A10 accumulates shikimate and traces of 3-dehydroshikimate in the culture broth. In shake flasks (batch cultures) titers up to 1-1.5 g/l of shikimic acid were obtained. Batch fermentations in 30-l fermenters using an industrial production medium yielded titers up to 2 g/l of shikimic acid with a complete depletion of the glucose present in the medium (19). We have calculated that a strain optimization by blocking the rifamycin production of strain A10 directly after the biosynthetic branch point for the C_7N-unit would channel all the substrate flux for the shikimate pathway into the production of shikimate and result in titers up to 5 g/l.

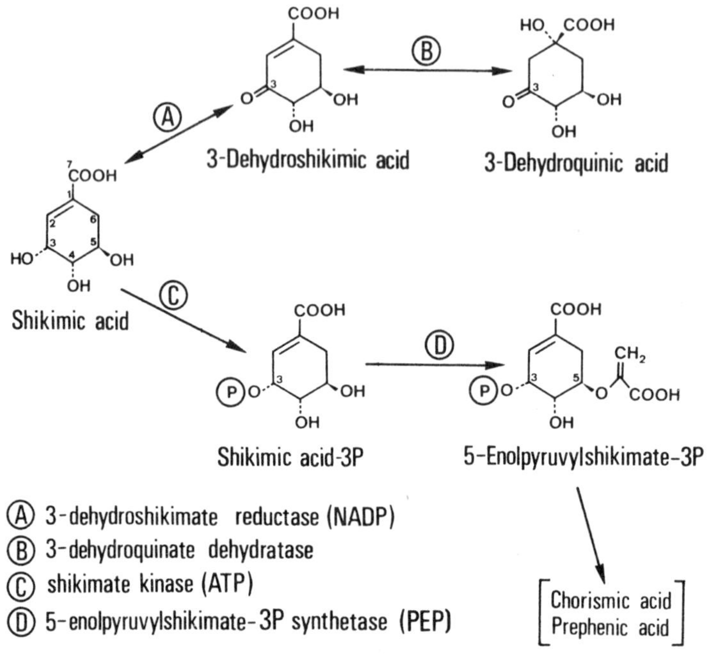

Figure 10. Enzymatic synthesis of important enzyme substrates starting from shikimic acid.

Although shikimic acid (Fig. 10) is an interesting chiral compound, it is, like ribulose, rarely used for chemical synthesis, probably also due to its high price. However, it would be a good starting material for enzymatic transformations using enzymes from microorganisms (e.g. E. coli) or plants (e.g. Pisum sativum). The products available by simple transformations are (Fig. 10): 3-dehydroshikimic acid, 3-dehydroquinic acid, shikimate-3-phosphate, 5-enolpyruvyl-shikimate-3-phosphate, and possibly also chorismate and prephenate. All these compounds are interesting enzyme substrates and intermediates for biochemical research. A number of biologically active shikimate derivatives have been isolated and described so far. Specifically labelled ^{13}C or ^{14}C substrates could be prepared starting from labelled shikimate produced by fermentation.

8. CONCLUSIONS

Methylotrophs isolated for specific waste degradation processes were found to be an excellent source for novel useful enzymes. Biomass of methylotrophs (single cell protein) can easily be produced on very cheap substrates, even waste substrates, in large quantities. The PQQ-dependent methanol dehydrogenase and related enzymes deserve further investigation. PQQ is also investigated as catalyst in the absence of an enzyme by Japanese workers (20-22). PQQ can be used as a catalyst for the oxidation of amines (20), for oxidative decarboxylation of amino acids (21) and for the oxidation of thiols (22). Auxotrophic mutants of Nocardia mediterranei can be applied for the production of expensive biochemicals.

REFERENCES

1) O. Ghisalba and M. Küenzi:'Biodegradation and utilization of monomethyl sulfate by specialized methylotrophs' Experientia **39**, 1257 (1983).
2) O. Ghisalba and M. Küenzi: 'Biodegradation and utilization of quaternary alkylammonium compounds by specialized methylotrophs' Experientia **39**, 1264 (1983).
3) O. Ghisalba, P. Cevey, M. Küenzi and H.P. Schär: ' Biodegradation of chemical waste by specialized methylotrophs, an alternative to physical methods of waste disposal' in: IWTUS-3 Third International Waste Treatment and Utilization Symposium 1984, Pergamon Press in print.
4) C. Anthony: The Biochemistry of Methylotrophs, Academic Press (1982).
5) C. Fromageot: 'Sulfatases and desulfinases' Meth. Enzymol. **2**, 324 (1955).
6) E.C. Sandberg and R.C. Jenkins: 'Hydrolysis of the sulfuric acid esters of steroid compounds' Meth. Enzymol. **15**, 684 (1966).
7) S.G. Hales, K.S. Dodgson, D.F. White, N. Jones and G.K. Watson:'Initial stages in the biodegradation of the surfactant sodium dodecyltriethoxy sulfate by Pseudomonas sp. strain DES1' Appl. Environ. Microbiol **44**, 790 (1982).
8) J.M. Cloves, K.S. Dodgson, G.F. White and J.W. Fitzgerald:'Purification and properties of the P2 primary alkylsulphohydrolase of the detergent-

degrading bacterium Pseudomonas C12B' Biochem. J. 185, 23 (1980).

9) H.P. Schär and O. Ghisalba: 'Hyphomicrobium bacterial electrode for determination of monomethyl sulfate' Biotechnol. Bioeng. 26, 897-901 (1985).

10) M. Gisin, S. Rösli, H.P. Schär and O. Ghisalba: unpublished results

11) H.P. Schär, P. Chemla and O. Ghisalba: 'Methanol dehydrogenase from Hyphomicrobium MS 223' FEMS Microbiol. Lett. 26, 117 (1985).

12) J.A. Duine, J. Frank and J. Westerling:'Purification and properties of methanol dehydrogenase from Hyphomicrobium X' Biochim. Biophys. Acta 524, 277 (1978).

13) J.A. Duine and J. Frank: 'The prosthetic group of methanol dehydrogenase' Biochem. J. 187, 221 (1980).

14) M. Ameyama, M. Hayashi. K. Matsushita, E. Shinagawa and O. Adachi: 'Microbial production of pyrrolo-quinoline quinone' Agric. Biol. Chem. 48, 561 (1984).

15) N. Gruenfeld, J.L. Stanton, A.M. Yuan, F.H. Ebetino, L.J. Browne, C. Gude and C.F. Huebner: ' Angiotension converting enzyme inhibitors: 1-glutarylindoline-2-carboxylic acid derivatives' J. Med. Chem. 26, 1277 (1983).

16) G.M. Ramos Tombo, H.P. Schär, W. Zimmermann and O. Ghisalba: in preparation.

17) C.S. Chen, Y. Fujimoto and C.J. Sih: 'Bifunctional chiral synthons via microbiological methods. 1. Optically active 2,4-dimethylglutaric acid monomethyl esters' J. Am. Chem. Soc. 103, 3580 (1981).

18) O. Ghisalba, J.A.L. Auden, T. Schupp and J. Nüesch: 'The rifamycins: properties, biosynthesis, and fermentation' in: E.J. Vandamme: Biotechnology of Industrial Antibiotics, Marcel Dekker, New York (1984),p.281.

19) O. Ghisalba: 'Biosynthesis of rifamycins (ansamycins) and microbial production of shikimate pathway precursors, intermediates and metabolites' Chimia 39, 79 (1985).

20) Y. Ohshiro, S. Itoh, K. Kurokawa, J. Kato, T. Hirao and T. Agawa: 'Micelle enhanced oxidation of amines by coenzyme PQQ' Tetrahedron Lett. 24, 3465 (1983).

21) S. Itoh, N. Kato, Y. Ohshiro and T. Agawa: 'Oxidative decarboxylation of α-amino acids with coenzyme PQQ' Tetrahedron Lett. 25, 4753 (1984).

22) S. Itoh, N. Kato, Y.Ohshiro and T. Agawa: 'Catalytic oxidation of thiols by coenzyme PQQ' Chemistry Lett. (J.C.S. Japan) 135 (1985).

SYNTHESIS OF CHIRALLY LABELLED SUBSTRATES USING ENZYMES

Douglas W. Young
School of Chemistry and Molecular Sciences
University of Sussex
Falmer
Brighton BN1 9QJ
U.K.

ABSTRACT. Samples of aspartic acid which are stereospecifically labelled with deuterium at C-3 have been prepared using the enzyme aspartase. These have been used as starting materials in syntheses of stereospecifically labelled samples of serine, glutamic acid, 2-aminoethanol, β-alanine and 2-fluoropropanoic acid. The latter compounds have been used to study the stereochemistry of some enzyme catalysed reactions. The enzymes glutamate pyruvate transaminase and β-methylaspartase have also been used synthetically in these studies.

1 INTRODUCTION

In studies of the interaction of enzymes with their natural substrates and with inhibitors, it is often necessary to prepare substrates which are labelled stereospecifically with isotopes such as ^{2}H, ^{3}H or ^{13}C. For some years we have been interested in the biosynthesis and catabolism of the DNA base thymidine (2) as summarised in Scheme 1. The biosynthesis is a cyclic process which involves three enzyme catalysed reactions. Two of the enzymes involved in the process are targets for anti-cancer drugs.

Our first investigations[1] in this area concerned the enzyme dihydrofolate reductase, and we were able to show that the 4-pro R hydrogen of NADPH (8) is transferred to the same face (the re face with respect to C-6 and the si face with respect to C-7) of the pteridine ring of the vitamin folic acid (9) when it is reduced to the coenzyme tetrahydrofolic acid (5). An X-ray study[2] of a ternary complex of dihydrofolic reductase with the coenzyme NADPH and the anticancer drug methotrexate (10) has shown that, although it is 4-pro R hydrogen of NADPH which approaches the pteridine ring, approach is in fact made to the opposite face of methotrexate (10) from that face of the pteridine ring which we have shown to be attacked in the vitamin folic acid (9).

M. P. Schneider (ed.), Enzymes as Catalysts in Organic Synthesis, 251–262.

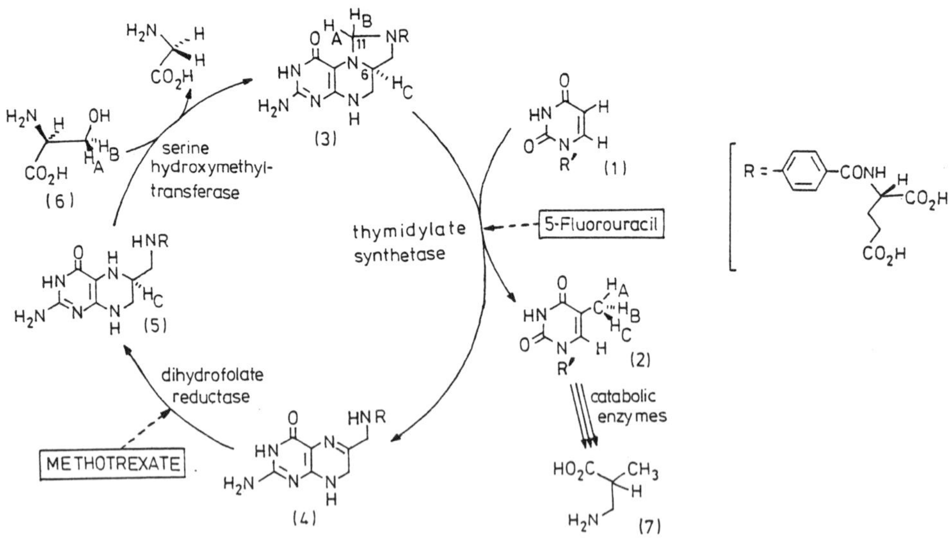

Scheme 1 - Biosynthesis and Catabolism of Thymidine

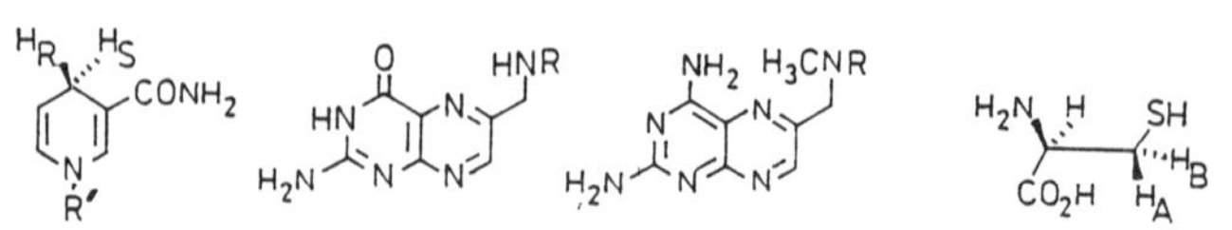

(8) (9) (10) (11)

The next phase of our work was to investigate the overall stereochemistry of the one carbon transfer required to convert the coenzyme tetrahydrofolic acid (5) to the 5,10-methylene derivative (3). This compound would act as the source of the methyl group of thymidine (2) in a process involving the unique transfer of H_C of the coenzyme together with the CH_2 moiety, so that the coenzyme would be oxidised in the process. To pursue this work we required samples of serine (6) stereospecifically labelled at C-3 with one of the isotopes of hydrogen.

2 SYNTHESIS OF STEREOSPECIFICALLY LABELLED L-SERINE and L-GLUTAMIC ACID

Some years ago in a study of the biosynthesis of the antibiotics penicillin G and cephalosporin C, we synthesised[3,4] samples of L-cysteine (11) stereospecifically labelled at C-3 with tritium. These were used to deduce the stereochemistry of the β-lactam ring closure step in the biosynthesis of penicillin G[3,4] and of cephalosporin C.[5] The samples of cysteine (11) were prepared by a totally synthetic route involving a time-consuming resolution step. This had to be performed not only on the unlabelled material but also on each of the two diastereotopically labelled samples. It was evident therefore that, if we wished to prepare other labelled compounds for enzyme studies, an improved synthesis would be required. The use of enzymes in a key step of the synthesis seemed especially attractive and we turned to the commercially available enzyme aspartase.

Aspartase is known[6] to add ammonia to the double bond of fumaric acid (12) with *trans* stereospecificity so that, in one step, we should be able to obtain aspartic acid as the (2S)-enantiomer and with a chiral label at C-3. Using [2,3-$^{2}H_{2}$]-fumaric acid (12, $H_A = {}^{2}H$), (2S,3S)[2,3-$^{2}H_{2}$]-aspartic acid (13, $H_A = {}^{2}H$) could be obtained, whilst use of unlabelled fumaric acid and $^{2}H_{2}O$ in the incubation gave (2S,3R)[3-$^{2}H_{1}$]-aspartic acid (13, $H_B = {}^{2}H$). The process proved amenable to the preparation of large amounts of the labelled aspartates and so the remaining problem was to use these as starting materials in the synthesis of labelled serine and other compounds which we were interested in using as substrates for enzymes whose stereochemistry we wished to study.

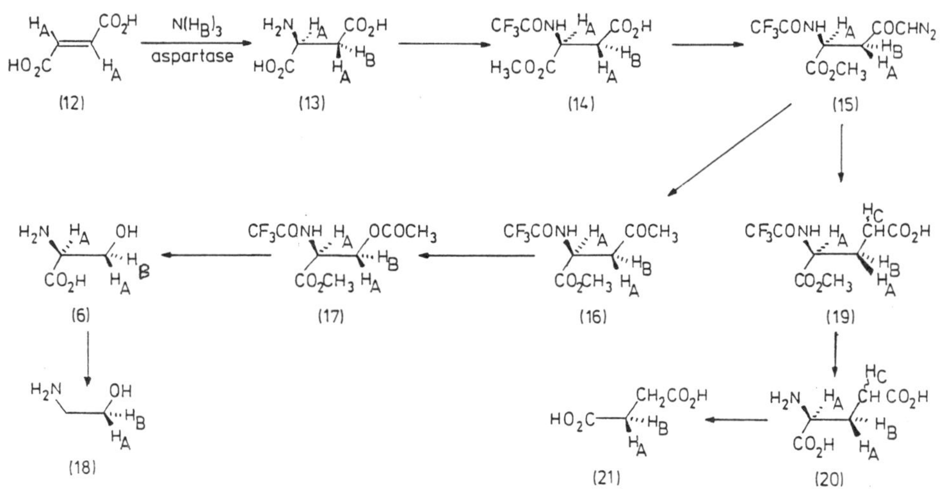

Scheme 2 - Synthesis of L-Serine and L-Glutamic Acid Stereospecifically Labelled at C-3 with Deuterium

Reaction of the labelled aspartic acids (13) with trifluoroacetic anhydride gave the intermediate N-trifluoroacetylanhydrides which, on reaction with methanol, yielded a mixture containing 80% of the α-monomethyl esters (14) and 20% of the regioisomeric β-monomethyl esters. The mixtures of esters were separately converted to mixtures of acid chlorides from which the β-acid chlorides could be obtained by recrystallisation from benzene. Reaction of these acid chlorides separately with diazomethane gave the corresponding diazoketones (15). These proved to be key intermediates for the synthesis of stereospecifically labelled serines (6) and glutamic acids (20).

In the synthesis of the labelled serines, the diazoketones (15) were separately reduced with hydrogen iodide in chloroform to give the methyl ketones (16). Baeyer Villiger oxidation and hydrolysis gave mixtures from which the labelled serines (6) (19%) and aspartic acids (13) (32%) could be isolated by ion-exchange chromatography. The key step in this synthesis was the Baeyer Villiger reaction which would be expected[7] to occur with retention of stereochemistry at the labelled chiral migrating group. The ^{1}H-n.m.r. spectra of the resultant labelled samples of serine indicated that migration of the primary group had occurred stereospecifically but the yield of serine was low due to a lack of regiospecificity in the Baeyer Villiger step. We had expected migration of the labelled centre rather than of the methyl group because of the well-known low migratory aptitude of methyl groups.[7] It would seem, however, that the effect of distant electron withdrawing groups, noted recently in some other work[8,9] had decreased the migratory aptitude of the methylene moiety to a level lower than that of the methyl group.

Although the yield in the last step was low, the availability of the labelled samples of aspartic acid (13, $H_A={}^2H$) and (13, $H_B={}^2H$) allowed us to prepare approaching 1 gm. of each of the samples of labelled serine (6, $H_A={}^2H$) and (6, $H_B={}^2H$).[10] This is ample for enzymic studies and we are currently investigating enzymic one-carbon transfer processes using these compounds. Decarboxylation of the labelled samples of serine has allowed us to obtain samples of (1S,2RS)[1,2-$^2H_2$]- and (1R)[1-$^2H_1$]-2-aminoethanol [(18, $H_A={}^2H$) and (18, $H_B={}^2H$) respectively]. We have used these compounds to show[11] that, in the vitamin B_{12}-mediated rearrangement of 2-aminoethanol (18) to acetaldehyde (22) using the enzyme <u>ethanolamine ammonia lyase</u> (see Scheme 3A), it is the (1S) hydrogen which migrates from C-1 to C-2.

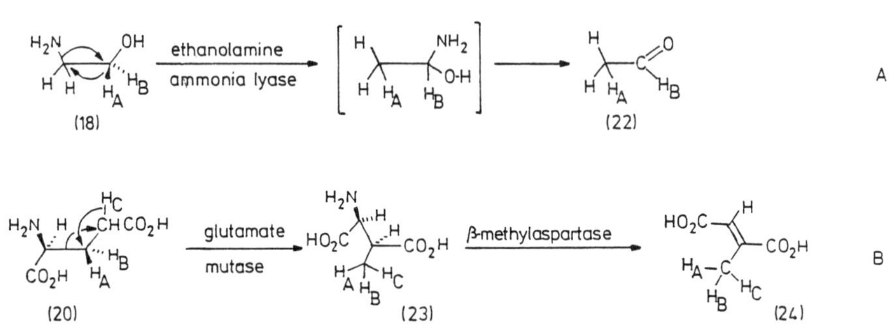

<u>Scheme 3</u> - Reactions Catalysed by the Vitamin B_{12} Dependent Enzymes (A) Ethanolamine Ammonia Lyase and (B) Glutamate Mutase

Because of our interest in the stereochemistry of a second vitamin B_{12}-mediated rearrangement catalysed by the enzyme glutamate mutase (Scheme 3B), we wished to synthesise samples of glutamic acid stereospecifically labelled at C-3. This could be achieved by photolytic Wolff rearrangement of the diazoketones (15) prepared in our serine synthesis. Use of ^2H_2O or ^3H_2O in the photolysis allowed non-stereospecific labelling at C-4. Hydrolysis of the product gave the appropriately labelled samples of glutamic acid (20). Although Wolff rearrangements were expected[12] to occur with retention of stereochemistry at the migrating centre, there had, to our knowledge, been no examples at that time of Wolff rearrangements which involved migration of a primary chiral centre. It was necessary therefore to confirm the absolute stereochemistry of the two samples of glutamic acid (20, H_A=^2H) and (20, H_B=^2H). These were degraded by reaction with chloramine T followed by acid hydrolysis to yield samples of (2R)- and (2S)-succinic acid [(21, H_A=^2H) and (21, H_B=^2H) respectively] which had o.r.d.[13,14] and c.d.[14] spectra in keeping with the literature. The absolute stereochemistry of the samples of (2S,3S)[2,3-$^2H_2$]- and (2S,3R)[3-$^2H_1$]-glutamic acid [(20, H_A=^2H) and (20, H_B=^2H) respectively] was therefore confirmed and it was evident that the Wolff rearrangement had occurred with retention of stereochemistry at the migrating primary chiral centre.[15]

Our studies on the sterochemistry of the glutamate mutase catalysed rearrangement using a mixed enzyme system from *Clostridium tetanomorphum* containing glutamate mutase and β-methyl aspartase (see Scheme 3B) were at an advanced stage[16] when it was shown[17] that the rearrangement proceeds with complete racemisation at the methyl group.

2 CATABOLISM OF URACIL TO β-ALANINE

Returning to our interest in the synthesis and catabolism of thymidine we noted that not only did the catabolic enzymes degrade thymine but that they were also able to degrade uracil (1, R=H) and 5-fluorouracil (28) as shown in Scheme 4 below. In view of the interesting differences in binding of folate and methotrexate observed with dihydrofolate reductase, it was of interest to compare the stereochemistry of the catabolism of these three substrates.

We have been able to prepare a partially purified enzyme system from beef liver which will perform the catabolic processes summarised in Scheme 4. The first and rate limiting step in these processes, the reduction of the pyrimidines, is catalysed by dihydrothymine dehydrogenase, an enzyme which has been shown to contain FAD[18]. The FAD is reduced to $FADH_2$ by the coenzyme NADPH, and we used glucose-6-phosphate dehydrogenase to recycle the NADPH by the oxidation of glucose-6-phosphate to 6-phosphogluconate. The catabolic experiments were performed in parallel, one set of reactions using labelled substrates where necessary and aqueous medium, and the other set using unlabelled substrates and ^2H_2O. The involvement of $FADH_2$ in the process meant that labelling could be achieved using ^2H_2O although there was an isotope effect of *ca.* 4 in these experiments.

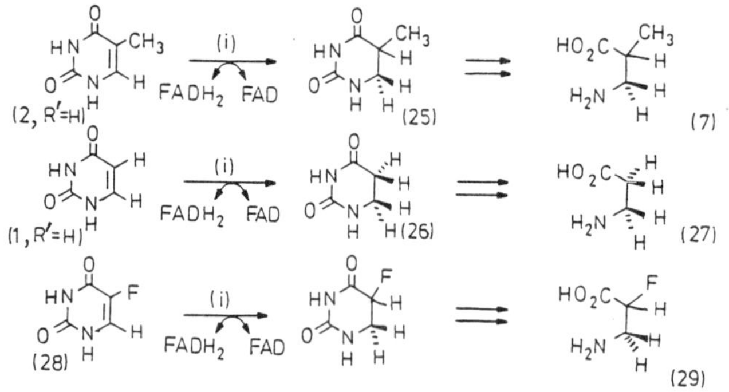

Scheme 4 - Catabolism of Uracil, Thymine and 5-Fluorouracil

The products from the catabolism of [5-$^2H_1$]- and [6-$^2H_1$]-uracil in H_2O and of uracil in ^2H_2O were samples of β-alanine stereospecifically labelled with deuterium and it was necessary to have an assay for the absolute stereochemistry of these compounds. This was achieved by first synthesising samples of β-alanine which were unambiguously stereospecifically labelled with deuterium.

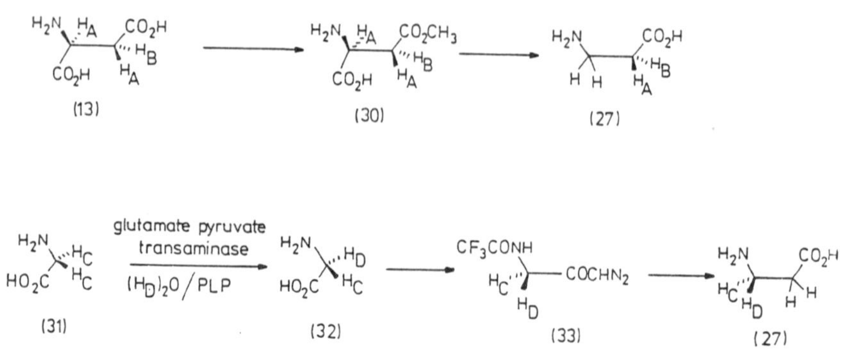

Scheme 5 - Synthesis of Samples of β-Alanine Stereospecifically Labelled in Each of the C-H Bonds with Deuterium

(2S) and (2R)[2-$^2H_1$]-β-Alanines [(27, H_A=^{2}H) and (27, H_B=^{2}H)] were readily prepared from our samples of labelled aspartic acid [(13, H_A=^{2}H) and (13, H_B=^{2}H) respectively] by first preparing the β-methyl esters (30) using thionyl chloride and methanol. Decarboxylation by heating in the melt with *para*-methoxyacetophenone followed by acid hydrolysis gave (2S) and (2R)[2-$^2H_1$]-β-alanines.

For synthesis of the C-3 labelled samples we used the enzyme *glutamate pyruvate transaminase*, known[19] to exchange the *2-pro-R* hydrogen of glycine, to prepare (2R)[2-$^2H_1$]-glycine (32, H_D=^{2}H) from glycine in ^2H_2O and (2S)[2-$^2H_1$]-glycine (32, H_C=^2H) from [2,2-$^2H_2$]-glycine in H_2O. These were converted to the N-trifluoroacetamides with trifluoroacetic anhydride and thence, *via* the corresponding acid chlorides, to the diazoketones (33). Wolff rearrangement and hydrolysis yielded (3R)[3-$^2H_1$]- and (3S)[3-$^2H_1$]-β-alanines.

Having synthesised four samples of β-alanine labelled unambiguously in each of the four C-H groups with deuterium, we were in a position to devise an assay for the metabolic experiments. This was readily achieved for the [3-$^2H_1$]-β-alanines since the camphanoyl amides had 360 MHz ^1H nmr spectra in which these hydrogens could be differentiated (Figure 1). Having assigned these hydrogen atoms it was evident from the spectra from the catabolic experiments (Figure 1) that the dehydrogenase had added hydrogen from the *si* face at C-6 of uracil (1, R=H).

Assignment of the stereochemistry to the [2-$^2H_1$]-β-alanines proved more difficult since nmr methods failed to differentiate these hydrogens in a variety of derivatives and under a variety of conditions. Eventually the problem was resolved by converting the compounds to samples of 2-aminoethanol (18) as in Scheme 6 by first preparing the N-trifluoroacetates and converting them, *via* the acid chlorides and diazoketones, to the methyl ketones (34). Baeyer Villiger reaction and hydrolysis gave samples of labelled 2-aminoethanol which were converted to the N,O-dicamphanoates. These had ^1H-nmr spectra (Figure 2) in good agreement with the spectra of samples of (1S) and (1R)-[1-$^2H_1$]-2-aminoethanol which we had previously prepared by decarboxylation of serine,[11] and a sample of β-alanine from the catabolism of [5-$^2H_1$]-uracil gave N,O-dicamphanoyl-2-aminoethanol, the ^1H nmr spectrum (Figure 2) of which indicated that the dehydrogenase had added hydrogen from the *si* face at C-5 of uracil. The dehydrogenase had therefore added hydrogen in an *anti* manner at C5 and C6 of the pyrimidine.[20]

3 CATABOLISM OF THYMINE TO 3-AMINO-2-METHYLPROPANOIC ACID

Catabolism of thymine (2, R'=H) yields 3-amino-2-methylpropanoic acid (7) and, since the 2R-isomer (7a) of this acid is excreted in human urine,[21] it is likely that the metabolic product has this configuration. The 2S-enantiomer (7b) is, however, found in some peptides[22] and it has been reported that the 2S-isomer is itself metabolised whilst the 2R-isomer is not.[23] It was therefore essential that the stereochemistry of the metabolic product be confirmed. From our earlier work on the enzyme glutamate mutase from *Clostridium tetanomorphum* we had a supply of the enzyme β-methylaspartase from this source (see Scheme 3B). We were therefore able to use mesaconic acid (24) as a substrate for this enzyme to prepare a sample of (2S,3S)-3-methylaspartic acid (23) of known[24] absolute stereochemistry. This compound could be decarboxylated by pyrolysis in a melt with *p*-methoxyacetophenone followed by hydrolysis in 3M aqueous hydrochloric acid. The (2S)-3-amino-2-methylpropanoic acid (7b) obtained in this experiment had $[\alpha]_D$ + 12.7° compared to $[\alpha]_D$ - 15.4° for the isomer obtained by catabolism of thymine using our bovine liver enzyme system.

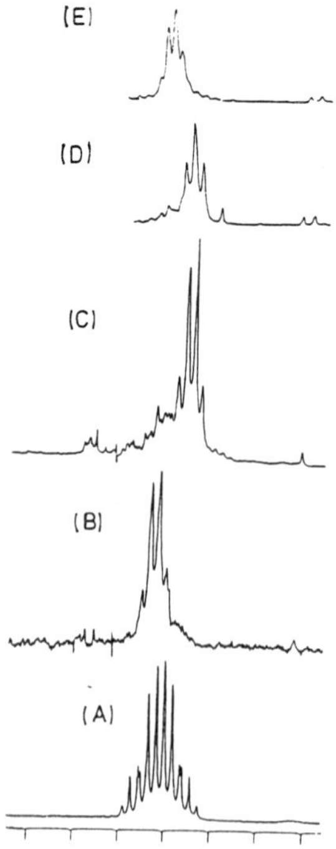

Figure 1 - 360 MHz ^1H nmr spectra ($CDCl_3$) of the β-alanine C-3 protons in the camphanoyl amide of (A) β-alanine (27); (B) synthetic (27, H_D=^2H); (C) synthetic (27, H_C=^2H); (D) (27) from catabolism of uracil in ^2H_2O; (E) (27) from catabolism of [6-^{2}H]-uracil in H_2O

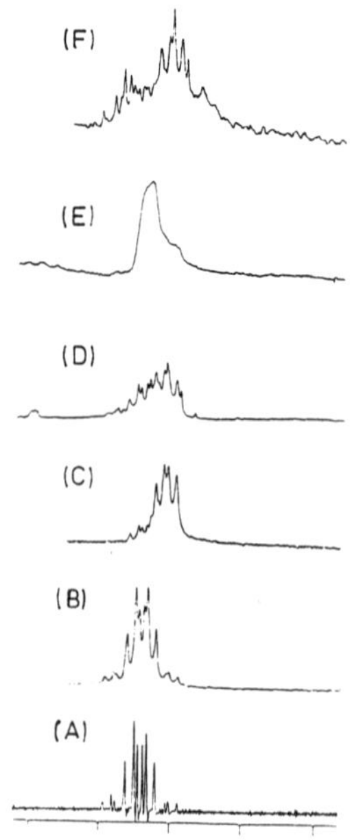

Figure 2 - 360 MHz ^1H nmr spectrum ($CDCl_3$) of N,O-dicamphanoyl-2-aminoethanol (A) unlabelled compound with line narrowing programme; (B) unlabelled compound without line narrowing; (C) authentic (1R)[1-$^2H_1$]-compound; (D) compound from degradation of synthetic (2R)-[2-$^2H_1$]-β-alanine; (E) compound from degradation of synthetic (2S,3RS)[2,3-$^2H_2$]-β-alanine; (F) compound from degradation of the β-alanine obtained by catabolism of [5-^{2}H] uracil in H_2O

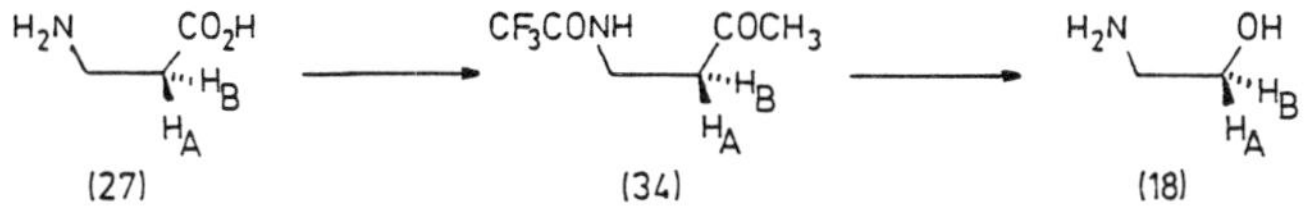

Scheme 6 - Degradation of Labelled β-Alanines to 2-Aminoethanols

That these compounds were enantiomeric could be further proved by preparing their camphanic acid amides and that of commercial (2RS)-3-amino-2-methylpropanoic acid. The 360 MHz ^{1}H-nmr spectra of these compounds showed clearly that the C-3 protons had different chemical shifts in the 2R- and 2S-isomers and that, although decarboxylation of (2S,3S)-3-methylaspartic acid (23) had been accompanied by a small amount of racemisation at C-2, the product from catabolism of thymine was clearly (2R)-3-amino-2-methylpropanoic acid (7a). This would imply that the product of the first step in thymine catabolism was (5R)-dihydrothymine and that reduction of thymine by dihydrothymine dehydrogenase had occurred from the *si*-face at C-5.

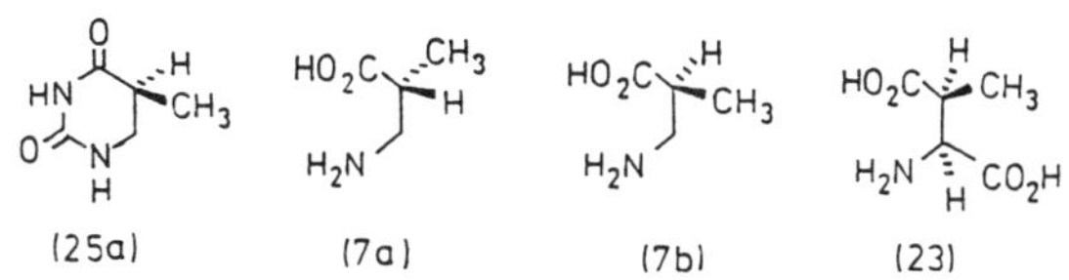

In order to assess the steric course of the reduction of thymine at C-6, it was necessary to have a sample of (2R)- or (2RS)-3-amino-2-methylpropanoic acid which was unambiguously stereospecifically labelled at C-3 with deuterium. Since catalytic hydrogenation of thymidine (2, R' = deoxyribose) takes place with asymmetric induction so that, on hydrolysis, (5S)-dihydrothymine (25a) is produced,[25] we used $^{2}H_2$ and $^{2}H_2O$ in the reduction step and hydrolysed the resultant dihydrothymine with hydrochloric acid, known[26] to result in racemisation at C-2, to obtain (2RS,3S)[3-$^{2}H_1$]-3-amino-2-methylpropanoic acid. The ^{1}H nmr spectrum of the camphanoate of this compound confirmed the stereochemical purity of the 3S label in both the 2S- and 2R-isomers of 3-amino-2-methylpropanoic acid and provided an assay for C-3 of the products from the catabolism of thymine. The ^{1}H nmr spectra clearly showed that enzymic reduction of thymine must have occurred from the *si*-face at C-6. The reduction process therefore involved *anti* addition of hydrogen at C-5 and C-6.[27]

4 CATABOLISM OF 5-FLUOROURACIL TO 3-AMINO-2-FLUOROPROPANOIC ACID

Catabolism of 5-fluorouracil (28) yields 3-amino-2-fluoropropanoic acid (29). One stereoisomer of this compound has been prepared[28] by reaction of N,N-dibenzyl-L-serine benzyl ester (35,R'=CH_2Ph) with diethylaminosulphur trifluoride (DAST) followed by hydrogenolysis. When we prepared the methyl ester (35,R'=CH_3) and treated it with DAST followed by hydrogenolysis and hydrolysis, we obtained 3-amino-2-fluoropropanoic acid (29), $[\alpha]_D = 28.4^o$. This was identical to a sample which we obtained by catabolism of 5-fluorouracil. Since the low anomalous dispersion of fluorine made use of Bijvoet's method difficult, we converted (29) into its methyl ester and thence into the camphanoyl amide (36). X-ray structure analysis of this compound showed it to be the (2R) isomer. The enzymic reduction of 5-fluorouracil had therefore occurred from the *si*-face at C-5.

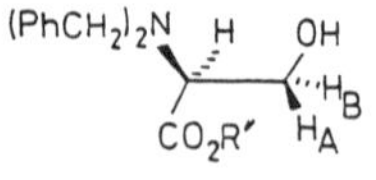

(35)

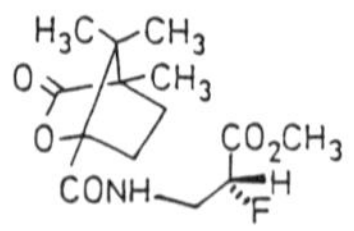

(36)

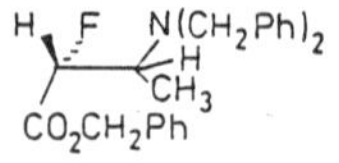

(37)

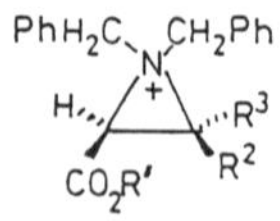

(38)

When L-threonine and L-allothreonine replaced L-serine in the DAST synthesis of α-fluoro-β-amino acids[28] the products were the *threo*- and *erythro*-isomers respectively of (37). Thus the *relative* stereochemistry was the same as in the starting amino acid. Our X-ray analysis of the L-serine product suggested that both centres would be inverted in the process which would be compatible with the intermediacy of an aziridinium ion (38). Since we had already prepared samples of L-serine stereospecifically labelled at C-3 using our aspartase synthesis we were in a position to synthesise samples of the catabolic product (29) stereospecifically labelled at C-3 with deuterium. The absolute stereochemistry at C-3 of these compounds would be epimeric with the stereochemistry at C-3 in the labelled samples of L-serine. When the samples of labelled L-serine (6) were converted to the deuteriated fluoro-β-alanines (29) and these were further converted to the N-camphanoyl methyl esters (36), the ^{2}H-nmr spectra allowed us to assign the absolute stereochemistry to the C-3 hydrogens so that the ^{2}H-nmr spectra of the derivatives (36) of the catabolic products of [6-^{2}H]-5-fluorouracil in H_2O and 5-fluorouracil in ^2H_2O showed that the enzymic reduction of 5-fluorouracil had occurred from the *si*-face at C-6.[27]

These experiments show that catabolism of the RNA base uracil, the DNA base thymine, and the anti-cancer drug 5-fluorouracil all involve *anti* addition of hydrogen to the pyrimidines at the *si*-face at C-5 and the *si*-face at C-6. This suggests that these compounds are bound similarly at the active site of the enzyme in contrast to the case of the substrate folic acid and the drug methotrexate which bind at the active site of *dihydrofolate reductase* with quite different geometries.

The work discussed in this lecture has been done at Sussex by my colleagues Peter Charlton, Steven Field, Vanessa Bertram, David Gani and Peter Hitchcock and I am grateful to them for their skill and energy in bringing it to a successful conclusion. I am also grateful to Drs. Feeney, Roberts and Birdsall at the National Institute for

Medical Research, Mill Hill, for their collaboration on part of the work and to the Science and Engineering Research Council (U.K.) for their financial support.

References

1. P.A. Charlton, D.W. Young, B. Birdsall, J. Feeney and G.C.K. Roberts, J. Chem. Soc., Chem. Commun., 1979, 922; ibid J. Chem. Soc., Perkin Trans. I, 1985, in press.
2. D.A. Matthews, R.A. Alden, J.T. Bolin, D.J. Filman, S.T. Freer, R. Hamlin, W.G.J. Hol, R.L. Kisliuk, E.J. Pastore, L.T. Plante, N. Xuong, and J. Kraut, J. Biol. Chem., 1978, 253, 6946.
3. D.J. Morecombe and D.W. Young, J. Chem. Soc., Chem. Commun., 1975, 198.
4. D.W. Young, D.J. Morecombe and P.K. Sen, Eur. J. Biochem., 1977, 75, 133.
5. J.A. Huddleston, E.P. Abraham, D.W. Young, D.J. Morecombe, and P.K. Sen, Biochem. J., 1978, 169, 705.
6. see D.W. Young in 'Isotopes in Organic Chemistry', eds. E. Buncel and C.C. Lee, Elsevier, Amsterdam, 1978, vol. 4, p.231.
7. P.A.S. Smith in 'Molecular Rearrangements', ed. P. de Mayo, Wiley Interscience, New York, 1967, part 1, p.577.
8. R. Noyori, T. Sato, and H. Kobayashi, Tetrahedron Lett., 1980, 21, 2569.
9. Z. Grudzinski, S.M. Roberts, C. Howard and R.F. Newton, J. Chem. Soc., Perkin Trans. I, 1978, 1182.
10. D. Gani and D.W. Young, J. Chem. Soc., Perkin Trans. I, 1983, 2393.
11. D. Gani, O.C. Wallis and D.W. Young, Eur. J. Biochem., 1983, 136, 303.
12. ref. 7 p.528.
13. J. Retey, J. Seibl, D. Arigoni, J.W. Cornforth, G. Ryback, W.P. Zeylemaker and C. Veeger, Eur. J. Biochem., 1970, 14, 232.
14. G. Ryback, unpublished observations; we thank Dr. George Ryback, Shell Biosciences Laboratory, Sittingbourne for running the o.r.d. and c.d. spectra of these compounds.
15. S.J. Field and D.W. Young, J. Chem. Soc., Perkin Trans. I, 1983, 2387.
16. S.J. Field, D.Phil. Thesis, Sussex 19
17. G. Hartrampf and W. Buckel, F.E.B.S. Letters, 1984, 171, 73.
18. T. Shiotani and G. Weber, J. Biol. Chem., 1981, 256, 219.
19. P. Besmer and D. Arigoni, Chimia, 1968, 22, 494.
20. D. Gani and D.W. Young, J. Chem. Soc., Chem. Commun., 1983, 576; ibid, J. Chem. Soc., Perkin Trans. I, 1985, in press.
21. K. Balenovic and N. Bregant, Tetrahedron, 1959, 5, 44.
22. Y. Kakimoto, A. Kanazawa and I. Sano, Biochim Biophys Acta, 1965, 97, 376.
23. M.D. Armstrong, K. Yates, Y. Kakimoto, K. Taniguchi and T. Kappe, J. Biol. Chem., 1963, 238, 1447.
24. M. Sprecher and D.B. Sprinson, J. Biol. Chem., 1966, 241, 868.
25. Y. Kondo and B. Witkop, J. Am. Chem. Soc., 1968, 90, 764.

26. V. Rachina and I. Blagoeva, Synthesis, 1982, 967, reference 12.
27. D. Gani, P.B. Hitchcock, and D.W. Young, J. Chem. Soc., Chem. Commun., 1983, 898; ibid, J. Chem. Soc., Perkin Trans. I, 1985, in press.
28. L. Somekh and A. Shanzer, J. Am. Chem. Soc., 1982, 104, 5836.

SYNTHESIS OF L-AMINO ACIDS BY ISOLATED ENZYMES AND MICROORGANISMS

C. Wandrey
Nuclear Research Center
Institute of Biotechnology
P.O. Box 1913
D-5170 Jülich
FRG

ABSTRACT

Some examples for the production of L-amino acids which were investigated in close cooperation with Gesellschaft für Biotechnologische Forschung, Braunschweig (Dr. Kula) and Degussa AG, Hanau (Dr. Leuchtenberger) are described. The paper is focussed on biochemical engineering aspects of continuous production. In all examples a re-use of the biocatalysts is achieved by membranes (enzyme-membrane-reactor; cell-membrane-reactor). The membrane reactor concept is an alternative to the use of carrier fixed enzymes respectively cells and probably the system of choice if continuous cofactor regeneration is attempted with isolated enzymes.

INTRODUCTION

Already for 1979 the world production of amino acids was estimated to be about 0.5 million tons ≙ 1.9 billion US-$ (1). Three amino acids (L-glutamic acid, D,L-methionine and L-lysine) dominate this field with respect to the output as well as to the value. These three amino acids are used for seasoning (L-glutamic acid) and as feed additives (D,L-methionine, L-lysine). In recent years there is an especially increasing use of amino acids in infusion solutions, as drugs, as starting material for agricultural chemicals and in the field of artificial sweeteners. Here normally only one optically active enantiomere is desired. Different accesses of enantiomeric compounds are shown in Fig. 1.

M. P. Schneider (ed.), Enzymes as Catalysts in Organic Synthesis, 263–284.

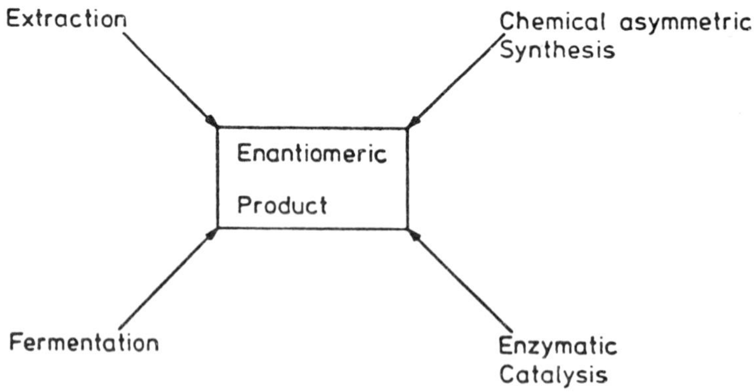

Fig. 1: Processes for the manufacture of enantiomeric compounds

Here mainly examples for enzymatic catalysis will be discribed. Using this route amino acid-derivatives or precursors can be used, so that well defined substrate solutions can be applied. Side reactions normally don't occur. The substrate is only used for biotransformations and not for growth and downstream processing is normally simpler and more effective than using other routes.

Fig. 2 shows different enzymatic routes.

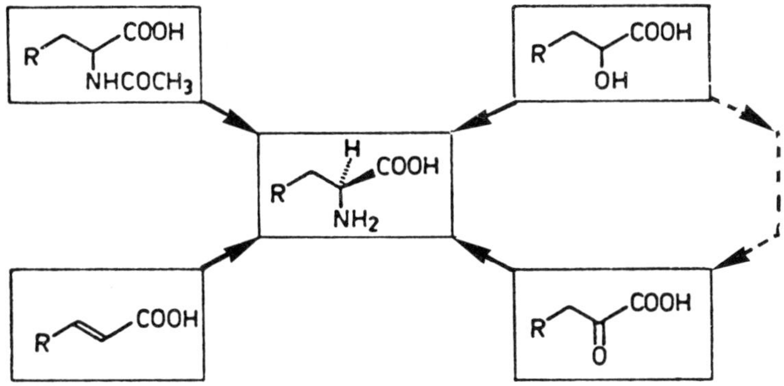

Fig. 2 Enzymatic routes to L-enantiomers of α-amino acids

Hydrolytic enzymes like acylases can be used, lyases which add ammonium to a double bond or transferases as well as amino acid dehydrogenases which use the corresponding α-keto acid as precursor. It is also possible to use racemic mixtures of α-hydroxy acids which are first converted to the α-keto acid and then stereospecificly to the L-amino acid. There is quite a number of other enzymatic routes to L-amino acids described in the literature (1 - 2).

For all enzymatic routes membrane reactors were used with native enzymes. This reactor type is compared to a fixed bed reactor with a carrier fixed enzyme in Fig. 3.

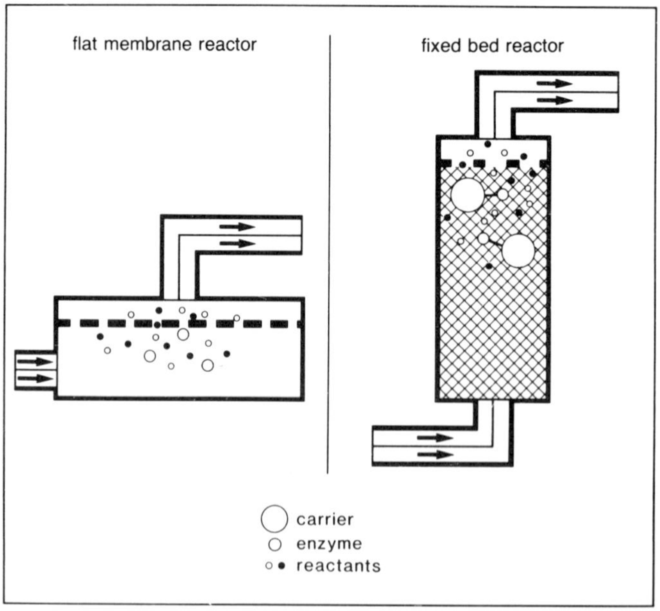

Fig. 3: Comparison of the membrane reactor with a native enzyme and a fixed bed reactor with a carrier fixed enzyme

For a continuous use of an enzyme a continuous separation from the reaction mixture must be achieved. This is possible with both reactor types. In principle there is only a difference in the pore size of the separation unit. For carrier fixed enzymes normal glass frits are sufficient, while for native enzymes ultrafiltration membranes have to be applied. Due to the difference in size between the catalyst and the product the latter alternative proved to be feasible. The best "membrane reactor" is obviously the microbial cell where many enzymes are used in "solution". In comparison to the microbial cell the technical membrane reactor exhibits on one hand poor separation capabilities, on the other hand (in contrast to the microbial cell) enforced flow can be used across the membrane. The advantages and disadvantages of an enzyme membrane reactor are summarized in Table 1.

Advantages	Disadvantages
- no transport limitations - high space time yield - no fixation procedure - no fixation chemicals - easy to sterilize - simple catalyst supplementation - constant space time yield possible - ultrafiltered product stream - useful for continuous cofactor regeneration	- substrate filtration required - higher investment cost for the reactor (in comparison to a fixed bed reactor) - many enzymes are less stable in solution than in the carrier fixed form - plug flow condition can only be approximated in a cascade of membrane reactors

Table I: Advantages and disadvantages of an enzyme membrane reactor

The membrane reactor provides the possibility for a continuous homogeneous catalysis without transport limitations, so that high space time yields can be achieved. At very high catalyst concentrations the effect of concentration polarization in front of the membrane may occur, which can be minimized by sufficient mixing. In comparison to carrier fixed enzymes or entire cells the volumetric activity can be at least one order of magnitude higher in a membrane reactor, before transport limitations occur. If the native enzyme can be used, of course, there is no fixation procedure which often exhibits some enzyme deactivation. Furthermore the catalyst is not contaminated with fixation chemicals. The native enzyme can be introduced in a steam sterilized membrane reactor across a sterile filter, so that aseptic conditions can be guaranteed. Similarly additional fresh enzyme can be supplemented across a sterile filter in order to maintain constant volume activity in spite of some enzyme deactivation. Thus it is easily possible to work at constant space time yield for long periods of operation. The obtained product stream is an ultrafiltered solution which is useful for downstream processing and product specification. For enzymatic steps depending on transport metabolites (coenzymes) only a homogeneous catalysis is applicable so that the membrane reactor seems to be the system of choice.

On the other hand there are some disadvantages. The substrate solution should be prefiltered in order to use the ultrafiltration membranes as long as possible. The investment cost for the reactor itself normally is higher than for a simple tubular reactor. Here one should keep in mind, that the capital involved to fill a tubular reactor with immobilized enzymes can be remarkable. The strongest empirical argument against membrane reactors is the fact, that up to now many enzymes are more stable in the carrier fixed state than in solution. But if this problem cannot be overcome, it is possible to use carriers of only some microns in size in a slurry type of a membrane reactor. From the chemical engineering point of view a membrane reactor behaves like a continuous stirred tank reactor, so that plug flow condition can only be approximated by means of a cascade of membrane reactors. In practice this is not a great problem since the km-values with respect to most of the substrates are comparitatively low. On the contrary, using technically relevant substrate concentrations a substrate surplus inhibition may occur, which may be overcome by a membrane reactor working under outlet conditions.

Examples

For the different routes from Fig. 2 some reaction systems will now be described. The common aspects of activity, efficiency and stability will be dealt with in the first example in some more detail because this concept is transferable to the other examples and also because commercialization of the membrane reactor was first reached with this example.

Acylase

Acylase catalyzes the hydrolysis of N-acetylated amino acids stereospecificly.

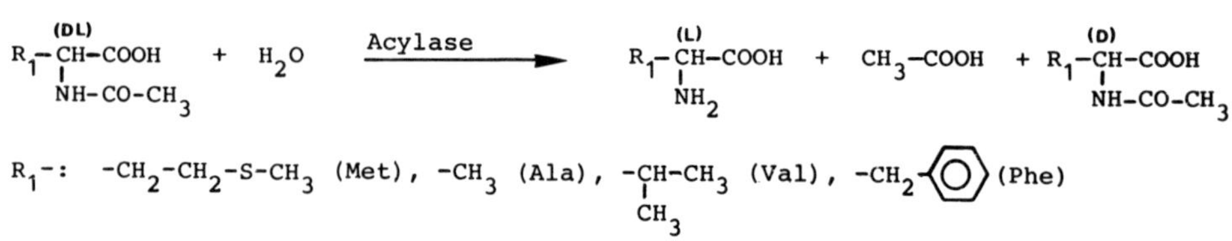

Thus it is possible to obtain the L-amino acid by separation from the non-reacted D-enantiomere. A similar approach is the use of esterases and amidases. Heterocyclic compounds can also be hydrolyzed stereospecificly. This is known for a number of hydantoinases (3), for L-α-amino-ε-caprolactame hydrolase (4) and 2-amino-Δ^2-thiazolin-4-carboxylate hydrolase (5).

In all these cases the product has to be separated from the non-reacted enantiomere, which has to be racemized in order to improve substrate utilization. There are enzymatic, chemical and thermal methods for racemization.

The formation of L-methionine by means of acylases was measured in detail. The hydrolysis of N-acetyl-L-methionine is an equilibrium reaction which can be followed polarimetrically. The achievable conversion decreases with increasing substrate concentration. This can also be seen from Fig. 4, where the reaction rate is given as function of the dimensionless substrate concentration for the acylase/L-methionine-example.

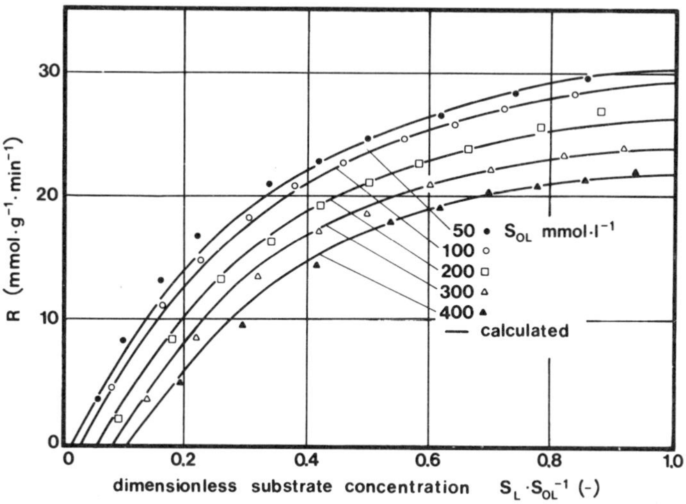

Fig. 4: Reaction rate (R) as function of dimensionless substrate concentration ($S_L \cdot S_{0L}^{-1}$); parameter: concentration of L-enantiomere in the racemic substrate (N-acetyl-D,L-methionine); pH= 7.0; T= 37°C

If the initial rates in Fig. 4 are compared one can realize that acylase shows a clear substrate excess inhibition. The combined influence of substrate and product results into pseudo Michaelis-Menten kinetics. The substrate excess inhibition is disguised by a strong product inhibition caused mainly by acetate (6).

Besides the activity of the catalyst, it is important how efficiently this activity can be used in a practical process. After carrier fixation of a native enzyme pore diffusion may decrease this efficiency in addition to activity loss during fixation. The use of native enzymes in membrane reactors offers the chance for a continuous homogeneous catalysis with theoreticly 100 % catalyst efficiency. Due to the enforced flow across the membrane concentration polarization with respect to the catalyst cannot be neglected completely. That means in an enzyme membrane reactor that at first part of the enzyme is concentrated near the membrane and also adsorbed. If the activity is measured in the bulk phase at the beginning - depending on enzyme concentration and flow conditions - only about 50 % of the total activity is found here.

Due to the limited stability of the native enzyme in solution one cannot keep constant conversion at constant residence time. Only by supplementation of fresh enzyme a constant activity can be maintained. With increasing operation time the adsorption capacity of the membrane is exhausted, so that the additional activity is found more and more in the bulk phase That means that after some time the catalyst efficiency in a membrane reactor is practically 100 % (Fig. 5).

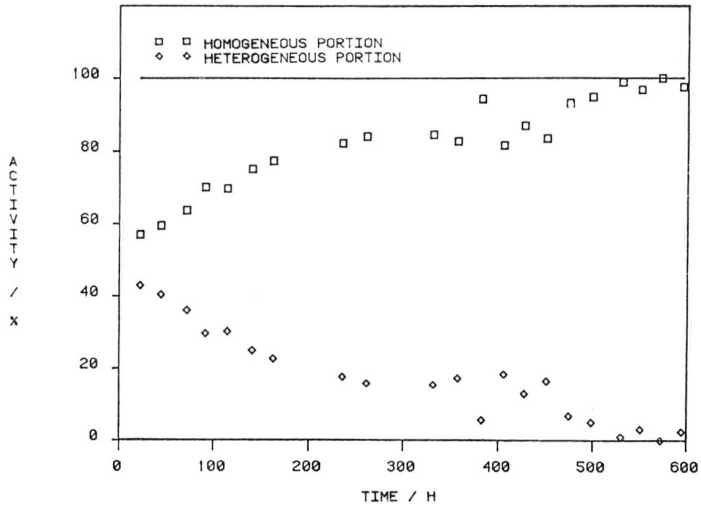

Fig. 5: Activity distribution as function of operating time at constant space time yield (achieved by catalyst supplementation) - samples were taken from the bulk phase to measure the homogeneous portion of the activity. The heterogeneous portion of the activity could be calculated from the known kinetics (7).

The operational stability of an enzyme can be measured very easily in a membrane reactor. For this purpose the appropriate amount of enzyme is added continuously to maintain constant conversion. Knowing the initial amount of catalyst for the desired conversion the deactivation constant under operational conditions can be determined as well as the enzyme consumption with respect to unit weight of product.

As can be seen from Fig. 6 this enzyme consumption increases with conversion because a higher catalyst concentration (at constant residence time) is needed to compensate for the lower rate if the state of equilibrium is approximated.

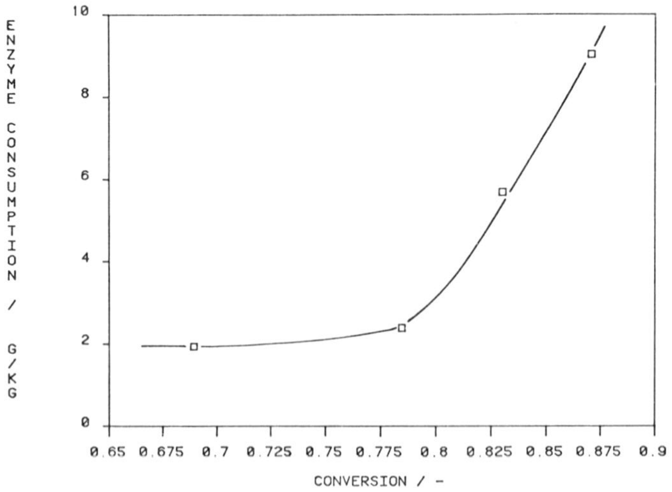

Fig. 6: Specific enzyme consumption as function of conversion measured at constant residence time and at different constant conversions achieved by continuous enzyme supplementation into a membrane reactor (7).

Since it is very difficult to predict enzyme stability under operational conditions it seems to be more sensible to measure the enzyme consumption and thus achieve information for process optimization.

An appropriate apparatus is shown in Fig. 7.

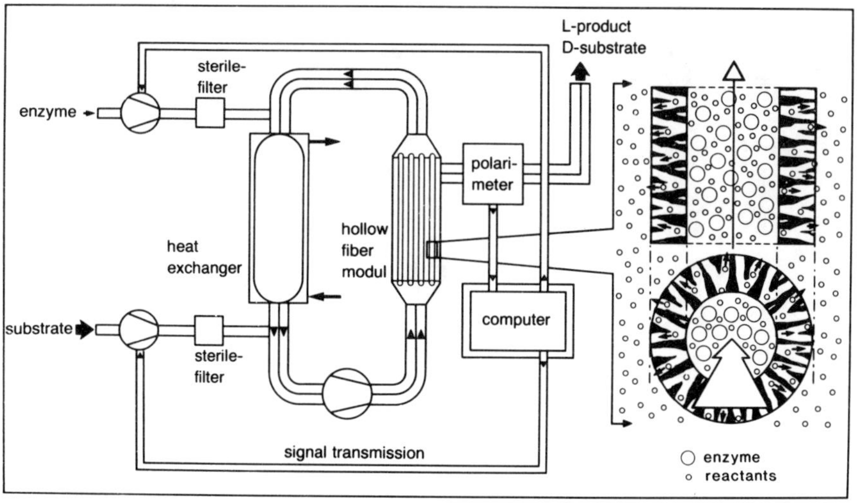

Fig. 7: Flow diagram of an enzyme membrane reactor controlled by a process computer. Product concentration is measured by a polarimeter. Constant conversion can be maintained by adjusting the speed of the substrate pump. Constant space time yield can be achieved by enzyme supplementation (8).

In the apparatus in Fig. 7 the ultrafiltration membrane is used in the form of a hollow fibre module. In this case backmixing is achieved by means of a recirculating pump so that concentration polarization can be avoided to a high extent. In Fig. 8 a series of hollow fibre modules in an industrial membrane reactor can be seen.

Fig. 8: Industrial enzyme membrane reactor for the production of L-amino acids, DEGUSSA AG, Konstanz

The plant in Fig. 8 has a production capacity of more than 200 tons of L-amino acids per year. The main products are L-methionine, L-alanine, L-phenylalanine and L-valine as well as D-valine (9).

Chymotrypsin

L-phenylalanine was produced from phenylalanine-D,L-methylester:

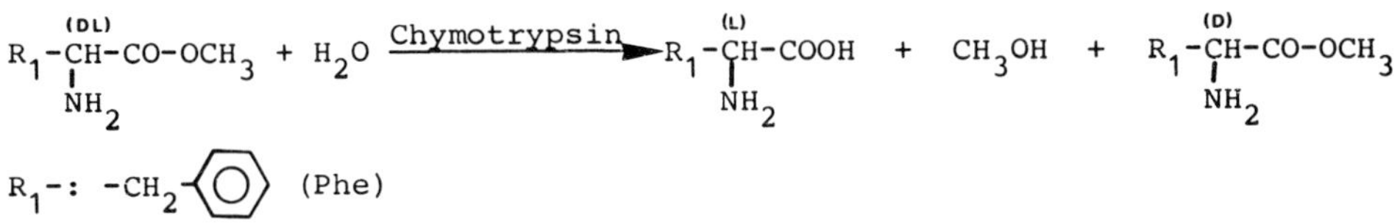

A membrane reactor as shown in Fig. 3 was used again (10). Besides the stereospecific hydrolysis of the L-ester by means of chymotrypsin a non-stereospecific spontaneous hydrolysis occurs, so that the maximal achievable catalyst concentration respectively the minimal possible residence time should be used in order to obtain maximal enantiomeric excess. With the system subtilisin/N-acetyl-D,L-methionine-methylester Schmidt-Kastner et al. (2) have reported residence times in a membrane reactor of less than 1 min and a space time yield with respect to N-acetyl-L-methionine to 27000 g/(l.d). Such values could never be achieved with carrier fixed enzymes or cells.

Aspartase

One example for ammonia lyases is aspartase which catalyzes the addition of ammonia to the double bond of fumarate:

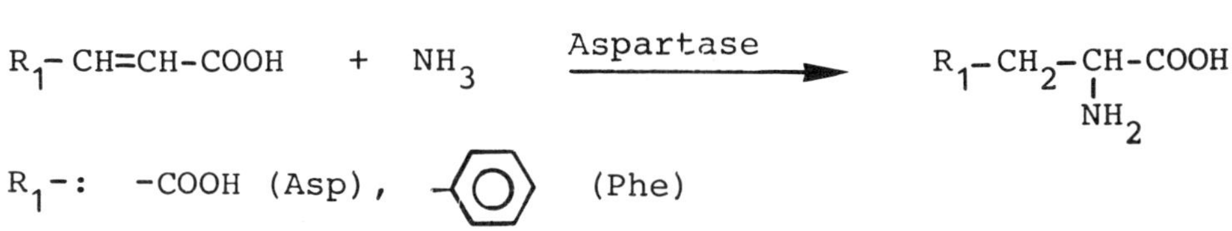

In a membrane reactor L-aspartic acid was produced via this route continuously for 2000 h - see Fig. 9 (11).

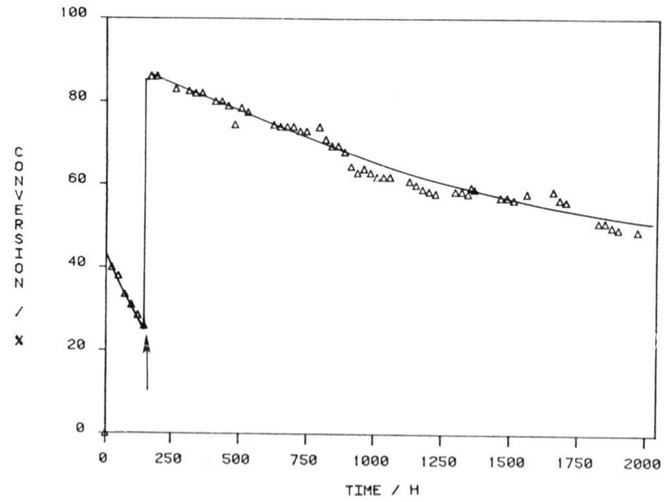

Fig. 9: Production of L-aspartic acid in an enzyme membrane reactor. Enzyme supplementation after 150 h. 400 mmol/l fumarate, 440 mmol/l ammonium formate, 1 mmol/l $MgCl_2$; pH= 8.5, T= 20°C, τ= 1.25 h

According to the raw material basis (fumaric acid and ammonium) L-aspartate is a comparitatively cheap L-amino acid and so can serve as amino donor in transaminations.

Amino transferases

In case of amino transferases the coenzyme pyridoxal phosphate/pyridoxamin normally is associated with the enzyme as prosthetic group so strongly, that a continuous homogeneous catalysis in a membrane reactor is possible without coenzyme supplementation:

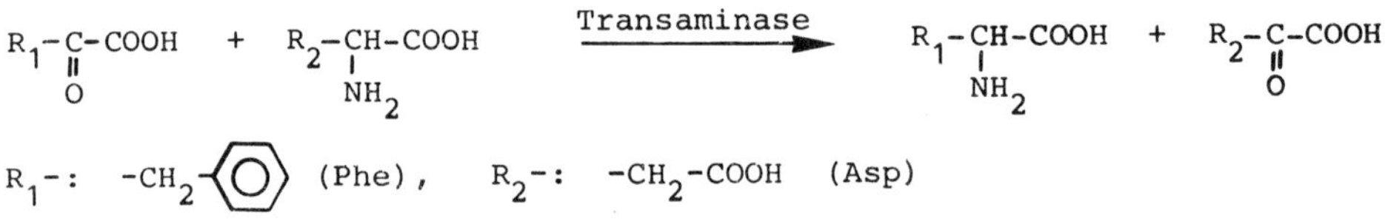

If the amino donor is cheap it can be used in excess in order to achieve a high conversion of the α-keto acid. Since such biotransformations are typical equilibrium reactions it is useful, if a product can be separated or is subject of a consecutive reaction. In case of oxalacetate originating from L-aspartic acid under operational conditions a decarboxylation to pyruvate was observed, so that practically quantitative conversion could be reached (12).

The decarboxylation of oxalacetate plays an important role when β-decarboxylase is used in principle as transaminase in order to transform L-aspartic acid to L-alanine:

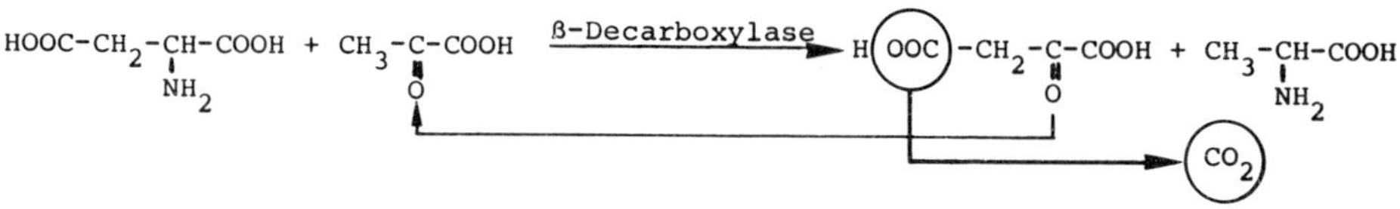

In this case pyruvate is a real second substrate for β-decarboxylase, so that at continuous operation in a membrane reactor the space time yield could be increased remarkably if a non-stoichiometric amount of pyruvate (5 mmol/l) was added to the substrate solution. Since the pH-optimum of β-decarboxylase is at pH 5.7, carbon dioxide - formed stoichiometrically - appears in a lot of gas bubbles. This causes an extreme deactivation of the enzyme at the gas/liquid interface (13). A gas phase can be avoided, if a sufficiently high system pressure is used (14).

Formate dehydrogenase/amino acid dehydrogenase

Instead of amino transferases also a combination of formate dehydrogenase and amino acid dehydrogenase can be used to obtain an L-amino acid from the corresponding α-keto acid by stereospecific reductive amination. In this case the cofactor NAD/NADH is a real transport metabolite: carbon dioxide once more formed on an equimolar basis is not a problem since a typical pH-optimum for such a system is pH 8.5:

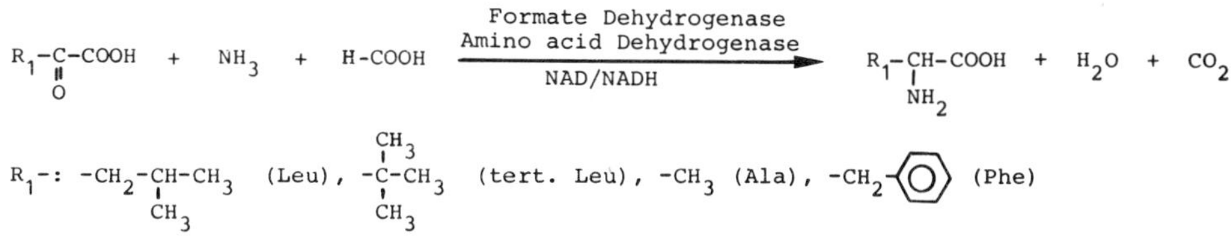

Continuous operation with native NAD/NADH cannot be achieved in a membrane reactor since the difference in size between the cofactor and the product molecule is not sufficient for an effective retention of the coenzyme. This problem could be solved when NAD was covalently bound to polyethylene gylcol 20000 (15). The resulting NAD-derivatives can be retained together with the enzymes in a membrane reactor as shown in Fig. 10:

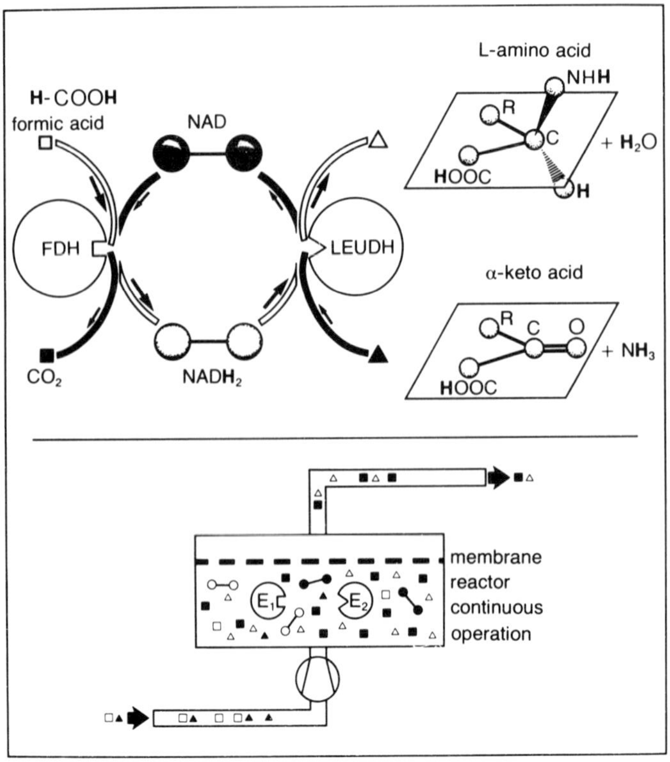

Fig. 10: Continuous cofactor regeneration in a membrane reactor with formate dehydrogenase (FDH, E_1) and leucine dehydrogenase (LeuDH, E_2)

The enlarged coenzyme is active, stabile and retainable, so that operation times of several months could be reached. During such a period of operation in average up to 130000 mole product could be produced per mole coenzyme (deactivated respectively lost across the membrane). That means, that the cost of the cofactor is no longer limiting. Quite remarkable space time yields are possible. In case of the production of L-tertiary leucine 640 g/(l.d) were reached (see also Table II).

Three enzyme systems

An especially interesting system for a membrane reactor is the combination of a L- and D-specific lactate dehydrogenase together with an amino acid dehydrogenase. Thus a racemic mixture of α-hydroxy acids can be transformed to an L-amino acid:

$$\underset{(DL)}{R_1\text{-CH(OH)-COOH}} + NH_3 \xrightarrow[\text{NAD/NADH}]{\text{3 Dehydrogenases}} \underset{(L)}{R_1\text{-CH(NH}_2\text{)-COOH}} + H_2O$$

R_1-: $-CH_2-CH(CH_3)-CH_3$ (Leu), $-CH_2-CH_2-S-CH_3$ (Met), $-CH_3$ (Ala)

How a membrane reactor is used in such a case on can seen in Fig. 11.

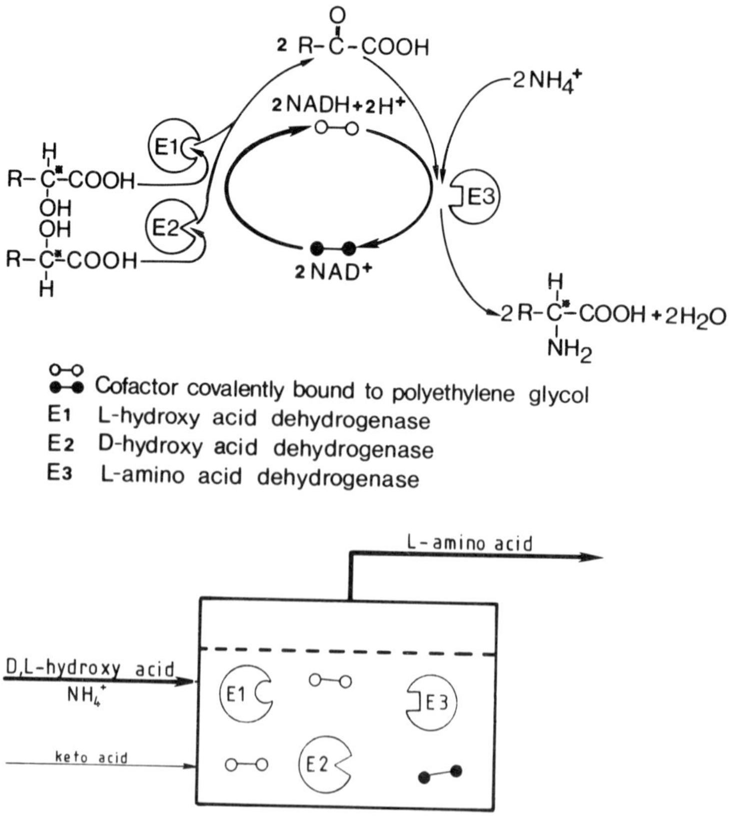

Fig. 11: Transformation of a racemic mixture of α-hydroxy acids to the corresponding L-amino acids in a membrane reactor. The substrate solution has to be supplemented with a non-stoichiometric amount of the α-keto acid, occuring as intermediate.

If we take the example that D,L-lactate should be transformed to L-alanine once more pyruvate is an intermediate product. Since pyruvate may be lost across the membrane before reductive amination can take place, NADH would pile up at the expense of NAD, if no pyruvate would be fed together with the substrate solution. As could be shown theoretically and in experiment there is a non-stoichiometric optimal concentration for pyruvate in the feed (16) depending on residence time and amino acid dehydrogenase concentration. Since, for example, alanine dehydrogenase has a small km-value for pyruvate, about 1 mmol/l pyruvate is sufficient, but without any pyruvate in the feed space time yield will finally become zero.

In Table II the enzymes and their sources are listed together with typical substrate and product concentrations. Next, the average conversion during continuous operation is given, the reactor volume, the space time yield, the enzyme consumption per unit weight of product and the cycle number (in cases where a cofactor is needed). Only bench scale data are listed. In case of the acylase system production scale was reached in 1981. In case of the cofactor regenerating systems first commercial products are expected for 1985.

Use of microorganisms

A great variety of microorganisms has been used for the production of L-amino acids. Wild strains, auxotrophic mutans as well as regulatory mutans have been applied (1).

In our laboratory some experiments using biosynthetic precursors were carried out. α-hydroxy and α-keto acids are not only useful precursors for enzymatic synthesis but also for microbial production. A product specific amino acid dehydrogenase may be used, but it is also possible to employ amino transferases:

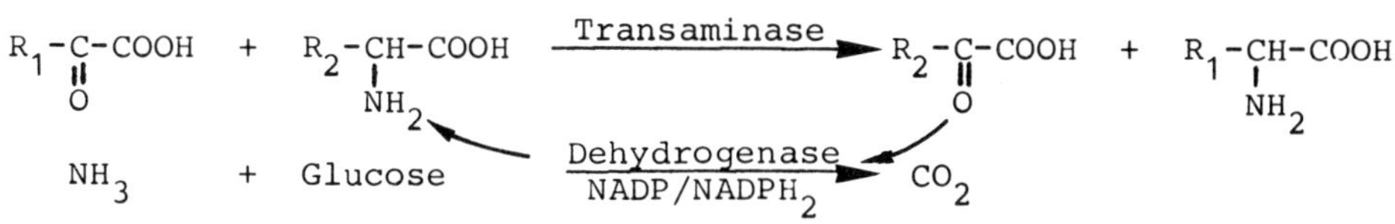

R_1-: CH_2-CH(CH_3)-CH_3 (Leu), R_2-: HOOC-CH_2-CH_2- (Glu)

In case of Corynebacterium glutamicum it is possible to combine the use of glutamate transaminase together with glutamate dehydrogenase. In such a case there is no direct reductive amination of the α-keto acid as precursor but via α-ketoglutarate. Ammonium is used as nitrogen source, while glucose serves as energy source as well as source to produce NADPH for reductive amination of α-ketoglutarate (17). In case of the production of L-leucine from α-ketoisocaproate up to 25 g/l product were reached at a substrate selectivity of more than 95 %. A somewhat similar approach leeds to L-isoleucine (18).

In Table II some biochemical engineering data for the use of enzymes in membrane reactors are summarized:

Enzyme	Substrate	Concentration mmol/l	Product	Concentration mmol/l	Conversion %	Reactor-volume l	Space time yield g/l·d	Enyzme consumption U/kg	Cycle number $\frac{mmol_{prod}}{mmol_{NAD}}$
Acylase (hog kidneys)	N-Acetyl-D,L-methionine	800	L-Methionine	340	85	1.0	1200	1000	
Acylase (Aspargillus orycae)	N-Acetyl-D,L-phenylalanine	400	L-Phenylalanine	168	84	0.18	400	1770	
Chymotrypsin (bovine pancreas)	D,L-Phenylalanine methylester	100	L-Phenylalanine	41	82	0.01	1400	-	
Aspartase (Escherichia coli)	Fumarate Ammonium formate	400 440	L-Aspartic acid	320	80	0.2	400	5000	
Transaminase (Pseudomonas putida)	Phenylpyruvate L-Aspartate	200 300	L-Phenylalanine	136	68	0.01	270	-	
ß-Decarboxylase (Pseudomonas dacurhae)	L-Aspartic acid Pyruvate	151 5	L-Alanine	112	74	0.18	53	140000	
Formate Dehydrogenase (Candida boidinii) Leucine Dehydrogenase (Bacillus cereus)	Ketoisocaproate Ammonium formate	100 400	L-Leucine	80	80	0.7	250	300 300	100000
Formate Dehydrogenase Leucine Dehydrogenase	Dimethylpyruvate Ammonium formate	500 1000	L-tert. Leucine	425	85	0.01	640	2000 1000	130000
Formate Dehydrogenase Phenylalanine Dehydrogenase (Brevibacterium sp.)	Phenylpyruvate Ammonium formate	20 2000	L-Phenylalanine	19	95	0.01	37	6000 6000	11500
L-Hydroxy isocaproic acid Dehydrogenase (Lactobacillus confusus) D-Hydroxy isocaproic acid Dehydrogenase (Lactobacillus casei) Leucine Dehydrogenase	D,L-Hydroxy isocaproic acid Ammonium sulfate Ketoisocaproic acid	100 250 1	L-Leucine	70	70	0.01	72	730 350 650	10100
L-Hydroxy isocaproic acid Dehydrogenase D-Hydroxy isocaproic acid Dehydrogenase Leucine Dehydrogenase	D,L-Hydroxy "methionine" Ammonium sulfate Keto-"methionine"	400 400 1	L-Methionine	240	60	0.01	143	- - -	8500
L-Lactate Dehydrogenase (rabbit muscle) D-Lactate Dehydrogenase (Lactobacillus confusus) Alanine Dehydrogenase (Bacillus subtilis)	D,L-Lactate Ammonium formate Pyruvate	400 800 1	L-Alanine	184	46	0.01	134	4700 2600 -	19200

Table II: Biochemical engineering data for the use of enzymes in membrane reactors

DISCUSSION

The use of isolated enzymes for L-amino acid production, even if a cofactor regeneration is required, is now technological feasable in many cases. The space time yields reached so far are superior to most fermentations. Product isolation, which has to be done continuously, is normally much easier than out of a fermentation broth. After enzymatic synthesis and product isolation the non-converted part of the substrate can be recycled to the reactor, so that a quantitative use of the substrate is possible. The costs for enzyme isolation seem to be tolerable in many cases since the enzyme is used in non-stoichiometric amounts.

With increasing demand for enzymes in reactors there is probably a great potential to obtain these enzymes more economically. The most important point is still the stability of enzymes in solution if continuous production periods of months are aimed for.

If a sufficient stability of an enzyme in an isolated form cannot be achieved or if a series of more than 2 or 3 enzymatic reactions is needed, entire microorganisms may be used for biotransformations. Also here biosynthetic precursors are useful for microbial production of L-amino acids, because thus one can use a better defined substrate and product solution taking advantage of the possibility, that many precursors can be synthesized by classical methods. If entire microorganisms are used, one has to make sure that the precursor is not used as energy or carbon source, since the precursor normally is more expensive than glucose, which may serve for maintainance. Because one is interested only in the catalytic capabilities of the cells and not in the cell mass itself excess cell mass formation should be avoided by means of limitation of some nutrients. For continuous operation, cells might be retained in a continuously operated reactor by means of a microfiltration module analogous to an ultrafiltration membrane in case of the use of isolated enzymes.

ACKNOWLEDGEMENT

Most of the enzymes as well as the coenzyme derivatives were obtained from Gesellschaft für Biotechnologische Forschung, Braunschweig (Dr. Kula). The precursors were provided by Degussa AG, Hanau (Dr. Leuchtenberger). This fruitful cooperation as well as the skilled technical assistance of Mrs. Mackfeld and Mrs. Haase is greatfully acknowledged.

REFERENCES

(1) K. Soda, H. Tanaka and N. Esaki (1983). In: Biotechnology (H.-J. Rehm and G. Reed, eds.), Verlag Chemie, Weinheim, Vol. 3, 479

(2) G. Schmidt-Kastner and P. Engerer (1984). In: Biotechnology (H.-J. Rehm and G. Reed, eds.), Verlag Chemie, Weinheim, Vol. 6a, 387

(3) H. Yamada, S. Shimizu, H. Shimada, Y. Tani, S. Takahashi and T. Ohashi (1980). Biochimie 62, 395

(4) T. Fukumura (1976). Agric. Biol. Chem. 40, 1687

(5) K. Sano, K. Yokozeki, F. Tamura, N. Yasuda, I. Noda and K. Mitsugi (1977). Appl. Environ. Microbiol. 34, 806

(6) C. Wandrey and E. Flaschel (1979). In: Advances in Biochemical Engineering (T.K. Ghose, A. Fiechter and N. Blakebrough, eds.) Springer Verlag, Berlin, 12, 147

(7) L. Jandel (1981). Dissertation, Technische Universität Clausthal

(8) C. Wandrey, R. Wichmann, M.-R. Kula and A. F. Bückmann, (1984). Umschau 3, 88

(9) W. Leuchtenberger, M. Karrenbauer and M. Plöcker (1984). Forum Mikrobiologie, Sonderheft, 40

(10) R. Weiß (1978). Dissertation, Technische Universität Hannover

(11) A. S. Jandel, H. Hustedt and C. Wandrey (1982). European J. Appl. Microbiol. Biotechnol. 15, 59

(12) H. Ziehr (1985). Dissertation, Technische Universität Braunschweig

(13) A. S. Jandel (1981). Dissertation, Technische Universität Clausthal

(14) T. Tosa, S. Takamatsu, M. Furui and I. Chibata (1984). Annals of the N. Y. Acad. of Sciences 434, 450

(15) A. F. Bückmann, M.-R. Kula, R. Wichmann and C. Wandrey (1981). J. Appl. Biochem. 3, 301

(16) E. Fiolitakis and C. Wandrey (1983). Proc. III. Rotenburg Fermentation Symp. Springer Verlag, Berlin, 273

(17) U. Groeger and H. Sahm (1984). Preprints of the 3rd Europ. Congress on Biotechnology, 1, 85

(18) I. Block and H. Sahm (1984). Preprints of the 3rd Europ. Congress on Biotechnology, 1, 45

DEVELOPMENT OF AN ENZYME REACTOR FOR FOOD CHEMISTRY

Ernst H. Reimerdes
Food Chemistry and Biotechnology
Bergische Universität - GHS Wuppertal
D-5600 Wuppertal 1

Abstract: The synthesis of spezific monosaccharides by hydrolysis of disaccharides and their isomerisation is performed by the use of biocatalysts. The development of this biotechnological procedure is described from the screening and isolation of the related enzymes, their characterization and immobilization as well as the development of a suitable reactor system and the processing characteristics. This system is applicable for the development of related systems for synthesis in food and organic chemistry.

1. Introduction

Biotechnology summarizes several different techniques as there are genetic engeneering, fermentation technology and enzyme technology with a wide range of applications. The first use of enzymatic reactions was during fermentation processes with whole cells or microorganisms. In these cases normally the final metabolic products are used, which are the results of multistep reactions. Among these examples the biosynthesis of vitamins, antibiotics or simple organic acids like citric acid and D- or L-lactic acid should be mentioned. Besides this all the fermented products e.g. in food production are the result of action of biocatalysts on natural chemical compounds resulting in numerous products as it can be demonstrated by the complex spectra of flavour compounds. All the products are related to biochemical pathways or bioorganic synthesis in nature.

Further approach is achieved by the use of isolated enzymes with special catalytic reactivities to perform specific reactions. This affords additional techniques concerning the screening, isolation and characterization of enzymes. On

M. P. Schneider (ed.), Enzymes as Catalysts in Organic Synthesis, 285–301.

This approach has the advantage that also intermediate metabolites can be produced. In the case of whole cells and related fermentation techniques these compounds are further metabolized.

Therefore the use of single biocatalysts gets more and more attractive for the synthesis of organic compounds, due to the high selectivity of these biocatalysts and there pronounced stereo-, regio- and enantiomeric specificity. Instead of this natural substrates homologue or similar compounds can be used due to bulk tolerances at the active sites of the biocatalysts.

The application of biocatalysts for the synthesis in food and organic chemistry affords special knowledge about the screening and isolation of enzymes related to specific chemical reactions, the characterization of the biocatalysts, the immobilization of enzymes, the development of suitable enzyme reactor systems as well as the definition of the processing conditions and product scale up.

During the early stages of bioorganic synthesis enzymes were usually employed, which were commercially available. These enzymes were characterized with natural substrates by typical biochemical procedures. On the basis of these results the enzymes are tested for their applicability to homologue substrates as required in each synthetic procedure. Due to the characteristics of the active site of an enzyme there are large differences in substrate specificity. These features can only be evaluated by the use of the knowledge from enzyme chemistry and by specific experiments. There are numerous results in biochemical literature on these special topics.

If suitable enzymes are not available, a screening set up has to be evaluated in order to find biocatalysts for special synthetic problems. In most cases microorganisms are used as enzyme sources. The enzymes can be found naturally or can be formed by adaptation to specific substrates or introduced by genetic engineering. The screening can be performed by chromogenic substrates related to the typical substrates and reactions wanted.

The isolation of enzymes is performed by typical biochemical procedures after solubilisation, e.g. ammonium sulfate or acetone precipitation and ion exchange or molecular sieve chromatography. The purification is characterized by activity/g of protein with specific substrates. Besides this the puritiy of the biocatalysts with respect to other enzymes is of importance, because these might cause side reactions of the substrates and the products.

Enzymes can be used in solution or in immobilized form. Immobilization has the advantages that the enzymes can be reused, which means large turnover rates can be achieved and the enzymes do not remain in the product. Immobilisation can be performed by membrane systems or by covalent or adsorptive binding to polymer matrices. Special immobilisation techniques have to be evaluated related to each enzyme system, due to the typical protein surface of the enzyme proteins and to the steric situation of the active site. Several procedures are available for these purposes.

The immobilized enzymes often have different kinetic properties, which means each enzyme system has to be characterized with respect to its specific synthetic properties. Often spacer molecules are used between enzyme and carrier in order to avoid sterical hindrance. Another important factor is the macrosurface of the carrier in order to get high amounts of activity bound to the system. Besides this good flow properties are necessary in order to pass large amounts of substrates through the system.

Enzyme reactors are directly related to the synthetic system. With low molecular weight substrates membrane reactors, enzyme entrappment and immobilization can be used, while large molecular weight substrates are not suitable for membrane sytems and entrapped enzymes. The processing parameters of the enzyme reactor depend on the enzyme beads, which means bound activity, macrosurface and flow properties. These factors influence the construction and dimensions of an reactor system. For this reason special polymer beads have been developed for the immobilisation of enzymes.

2. Lactose hydrolysis to glucose and galactose

Lactose is a typical disaccharide with the formula:

4-O-ß-D-Galactosyl-D-glucopyranose

which can be hydrolysed into the monosaccharides galactose and glucose. The glucose again can be isomerized to fructose. These applications are of interest in food chemistry because large amounts of lactose are obtained during cheese making in the resulting byproduct whey. By hydrolysis and isomerisation products like high glucose or high fructose sirups are produced which can be used in food products.

In general lactose can also be hydrolysed by mineral acids or acidic ion exchangers, if this disaccharide is available in relatively pure form. Unfortunately whey represents a highly heterogeneous substrate solution containing proteins, low molecular weight nitrogen substances and minerals among others. These compounds cause to a large extent side reactions during acid hydrolysis, resulting in a dark brown liquid and problems during scale up for further use (01).

On the other hand hydrolysis by a biocatalyst, in this case ß-1-4-galactosidases, which reacts highly selectively with lactose allows the specific formation of galactose and glucose in such heterogeneous substrate solutions without any side reactions. For these reasons the development of enzyme reactors for the hydrolysis of lactose in whey and milk serum has been investigated (02-04).

The application of biocatalysts to these problems and the development of the related technology requires the following steps (05-06):

- concept for lactose hydrolysis
- properties of the raw material
- enzyme requirements
- enzyme screening
- enzyme purification
- enzyme characterization
- immobilisation of the enzymes
- reactor concept for batchwise or continuous operation and additional technology
- evaluation of processing parameters
- product processing and application

This general scheme is applicable to any other use of biocatalysts and therefore can serve as model for the development of such biotechnological procedures.

3. Concept for lactose hydrolysis and properties of the raw materials

Substrate characterization basically means the chemical

structure of the starting material and the chemical reaction which has to be performed. In this case a disaccharide containing galactose and glucose with ß-1-4-linkage is to be hydrolysed into the two monosaccharides and the resulting glucose shall be isomerized to fructose. These parameters define the requirements for the related the biocatalysts.

In the case of heterogenous substrates as whey or milk serum further parameters have to be considered like the mineral content and the pH of the substrate solution, because these have influence on the enzyme activities . In the case of whey and milk serum the pH is 4-5 and 6-7 respectively and a relatively high mineral content especially calcium and phosphate ions are present in these substrates.

4. Enzyme requirements

Concerning the hydrolysis of lactose specific ß-1-4-galactosidases are necessary, which are able to remove and transfer the galactosyl residue. Further requirements are the pH-optima for their activities in the two relevant ranges of pH 4-5 and pH 6-7 as well as high tolerances for calcium and phosphate ions.

5. Enzyme screening

In order to find the required biocatalysts by systematic screening a simple and reproducible test system for the enzyme activity is essential. In most cases the natural substrates, in this case lactose are applied and the products analysed. In special cases like proteolytic enzymes and glycosidases synthetic, chromogenic substrates are available, e.g. nitroanilides. After incubation the release of 4-nitroanilin due to the hydrolytic reaction is easily detectable by photometric procedures.

For enzyme screening special knowledge about their availability is necessary. In most cases microorganisms are considered as the most promising enzyme sources. In the case of ß-1-4-galactosidases the screening can also be extended to mammalian and plant tissues because these types of enzymes are ubiquitous.

During the screening several sources have been evaluated. With respect to the above requirements the following enzymes have been considered to be suitable for the application to lactose hydrolysis in whey and milk serum:

Enzymes for lactose hydrolysis in whey and milk

	pH-Optimum	Temperature-optimum (oC)
Aspergillus orycae	4-5	45
Aspergillus niger	3-5	55
Saccharmyces fragilis	6-7	50
lactis	6.5-7.5	45
Kluyveromyces species	6-7	45
Escherichia coli	6-7	45
Bacillus subtilis	6-7	50
stearothermophilus	6-7	65

The list of the enzymes demonstrates that both pH-ranges with respect to the substrate conditions are covered. Beside this the enzyme activities are not reduced due to the high content of calcium and phosphate ions (05).

In the case of food related products it is further necessary that the enzymes and there sources are 'generally recognised as safe' for human consumption. The ß-galactosidase of E. coli does not fulfill this requirement but it was of special interest during the immobilization experiments.

5. Enzyme purification

In order to perform specific reactions with biocatalysts and for the immobilization the purification of the enzymes is necessary, especially since due to the enzyme sources a complex mixture of enzymes, proteins and other by-products are obtained from the biomass. Enzyme purification is well known from biochemical research even in large scale preparations(07). In the case of microbial sources it is of importance weatherthe enzyme is released to the culture broth or contained inside the cells. In the last case before purification of the enzymes the cells have to be destroyed This can be achieved by several techniques like freeze/thawing, ultrasonic treatment or desintegrating the cell walls by specific enzymes. If the enzymes are bound to the cell membranes special solubilization techniques have to be applied or the membrane fragments can be used directly.

From solutions enzymes are normally isolated by precipitation e.g. with ammonium sulfate or acetone resulting in crude enzyme preparations. In most cases these preparations can be used for preliminary experiments. If highly purified enzyme preparations are required e.g. for their immobilization further purification can be achieved by fractionated precipitation, ion exchange and molecular sieve chromatography or two phase extraction techniques (05,07).

The enzyme of Bacillus subtilis has been purified by the following steps: solubilisation by ultrasonic treatment, ammonium sulfate precipitation and molecular sieve chromatography, resulting in a highly purified enzyme preparation. Similar procedures are used for the purification of the other enzymes (08).

6. Enzyme characterization

Enzyme characterization has been performed with respect to typical biochemical parameters as there are Michaelis constant, turnover rates, substrate and product inhibition and pH-optimum, e.g.:

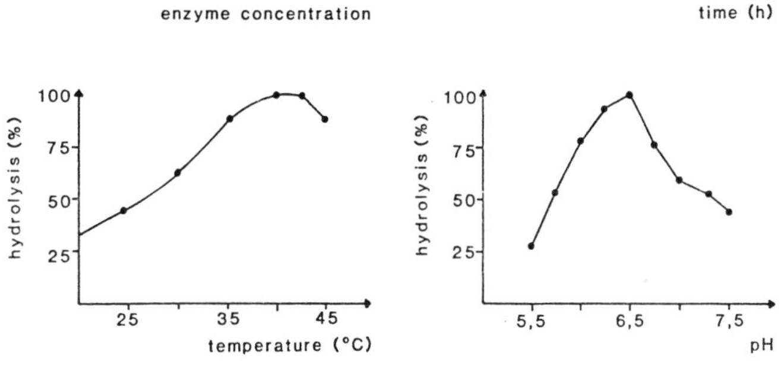

Characterization of β-galactosidase (s. fragilis) for lactose hydrolysis

With respect to the special requirements for the lactose hydrolysis in whey and milk serum also the influence of the related minerals has been evaluated.

With respect to the former parameters the above mentioned enzymes were suitable for the required lactose hydrolysis.

In the case of the enzymes of E. coli and Bacillus subtilis typical product inhibition has been found. This limits the concentration of the substrate during hydrolysis, because due to product formation the efficiency of the hydrolysis will be reduced.

7. Immobilization of the enzymes

The easiest way to perform bioorganic sythesis with enzymes is their addition to the substrate solutions but in this case the the loss of the biocatalysts and the contamination of the products by the enzyme proteins have to be considered (05).

These disadvantages can be avoided by immobilization techniques (09-12). The advantages of immobilized enzymes are summarized as follows:

- reuse of the enzymes
- no enzymes in the products
- often higher stability
- good technological properties
- exact defined turnover rates

There exist several possibilities for enzyme immobilization:

- entrappment of the enzymes
- adsorptive binding and
- covalent binding

Entrappment of enzymes can be performed in two major forms, by formation of polymer networks which engulf and immobilize enzymes or by membrane systems. Both types are useful only for reactions with low molecular substrates, which can penetrate the polymers or pass the membranes. Special cellulose acetate fibres (13) and protein or other polymer products have been used for the entrappment of enzymes. In the case of membrane systems typical membrane reactors have been developed which allow the passage of the substrates through the enzyme containing solution and retaining the enzymes (14). During the passage the biocatalysis is performed resulting in the related products.

For all types of substrates biocatalysts immobilized by adsorptive or covalent binding can be used. In general the enzyme molecules are adsorbed to the surface of special carriers e.g. glass, titanium dioxide, polyacrylamide or polyphenol beads (09-12). Binding is achieved by ionic, hydrophobic or hydrogen binding according to related functional properties of the carrier and the enzyme surfaces. One of the most succesful system, is the L-aminoacid acylase bound to ion exchange resins (15). The advantage of adsorptive binding is the reuse of the carrier after inactivation of the enzyme. The inactive enzyme molecules can be washed off with detergents and the carrier can be reloaded with new enzymes (16).

Unfortunately not all enzymes are suitable to form stable adsorptive bonds with the available carriers. For this reason several attempts have been made to modify the carrier e.g. by the introduction of hydrophobic groups.

A different approach has been applied to the ß-galactosidases from Saccharomyces species. For the first time the modification of the enzyme protein surface has been tried in order to introduce functional groups for adsorptive binding purposes (16-18). This has been performed by reductive alkylation via azomethine intermediates and reduction with sodium borohydride. In this way different hydrophobic groups have been introduced to the enzyme surface, e.g methyl- or isopropyl-groups using formaldehyde or acetone. Even mixed alkylation has been performed using both reagents. The alkylation takes place at the e-amino-groups of lysine residues at the enzyme surface.

By this modification the enzyme surface becomes more hydrophobic and binding studies with hydrophobic carriers demonstrated much higher binding efficiency and stability. The half-live of the immobilized enzymes increased up to twenty times (16-18).

Using this method it is very important that the enzyme activity is retained during the modification procedure. In the case of the ß-galactosidases of the Saccharomyces species this was successfully achieved, but it has to be evaluated for each special enzyme system.

The possibility to use these hydrophobic modified enzymes in organic solvents or in two phase systems is another important consequence of this approach.

The major approach for enzyme immobilization is covalent binding. Numerous procedures have been developed for this purpose (09-12). Frequently used are the binding via glutaraldehyde after functionalizing the carrier with amino groups, the coupling via epoxide groups, the cyanogen bromide procedure and binding via anhydride groups. A very special procedure is the binding via photoalkylation according to the Schenck procedure (19).

As shown in an investigation with trypsin covalent immobilization causes changes at the carrier and the enzyme surface in relation to the reacting functional groups. This has been shown to be of importance for the kinetic properties of the enzymes (20-23). Due to the reactions the charges at the carrier and the enzyme surface are changed as summarized in the following scheme for trypsin:

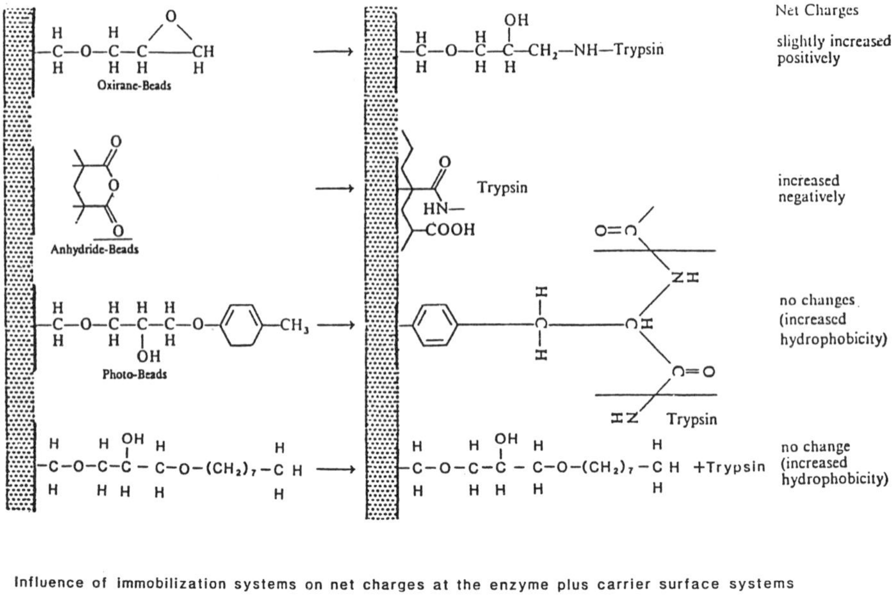

Influence of immobilization systems on net charges at the enzyme plus carrier surface systems

Besides the binding techniques special features concerning the carrier systems which are related to the technological properties are of importance. In the case of the Corning procedure macroporous glass or titanium dioxide beads are used which have been modified at the surface by introduction of amino groups. The enzymes are adsorbed to the surface and covalently bound by cross-linking with bifunctional aldehydes, e.g. glutaraldehyde. This procedure has also been applied successfully to ß-galactosidases(24-25).

Another system with highly interesting properties are the polyacrylic beads developed by Röhm (19). In this case beads are obtained with a large surface and a very rigid structure and good technological properties. Binding groups are: epoxide groups, anhydride groups and toluene groups for fotoalkylations. Via the epoxide groups it is easy to introduce spacers of different length in order to preven sterical hindrance concerning the enzyme substrate reactions.

Good technological properties of the enzyme beads means.

stable macroporous structure

high enzyme activity due to the number available binding groups

good flow properties

no sterical hindrance for the catalytic reactions

In order to prove the quality of the immobilized enzyme systems a similar characterization as for the enzymes is necessary concerning the kinetic parameters. Besides this temperature-time profiles have to be evaluated in order to find optimal processing conditions of long term application.

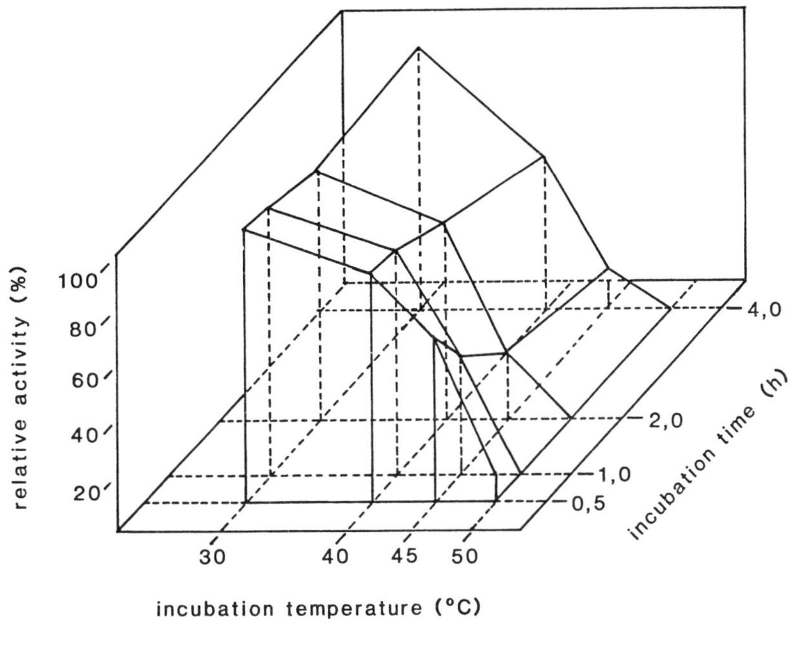

Temperature – time – activity – diagram of β-galactosidase immobilized by adsorptive binding after mixed reductive alkylation

The temperature stability of the enzymes varies to a large extent. Generally the temperature stability is in the range of 25-45 oC, but with the enzyme of Bacillus stearothermophilus obtained from hat springs the temperature optimum is found to be between 50 and 60 oC.

7. Reactor concept

Immobilized enzymes can be used in a batchwise or continuous manner. Batchwise operation means addition of the enzyme beads, operation under defined temperature-time conditions, separation of the enzyme beads by filtration or centrifugation and processing up of the resulting products(2,3,5,24,25).

For continuous operation special reactor systems are required. Concerning laboratory experiments simple equipment like the ones used for column chromatography can be utilized. Due to the size of the enzyme beads normal nylon gaze is suitable to retain of the enzyme beads in the column.

For the application of immobilized ß-galactosidases to whey and milk serum a special pilot reactor system has been developed as demonstrated in the following graph (2,3,5):

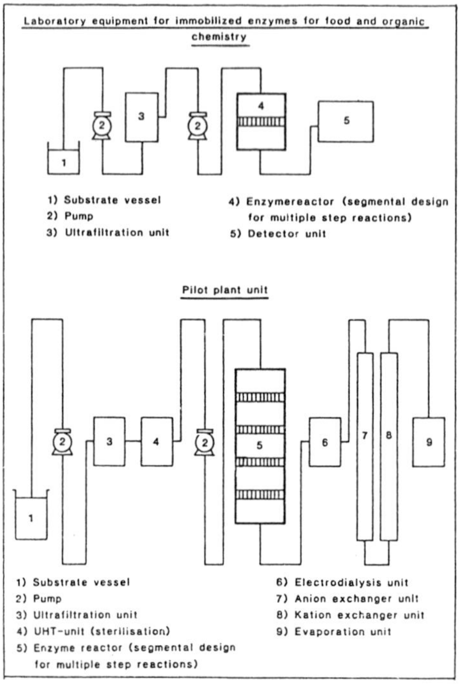

The system consists of an ultrafiltration and heating unit, the enzyme reactor with a pumping unit, a desalting system and an evaporator. The central facility is the enzyme reactor, which has been set up in segments, each of the segments being an independent unit. The dimensions of the units are 14.5 cm in height and 23.4 cm in diameter. This means the volume for the enzyme beads is 6.2 l. The number of segments is five. Due to this concept it is possible to use different amounts of enzyme beads in the range from 6.2 to 31.0 l. Besides this it is possible to use different enzymes in the single units in order to perform multi step reactions(05), e.g.:

$$\text{Lactose} \xrightarrow{\text{ß-galactosidase}} \begin{matrix}\text{glucose}\\ \text{galactose}\end{matrix} \xrightarrow[\text{isomerase}]{\text{glucose}} \begin{matrix}\text{fructose}\\ \text{galactose}\end{matrix}$$

The additional facilities are needed to handle the special substrates milk and whey. With ultrafiltration proteins which might contaminate and inactivate the enzyme beads are removed from the solutions. The heating system is applied to sterilize the substrates in order to avoid microbial contamination of the enzyme beads. The desalting unit, a combined system of electrodialysis and ion exchange treatment is used for further processing of the products as well as the evaporator.

With the above pilot reactor immobilized ß-galactosidases have been applied to the treatment of whey. Representative results are: 1000 l/h at 25-30 oC for 7 h followedby 1 h cleaning resulting in a hydrolysis rate of 85-90%. The hygienic control of the beads and the resulting products showed no contamination by microorganisms. Long term evaluation showed that the half life of the enzyme activity with this system is in the range of 6 month during continuous operation are obtained.

8. Product processing - further reactions

In the case of whey as a substrate the applications of the resulting products in foods afford a special processing in order to remove part of the by products like minerals and to concentrate the product solution to obtain sirups. This is performed by the desalting and evaporator units (02,03,.05,24-33).

Concerning the isomeration of glucose to fructose commercial immobilized glucose isomerase from Bacillus species is used.

In this case desalting of the substrate solution is also necessary, because this enyzme system is sensitive to calcium ions which strongly reduce the activity (34).

9. Further enzyme systems

The above model system is one representative example of the application of immobilized enzymes. Further systems of biocatalysts have been applied successfully for bioconversion in food chemistry(35), e.g.:

immobilized chymosin
pepsin
trypsin
papain
sulfhydryl oxidase
catalase
phosphatase
amino acid acylase

which demonstrates the range of possibilities by this bioorganic chemistry. There are many other examples of the application of biocatalysis in organic chemistry which offer special opportunities to perform complicated reactions due to the specificity of the enzymes.

10. Conclusions

With respect to the biochemical pathways and the high selectivity and specificity of biocatalysts the use of enzymes in chemistry offers special advantages which might be a special supplementation to classical organic chemistry.

The development of an enzyme reactor system for the preparation of glucose and galactose from lactose and the isomerization of the glucose to fructose using immobilized enzymes from laboratory scale up to a pilot system is shown. This includes all steps from substrate characterization to enzyme screening, isolation, immobilization and application. Additional requirements were necessary due to the use of the natural heterogenous substrate whey, e.g. ultrafiltration, sterilization and desalting equipment.

This system is in practical use in food chemistry and technology but might serve as a model to solve problems in organic chemistry with respect to the application of biocatalysts and their high stereo-, regio- and enantiomeric specicity and selective substrate affinity.

11. References

01. de Boer, R., Robertsen, T.: A purified hydrolized lactose sirup made from ultrafiltration permeate Neth. Milk Dairy J. 35: 95 (1981)

02. Reimerdes, E.H., Gottschick, W.: Neue Verwendungsmöglichkeiten für Milch, Molke und Permeate durch Milchzuckerspaltung (I) Lebensmitteltechnik 9: 402 (1981)

03. Reimerdes, E.H., Gottschick, W.: Neue Verwendungsmöglichkeiten für Milch, Molke und Permeate durch Milchzuckerspaltung (II) Lebensmitteltechnik 10: 466 (1981)

04. Klostermeyer, H., Herlitz, E., Jürgens, R.H., Reimerdes, E.H., Thomasow, J.: Lactasebehandlung von Magermilch zur Herstellung laktosereduzierten Milchpulvers Kieler Milchw. Forschungs-Ber. 30: 295 (1978)

05. Reimerdes, E.H.: Theoretical and practical aspects of lactose hydrolysis German-Norwegian Biotechnology Workshop, BMFT, Jülich 51 (1982)

06. Reimerdes, E.H.: Vom löslichen zum immobilisierten Enzym Int. Milchw. Congreß, Moskau (1982)

07. Kula, M.-R., Kroner, K.H., Hustedt, H.: Purification of enzymes by liquid-liquid extraction Adv. Biochem. Eng. Biotechnol. 26: 73 (1982)

08. Reimerdes, E.H.: unpublished results

09. Zaborski, O.R.: Immobilized Enzymes CRC Press, Chemical Rubber, Cleveland, USA (1974)

10. Manecke, G.: Immobilisierte Enzyme, Chimia, 28: 467, (1974)

11. Goldstein, L., Manecke, G.: The chemistry of enzyme immobilization. In Immobilized Enzyme Principles (Wingard, Jr. L.B., Katchalski-Katzir, E., Goldstein, L. (Eds.), Academic Press, New York, 23 (1976)

12. Messing, R.A.: Carrier for immobilized enzymes, Process Biochem. 9: 26 (1974)

13. Pastore, M., Morisi, F., Zacardelli, D.: Reduction of lactose content of milk using entrapped beta-glactosidase Insolubilzed enzymes, Salmona, M., Saronia, C., Garattini, G.,Raven Press, New York, 211 (1974)

14. Wandrey, C.H.: Bioorganic synthesis of L-amino acids by isolated enzymes and microorganisms
This volume

15. Chibata, I., Tosa, T., Mori, T., Matsuo, Y.: Preparation and industrial application of immobilized aminoacylase
Fermentation technology today, Proceedings of the 4th International Fermentation Symposium, Kyoto, Japan (1972)

16. Reimerdes, E.H., Gottschick, W., Scholz, N.: Chemical modification of the enzyme surface of beta-galalactosidase for adsorptive immobilization
Symp. Int.: Use of enzymes in food technology, Versailles 121 (1982)

17. Reimerdes, E.H., Scholz, N.: Die hydrophobe Modifizierung der beta-Galaactosidase von Saccharomyces lactis für die adsorptive Immobilisierung
Milchwissenschaft 36: 608 (1981)

18. Reimerdes, E.H., Gottschick, W., Scholz, N.: Charakterisierung und Anwendung hydrophob immobilisierter beta-galactosidase von Saccharomyces lactis
Milchwissenschaft 36: 664 (1981)

19. Krämer, D., Lehmann, K., Pennewiss, H., Pleiner, H.: Oxirane Acrylic Beads for Protein Immobilization : A Novel Matrix for Biocatalysis and Biospecific Adsorption
Röhm, Darmstadt (1979)

20. Reimerdes, E.H., Herlitz, E.: The properties of beta-galactosidase immobilized in various forms
Abstracts of the Fifth Int. Congreß of Food Science and Technology, IUFOST, Kyoto, Japan 224 (1978)

21. Reimerdes, E.H.: The evaluation of immobilized trypsin with respect to different binding systems and the catalytic properties
Abstracts of the First European Congreß of Biotechnology, Interlaken , Verlag Chemie 2 (1978)

22. Reimerdes, E.H.: The effect of binding systems on immobilized enzyme suitability
Food Process Engineering II: Enzyme Engineering in Food Processing, Linko, P., Lanrinkari, J., (Eds), Applied Science Publishers 69 (1983)

23. Reimerdes, E.H.: Model proteolysis of beta-casein with immobilized trypsin and its technological significance
Enzyme Engineering, Vol. 4, Brown, G.H., Manecke, G., Wingard,Jr. L.B. (Eds), 275 (1978)

24. Pitcher, W.H., Immobilized lactase
Milk Industry 78: 4 (1976)

25. Messing, R.A.: Immobilized Enzymes for industrial reactors
Academic Press, New York (1975)

26. Pye, E.K., Wingard, Jr., L.B.:
Enzyme Engineering Vol. 2 (1974), Plenum Press, New York

27. Shukla, T.P.: Beta-galactosidase technology: A solution to the lactose problem
CRC Critical Reviews in Food Technology 325 (1975)

28. Berghofer, E.: Anwendung immobilisierter Enzyme in der Lebensmitteltechnologie
Lebensmittel und Ernährung 29: 225 (1976)RR

29: Weetall, H.H.: Immobilized Enzymes: Some Applications to Food and Beverages, Part 2: Applications
Food Prod. Development 94: 94 (1973)

30. Reimerdes, E.H., Scholz, N.: Herstellung eines laktosereduzierten Kakaos
Deutsche Milchwirtschaft 34: 1280 (1981)

31. N.N.: Laktofruit: A new soft drink
Food Engineering 10 (1978)

32. Talley, E.A., Hunter, A.S.: Solubility of lactose and its hydrolytic products
Am. Chem. Soc. 42: 2789 (1952)

33. Aries, R.: Eiscreme und moderne Molkereitechnik
Noreuropaeisk Mejeri-Tidskrift 9: 297 (1977)

34: Novo, Kopenhagen

35. Reimerdes, E.H.: Moderne Lebensmitteltechnologie - Einsatz von Enzymen
Vorträge zur Hochschultagung, Kiel , Paul Parey-Verlag 223, (1982)

PREPARATION AND PROPERTIES OF SEMISYNTHETIC FLAVOENZYMES

Donald Hilvert and E.T. Kaiser
Laboratory of Bioorganic Chemistry and Biochemistry
The Rockefeller University
1230 York Avenue
New York, New York 10021

ABSTRACT. Derivatives of 10-methylisoalloxazine have been used to alkylate the active site sulfhydryl group of the protein templates papain and glyceraldehyde-3-phosphate dehydrogenase. The resulting semisynthetic flavoenzymes have contrasting substrate specificity and stereoselectivity in the oxidation of 1,4-dihydronicotinamides. Furthermore, rate accelerations of 100 to 1000-fold are observed in the reactions of each of these catalysts with their optimal substrates compared to the analogous reactions with non-enzymatic model flavins. The success of these first generation catalysts suggests that the "chemical mutation" of existing tertiary structures holds great promise for the development of effective artificial enzymes.

The practical application of enzymes to the production of drugs, foods and fuels has received, recently, increasing attention. This is easy to understand since few chemical catalysts can match the catalytic efficiency and specificity achieved by enzymes under mild aqueous conditions. The potential usefulness of protein catalysts could, however, be greatly extended if new active sites could be designed from first principles and constructed from the constituent amino acids. Effective artificial enzymes would be particularly advantageous, for instance, in the catalysis of chemical transformations unrelated to physiological processes.

Tertiary structures already present in naturally occurring enzymes provide excellent starting points for the construction of artificial enzymes, and we have explored an approach towards catalyst design involving "chemical mutation" of existing enzyme active sites (1). Introduction of foreign catalytic functionality into the binding pocket of a biological macromolecule, through covalent linkage to selected amino acid residues, can lead to semisynthetic enzymes which combine the characteristic reactivity of the prosthetic group with the binding specificity of the protein template.

Initially, we felt that it would be best to avoid systems in which a series of rapid proton transfers are required for effective catalysis to occur and have instead studied systems in which the catalytic effects

M. P. Schneider (ed.), Enzymes as Catalysts in Organic Synthesis, 303–310.

are likely to arise from holding a substrate in close proximity to the catalytic funtionality. In this context, cofactor molecules, including vitamins, which possess intrinsic catalytic activity of their own, seemed to us to have distinct advantages as potential prosthetic groups. For example, isoalloxazines (riboflavin analogs) promote many processes catalyzed by flavin-dependent enzymes (2), including thiol and 1,4-dihydronicotinamide oxidation and various monooxygenation reactions, albeit at vastly lower rates and with less selectivity than the holoenzyme systems. Incorporation of flavin cofactors into the active sites of specific protein templates allows us to enhance the reactivity and fine-tune the substrate specificity of these simple catalysts. We have therefore prepared a series of 10-methylisoalloxazine derivatives that contain reactive sites that can alkylate nucleophilic amino acid residues (1).

In our work to date we have developed successful semisynthetic enzymes using two naturally-occurring enzyme systems, papain and glyceraldehyde-3 phosphate dehydrogenase. Papain is a cysteine protease whose structure has been determined by x-ray diffraction to 2.8 Å resolution (3). The active site consists of a 25 Å long, hydrophobic groove containing the catalytically essential sulfhydryl group of Cys-25. This residue is transiently acylated during the enzyme-catalyzed hydrolysis of peptide and ester substrates, so that alkylation causes the loss of native enzymatic activity. However, modification of Cys-25 often does not prevent substrate binding (4), and model building convinced us that a flavin moiety could be introduced at this position without blocking the binding site.

Alkylation of the Cys-25 sulfhydryl of papain with the 10-methylisoalloxazine derivatives we have synthesized has yielded a family of "flavopapains" with widely varying catalytic efficiencies (1). The most successful of these semisynthetic enzymes is obtained from the reaction of 8α-bromoacetyl-10-methylisoalloxazine with the enzyme (5) (Fig. 1). This artificial flavoenzyme actually rivals many naturally occurring flavoproteins with respect to the rate effects and substrate specificity it achieves.

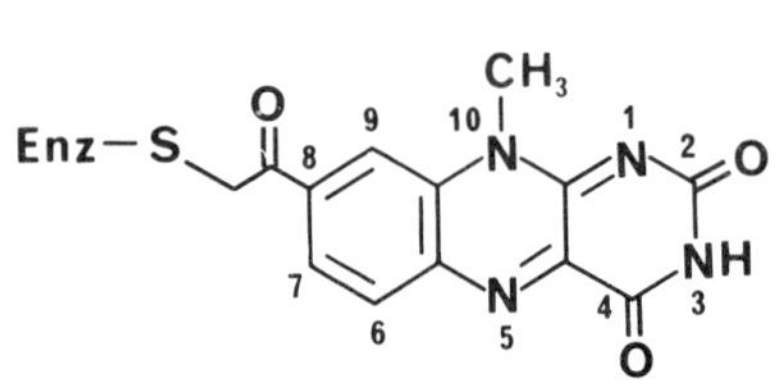

Fig. 1. 8-acetylflavoprotein

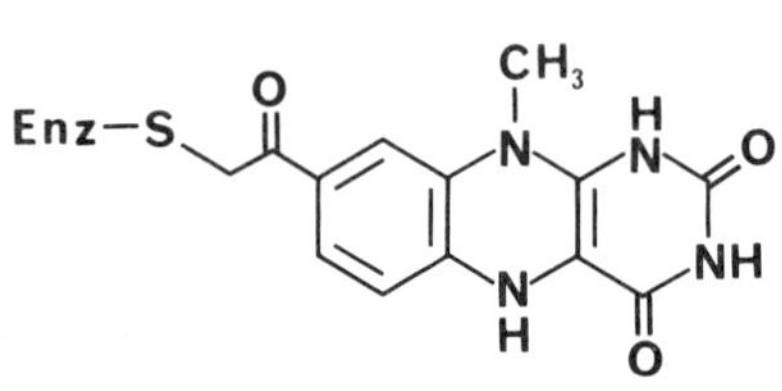

Fig. 2. reduced flavoprotein

We have evaluated the activity of 8-acetylflavopapain by studying its ability to oxidize N-alkyl-1,4-dihydronicotinamides (5). The kinetics of the enzymatic reaction are characterized by saturation behavior, as expected if the substrate binds to the active site before it is oxidized. Furthermore, in the presence of molecular oxygen, dihydronicotinamide oxidation is catalytic in flavin. In the first half-reaction, isoalloxazine is reduced at the active site (Fig. 2) by transfer of a hydride equivalent from the substrate. The resulting dihydroflavin species subsequently reacts with O_2 to regenerate the oxidized flavoenzyme. As illustrated in Equation 1, however, flavin reoxidation can conceivably occur either in the presence or absence of

$$E_{ox} + NRNH \underset{K_s}{\rightleftharpoons} (E_{ox}\bullet NRNH) \xrightarrow{k_2} (EH_2\bullet NRN) \longrightarrow EH_2 + NRN \qquad (1a)$$

$$(EH_2\bullet NRN) \xrightarrow{k_{ox}'[O_2]} (E_{ox}\bullet NRN) \longrightarrow E_{ox} + NRN \qquad (1b)$$

$$EH_2 \xrightarrow{k_{ox}[O_2]} E_{ox} \qquad (1c)$$

bound substrate or product. In this scheme E_{ox} and EH_2 are the oxidized and reduced forms of flavopapain, E•NRN(H) are the Michaelis complexes, and NRNH and NRN are the reduced and oxidized forms of the substrate nicotinamide. The rate equation which describes this complex mechanistic behavior is given in Equation 2, where $(\phi n)^{-1}$ are the steady state rate constants. As shown, this equation is related to the standard Michaelis-Menten formulation familiar to enzymologists, although (kcat)app and (Km)app here are complex kinetic parameters.

$$\frac{1}{k_{obs}} = \left[\phi_0 + \frac{\phi_2}{[O_2]}\right] + \frac{1}{[NRNH]}\left[\phi_1 + \frac{\phi_3}{[O_2]}\right] = \frac{1}{(k_{cat})_{app}} + \frac{1}{(k_{cat}/K_m)_{app}[NRNH]} \qquad (2)$$

Flavopapain derived from 8-acetyl-10-methylisoalloxazine is a very efficient catalyst for the oxidation of N-alkyl-1,4-dihydronicotinamides (5). In air-saturated buffer (pH 7.5, 25°C), for example, (kcat/Km)app for the oxidation of N-hexyl-1,4-dihydronicotinamide by this catalyst is in the vicinity of 10^6 $M^{-1}s^{-1}$. This corresponds to a ca. 600-fold rate acceleration relative to the model flavin, 8-acetyl-10-methylisoalloxazine. On the other hand, NADH itself is a relatively poor substrate for the enzyme giving only a modest 4-fold rate enhancement over the analogous non-enzymatic model reaction. This finding is readily rationalized in terms of the hydrophobic nature of the substrate binding site which should greatly favor the hydrophobic N-alkyl-1,4-dihydronicotinamides as substrates over the bulky, hydrophilic NADH molecule.

The reduction of 8-acetylflavopapain by N-alkyl-1,4-dihydronicotinamides is, in fact, so rapid that the air-oxidation of the dihydroflavin produced is typically rate-limiting. A rapid build-up of the reduced flavoenzyme with its characteristic absorbance at 530 nm can be observed when the enzyme is mixed with a moderate excess of the various dihydronicotinamides. Furthermore, as expected if Equation 2 holds, the values of (kcat)app and (Km)app determined experimentally depend on the concentration of molecular oxygen in the reaction medium. The oxygen effect seems to be roughly the same on both parameters, since (kcat/Km)app is not significantly changed at two different oxygen concentrations. This observation indicates that the $\Phi_3/[O_2]$ term in Equation 2 is negligibly small and suggests that a ternary complex is not kinetically important in the reoxidation of reduced 8-acetylflavopapain. We therefore concluded that the apparent second-order rate constant can be used as a legitimate reactivity index for flavopapain (5).

In order to determine the microscopic rate constants k_2 and Ks for the oxidations catalyzed by 8-acetylflavopapain, anaerobic studies have been carried out by Dr. C. Radziejewski in our group in collaboration with Dr. D. Ballou of the University of Michigan using stopped-flow spectroscopy (6). In the case of the N-hexyl-1,4-dihydronicotinamide, the best substrate studied, k_2 for the reduction step is 370 ± 70 s^{-1} and Ks for formation of the Michaelis complex is 230 ± 50 μM at pH 7.6 and 24.5°C. The bimolecular rate constant k_2/Ks for this reaction (1.6 x 10^6 $M^{-1}s^{-1}$) is similar in magnitude to the kcat/Km value determined under aerobic conditions, providing further support for the conclusion that the latter parameter is a valid measure of flavopapain's catalytic efficiency. The model reactions, on the other hand, did not show saturation under anaerobic conditions up to concentrations of the dihydronicotinamides of 400 μM.

An additional series of stopped-flow experiments (6) conclusively demonstrate that the low turnover numbers [(kcat)app] observed under aerobic conditions are due entirely to the slow rate of dihydroflavin reoxidation. The reaction of molecular oxygen with the reduced form of flavopapain is a complex, autocatalytic process which is roughly 7 to 8 times slower than the corresponding reaction of oxygen with the model compound, 1,5-dihydro-8-acetyl-10-methylisoalloxazine. Other oxidants, including dichloroindophenol (DCIP) and 3-(4',5'-dimethylthiazol-2-yl)-2,4-diphenyltetrazolium bromide (MTT), have been found (6) to be far superior to molecular oxygen as electron acceptors in this reaction. Indeed, MTT has been successfully used under turnover conditions as the terminal electron acceptor in the oxidation of N-benzyl- and N-propyl-1,4-dihydronicotinamide by the enzyme. The kcat values that are obtained in these reactions correspond well with the k_2 values determined anaerobically. Thus, in the presence of an appropriate electron acceptor, flavopapain can be an extremely efficient redox catalyst marked simultaneously by high reactivity (kcat/Km) and favorable turnover rates (kcat).

We have shown that it is possible to take a hydrolytic enzyme like papain and convert it, through covalent modification with a flavin analog, into a semisynthetic flavoenzyme with catalytic activity comparable to that of naturally occurring enzymes. More recently we have been involved in the development of a series of artificial flavoenzymes that have substrate specificity and stereoselectivity opposite that of the flavopapains. For these investigations we have chosen a new protein template, glyceraldehyde-3-phosphate dehydrogenase (GAPDH). The native enzyme (7), which is a tetramer of identical subunits, catalyzes the oxidation of glyceraldehyde-3-phosphate (GAP) with concommitant reduction of NAD^+. Like papain, GAPDH has a catalytically essential sulfhydryl group, Cys-149, that can be readily alkylated (7). From x-ray diffraction data it is known that this Cys residue is located near the pyridinium moiety of the bound nicotinamide cofactor (8). However, alkylation of this group apparently does not block the binding site for NAD^+/NADH (9). We were therefore encouraged to test the possiblity that the modification of Cys-149 with our flavin derivatives might result in a flavo-GAPDH species that would selectively oxidize NADH with the *si*-face stereoselectivity (7) exhibited by the native enzyme.

GAPDH has been isolated from a range of sources extending from bacterial cells to animal tissue (7). In our preliminary experiments, however, we have used commercially available rabbit muscle enzyme. The x-ray structure of this protein has not been determined, but that of closely related forms of the enzyme are available (8,10). This allows us to consider in detail the steric environment of the flavins at the active site and how substrate molecules may be brought into proximity to them. Such considerations convinced us that alkylation of the active site sulfhydryl group with 7α-bromoacetyl- and 8α-bromoacetyl-10-methylisoalloxazine (5,11), which had been developed for study with flavopapain, might well yield NADH-specific oxido-reductases (Fig. 1 and Fig. 3). Treatment of apo-GAPDH (NAD^+-free) from rabbit muscle with these flavin derivatives results in incorporation of approximately one flavin per active site (12). The extent of modification can be easily estimated from anaerobic titration of the flavoenzyme with dithionite and Biuret protein determination.

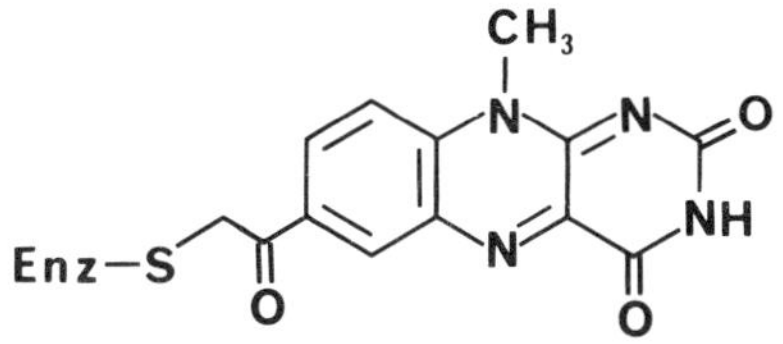

Fig. 3. 7-acetylflavoprotein

We have studied (12) the catalytic potential of the flavo-GAPDH species by examining their ability to oxidize various 1,4-dihydronicotinamide derivatives, as well. In air-saturated aqueous buffer (50 mM EPPS, pH 8.0, and 25.0°C) both 7- and 8-acetylflavo-GAPDH are effective oxidants of NADH itself but not of the hydrophobic NADH analogues. The apparent second-order rate constant kcat/Km for the oxidation of NADH by 7-acetylflavo-GAPDH is 1067 $M^{-1}s^{-1}$, while that for the 8-acetyl derivative is 475 $M^{-1}s^{-1}$. In each case this parameter is roughly 85 times larger than the second-order rate constant obtained under identical conditions with the corresponding model compound, 7- or 8-acetyl-10-methylisoalloxazine. Furthermore, saturation behavior is observed in these reactions at relatively low concentrations of NADH, with apparent Km values of roughly 25 μM measured for each enzyme derivative. In contrast, 10 to 20-fold larger apparent Km values and rate accelerations of at most 2 to 3-fold are obtained in the oxidation of the simple N-alkyl-1,4-dihydronicotinamides.

The clear preference for NADH as substrate over the hydrophobic NADH analogues provides a point of contrast with the flavopapains and is presumably due to the polar nature of GAPDH's active site. In addition, NADPH, which is not oxidized by the native enzyme, is also a poor substrate for flavo-GAPDH, indicating that template specificity is preserved subsequent to modification. Appropriate choice of a protein template thus allows us to alter the substrate specificity of our semisynthetic enzymes in a predictable manner.

The catalytic efficiency of the flavo-GAPDH enzymes, like that of flavopapain, seems to be limited by the rate of dihydroflavin reoxidation. However, increasing the oxygen concentration in the medium augments both kcat <u>and</u> kcat/Km in the case of flavo-GAPDH (12), whereas only kcat is affected for flavopapain (6). When the oxidation of NADH by 7-acetylflavo-GAPDH is carried out in an oxygen saturated medium, for example, kcat and kcat/Km increase 1.7 and 1.5-fold relative to their respective values determined in air-saturated buffer. Reoxidation of the prosthetic group in the flavo-GAPDH enzymes apparently occurs with substrate/product bound at the active site. The product NAD^+ does have some affinity for the binding pocket as evidenced by the fact that it is a competitive inhibitor of NADH oxidation with a Ki of roughly 330 μM (12).

Although the observed rate acceleration for NADH oxidation catalyzed by flavo-GAPDH is substantial, it is due primarily to the low value of (Km)app measured. As with flavopapain, the apparent kcat value is small under aerobic turnover conditions (ca. 0.03 s^{-1}). This parameter presumably does not, however, reflect the actual catalytic efficiency of the reduction step, since molecular oxygen is such a poor oxidant for the reduced flavoenzyme. It is therefore desirable to measure the microscopic rate constants. Unfortunately, various electron accepting dyes like DCIP and MTT are actually less efficient oxidants of dihydroflavo-GAPDH than molecular oxygen, perhaps because bound product inhibits binding of the bulky dye molecules to the active site. Furthermore, the instability of the rabbit muscle enzyme modified with our flavin derivatives has precluded measurement of k_2 and Ks by stopped-flow techniques.

The instability of rabbit muscle apo-GAPDH (NAD^+) is well documented in the literature (7). In contrast, the same enzyme isolated from the thermophilic bacterium *Bacillus stearothermophilus* is exceptionally stable (13,14). Together with Dr. Y. Hatanaka in our group we have recently isolated the bacterial protein and carried out a preliminary investigation of the semisynthetic enzyme resulting from modification of its active site sulfhydryl group with 7α-bromoacetyl-10-methylisoalloxazine. One flavin can be readily introduced per active site, and alkylation of the enzyme leads to the loss of the four rapidly titrating thiols of the native tetramer. The resulting species, like its muscle enzyme counterpart, is a good catalyst for NADH oxidation.

In air-saturated buffer (50 mM EPPS, pH 8.0, 25°) the apparent kcat/Km value is 4952 $M^{-1}s^{-1}$, which corresponds roughly to a 400-fold rate acceleration compared to the non-enzymatic model reaction. Although the apparent bimolecular rate constant for bacterial flavo-GAPDH is almost five times larger than the value obtained with the muscle enzyme, (kcat)app is only slightly enhanced (0.036 s^{-1}). The observed increase in rate acceleration relative to the non-enzymatic model is therefore due almost entirely to the lowering of the Km parameter. Nevertheless, the magnitude of the rate effect is clearly similar to that observed with flavopapain and its optimal substrate.

As anticipated, the thermophilic semisynthetic enzyme is much more stable than the analogous muscle enzyme derivative. We find, for example, that the activity of bacterial flavo-GAPDH which has been precipitated with ammonium sulfate does not appreciably change after storage at 4°C for one month. Furthermore, this enzyme functions well at 55°C. The apparent kcat value at this temperature for oxidation of NADH is 0.166 s^{-1}, while (kcat/Km)app is 1.30 x 10^4 $M^{-1}s^{-1}$. These parameters are respectively 5 and 2.5 times larger than the corresponding values measured at 25°C. The stability of the thermophilic enzyme should facilitate its kinetic characterization, and we are currently elucidating the microscopic rate constants by stopped-flow spectroscopy.

In order to characterize the enzymatic behavior of flavo-GAPDH further, we have investigated the stereochemistry of hydrogen transfer from the dihydronicotinamide to the flavoprotein. Native GAPDH distinguishes the diastereotopic methylene protons at the C(4) position of NADH and exclusively removes the pro-S hydrogen during oxidation of this substrate (7). In their reaction with the chirally deuterated NADH derivatives, (4R-D)NADH and (4S-D)NADH, we find that 7-acetyl- and 8-acetylflavo-GAPDH from rabbit muscle also exhibit substantial si-face stereoselectivity (12). We have examined the product ratios (15) and rate behavior of the oxidation of these substrates and have estimated (12) from this data a 7:1 preference for transfer of the pro-S hydrogen in the case of 7-acetylflavo-GAPDH by assuming, as a first approximation, that secondary isotope effects are negligible and by ignoring the oxygen dependence of kcat/Km. For the non-enzymatic models, on the other hand, a 3:1 preference for transfer of the pro-R hydrogen is found (12). Removal of the pro-R hydrogen of NADH is similarly favored by the flavopapain derived from 7α-bromoacetyl-10-methylisoalloxazine (16). We are thus able to select the stereochemistry of hydride transfer in a systematic manner by simply choosing the proper protein template.

The design and construction of protein catalysts from the constituent amino acids is still a distant goal. Nevertheless, we have shown that it is possible to modify existing tertiary structures with coenzyme analogs to obtain semisynthetic enzymes with catalytic activities very different than those of the corresponding native enzymes. The large rate accelerations and the predictable substrate specificity exhibited by the artificial flavoproteins, 8-acetylflavopapain and 7-acetylflavo-GAPDH, in the oxidation of dihydronicotinamides are particularly noteworthy. We believe that the "chemical mutation" strategy thus holds great promise for the rational design of artificial enzymes and we anticipate that this methodology, possibly in conjunction with recombinant DNA technology, will be applicable in the future to the development of even more complex and finely tuned catalysts.

ACKNOWLEDGEMENTS

The partial support of this research by NIH Postdoctoral Fellowship AM-07232 (D.H.) and NSF grant CHE-8218637 (E.T.K) is gratefully acknowledged.

REFERENCES

(1) Kaiser, E.T.; Lawrence, D.S. Science 1984, (226), 505-511.
(2) Walsh, C. Acc. Chem. Res. 1980, (13), 148-155.
(3) Kamphuis, I.G.; Kalk, I.G.; Swarte, M.B.A.; Drenth, J. J. Mol.Biol. 1984, (179), 233-256.
(4) Lowbridge, J.; Fruton, J.S. J. Biol. Chem. 1974, (249), 6754.
(5) Slama, J.T.; Radziejewski, C.; Oruganti, S.R.; Kaiser, E.T. J. Amer. Chem. Soc. 1984, (106), 6778-6785.
(6) Radziejewski, C.; Ballou, D.; Kaiser, E.T., J. Amer. Chem. Soc., in press.
(7) Harris, J.I.; Waters, M. in "The Enzymes"; Boyer, P., Ed.; Academic Press: New York, 1976; vol. XIII, part C, pp 1-49.
(8) Moras, D.; Olsen, K.W.; Sabesan, M.N.; Buehner, M.; Ford, G.C.; Rossmann, M.G. J. Biol. Chem. 1975, (250), 9137-9162.
(9) Rafter, G.W.; Colowick, S.P. Arch. Biochem. Biophys. 1957, (66), 190-207.
(10) Biesecker, G.; Harris, J.I.; Thierry, J.C.; Walker, J.E.; Wonacott, A.J. Nature (London) 1977, (266), 328-333.
(11) Levine, H.L.; Kaiser, E.T. J. Amer. Chem. Soc. 1978, (100), 7670-7677.
(12) Hilvert, D.; Kaiser, E.T. manuscript submitted.
(13) Amelunxen, R.E. Biochim. Biophys. Acta 1967, (139), 24-32.
(14) Suzuki, K.; Imahori, K. J. Biochem. 1973, (74) 955-970.
(15) Arnold, L.J., Jr.; You, K.; Allison, W.S.; Kaplan, N.O. Biochemistry 1976, (15), 4844-4849.
(16) Levine, H.L.; Kaiser, E.T. J. Amer. Chem. Soc. 1980, (102), 343-345.

THE STUDY AND REDESIGN OF ENZYMES BY PROTEIN ENGINEERING

Robin J. Leatherbarrow, Tim N.C. Wells & Alan R. Fersht
Department of Chemistry,
Imperial College of Science & Technology,
London SW7 2AY
U.K.

ABSTRACT. Protein engineering offers an approach to the modification of enzymes in terms of alteration of specificity, catalytic efficiency, pH dependence and stability. We describe how site-directed mutagenesis has been used on the tyrosyl-tRNA synthetase of *Bacillus stearothermophilus* to provide details of the energetics of individual hydrogen bonded interactions between the enzyme and its substrates. The importance of such studies is twofold. First, they give information on the mechanism of catalysis by this enzyme, and second, they indicate quantitatively how alteration of individual interactions may be used to manipulate binding and kinetic constants.

1. INTRODUCTION

1.1. The Involvement of Amino Acid Side Chains in Catalysis

Enzymic catalysis is achieved by interactions between sites on the enzyme and the molecules involved in the reaction. These interactions are used to alter the energetic states of bound substrates, products, and most importantly, intermediates in the reaction pathway. By these means the rate of the reaction is considerably enhanced compared with that in free solution. Interactions with the enzyme determine the energetic levels of the various states ES, $ES^{\ddagger}$ and EP (Fig. 1). The rate of the reaction depends primarily upon the energy level of $ES^{\ddagger}$, and therefore specific interactions between the enzyme and this transition state may be used to lower this energy barrier and so enhance rate. Lowering of the ES energy i.e. specific interaction with the substrate, may be used to lower the dissociation constant of the ES complex and hence the K_M of the enzyme. It should be noted that if the energy level of the ES complex is too low, the reaction will pass through a large 'thermodynamic pit' which will lead to decreased rate. It can therefore be advantageous, in rate terms, to destabilize the ES complex.

The interactions between the enzyme and the reactant states are typically non-covalent and will be mediated principally via the amino acid side chains of the active site. The remarkable selectivity and

M. P. Schneider (ed.), Enzymes as Catalysts in Organic Synthesis, 311–323.

Figure 1. Schematic Gibbs energy profile for enzymic catalysis of the reaction S = P.
S, substrate; E, enzyme; ES, enzyme-substrate complex; $ES^{\ddagger}$, enzyme-bound transition state; EP, enzyme-product complex. $\underline{K}_S$, the dissociation constant of the substrate is proportional to ΔG_S; the rate constant, $\underline{k}$, for the reaction is proportional to $\Delta G^{\ddagger}$ [14].

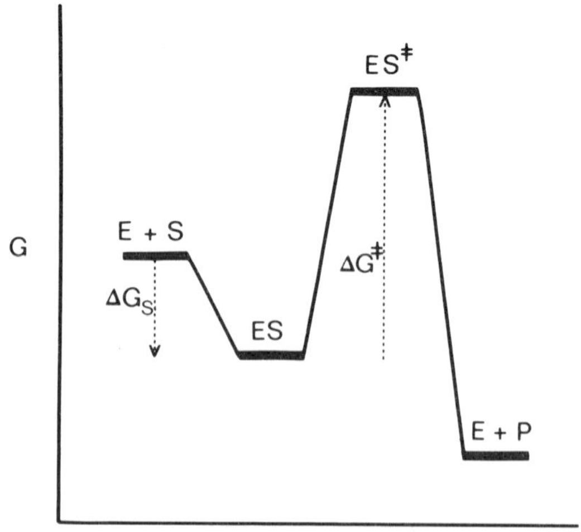

efficiency of enzymic catalysis is therefore a consequence of the topological arrangement of the interacting amino acids. The specificity, and the extent of rate enhancement depend on the reaction free energies of the various states ES, $ES^{\ddagger}$ and EP. These energies in turn depend simply on the sum of the energetic contributions of interactions to each of these various states.

Altering the constituent amino acids of an enzyme active site therefore provides both a means to manipulate catalytic properties, and a method of quantitating the interaction made with any single side chain. At the simplest level, disruption of interactions between enzyme and substrate will affect binding whereas altered transition state interactions will affect rate. Given the structural details of an enzyme as obtained by X-ray crystallography, it may be envisaged how rational alteration of enzyme properties may be achieved by the replacement of one or more interacting amino acid side chains. The techniques used for such manipulations of protein structure stem from recent advances in molecular biology. Alterations are made at the level of DNA, which result in production of enzymes containing specific single or multiple amino acid replacements. The potential uses of such engineered proteins include tailored alterations in the kinetic constants $\underline{K}_M$ and $\underline{k}_{cat}$, substrate specificity, pH optima and stability. In addition to such potentially commercial applications, kinetic studies on mutant proteins may be used to give extensive information on the nature of the catalytic process. These studies also give quantitative data on the energetics of interactions between individual amino acid side chains and each of the various reactant states. As described earlier, the energetics of these interactions are vital to the understanding of how the enzyme catalyses the reaction, and are not forthcoming from structural studies alone.

1.2. System of study

We are utilising protein engineering to perform a systematic study of the reactions catalysed by the enzyme tyrosyl-tRNA synthetase (TyrTS) from _Bacillus stearothermophilus_ [reviewed in 1]. This enzyme is a typical aminoacyl-tRNA synthetase, enzymes which are central to protein biosynthesis, and catalyses the addition of tyrosine to its acceptor tRNA. The reaction proceeds in two steps [2]. First, the tyrosine is activated by reaction with ATP to form an enzyme-bound mixed anhydride and liberate pyrophosphate. In the absence of pyrophosphate and tRNA the bound tyrosyl adenylate is stable, and its accumulation allows a single turnover of the enzyme to be monitored. In the presence of tRNA the second stage of the process occurs, and tyrosine is transfered to the tRNA.

$$\text{Tyrosine + ATP + Enzyme} \longrightarrow \text{Enzyme.Tyrosyl adenylate + PPi} \quad (1)$$

$$\text{Enzyme.Tyrosyl adenylate + tRNA}^{tyr} \longrightarrow \text{Tyrosyl-tRNA}^{tyr} \text{ + AMP} \quad (2)$$

The first of these reactions is particularly amenable to study by protein engineering for several reasons. First, the structure of the enzyme has been determined by X-ray crystallography to 0.21 nm [3], and in addition the structures of enzyme-bound tyrosine and enzyme-bound tyrosyl adenylate have been solved. Second, the stability of the enzyme.tyrosyl adenylate complex facilitates kinetic studies by allowing both a single turnover to be monitored, and by providing a stoichiometric titration for the number of active sites present in a solution. Third, the gene encoding this protein has been cloned into the _E. coli_ bacteriophage M13 in a manner which enables its expression in infected cells. Fourth, proteins from the thermophilic organism _B. stearothermophilus_ are thermostable, allowing a heating step to be included in the enzyme purification in order to remove any endogenous tyrosyl-tRNA synthetase present from the host _E. coli_.

1.3. Site-directed Mutagenesis

Site-directed mutagenesis is a technique whereby single or multiple amino acid replacements, insertions or deletions may be engineered into a protein [4,5]. The techniques require that the gene coding for the protein of interest be cloned into a suitable vector which will allow the necessary manipulations and also the expression of the encoded protein. In its simplest form (Fig. 2) the gene is inserted into a bacteriophage which will produce single-stranded DNA. A chemically synthesized oligonucleotide of around 17 nucleotides is constructed to be complementary to the region of interest except for one or more internal mismatched bases. The oligonucleotide is annealed to the

Figure 2. Construction of a mutation by use of site-directed mutagenesis. The synthetic oligonucleotide containing the mismatch (*) is annealed to the single stranded DNA template and then extended to produce closed circular DNA. When used to infect E. coli this duplex will give rise to both wild-type and mutant progeny. Selection of the mutant progeny allows production of the mutant progeny.

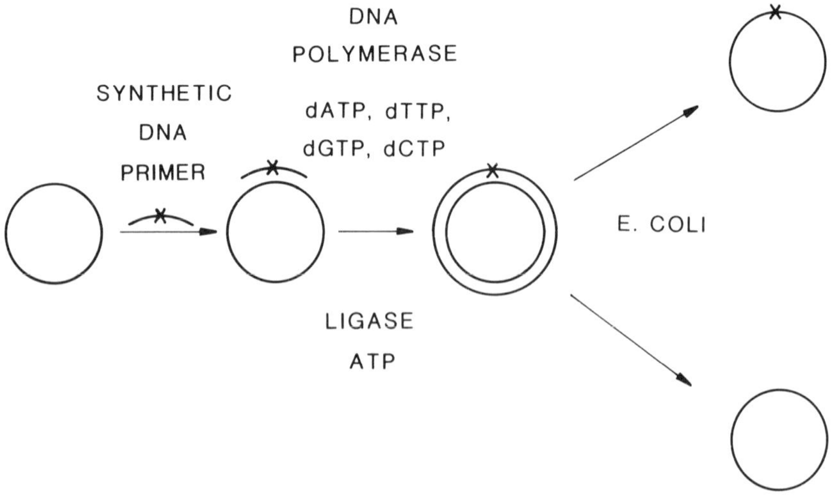

single-stranded DNA and used as a primer for extention around the genome using a DNA polymerase and a mixture of the 4 nucleotide triphosphates. The product is finally ligated with DNA ligase to generate double-stranded closed circular DNA. Due to the fidelity of DNA polymerase the newly synthesized strand is perfectly complementary to the original, however the mismatches placed in the oligonucleotide will lead to a mutation when this DNA is used to transfect E. coli. By use of appropriate oligonucleotide primers it is possible to direct any sequence alteration in the encoded protein.

2. RESULTS AND DISCUSSION

2.1. Energetic Contributions of Hydrogen Bonds in an Enzyme

Many of the important interactions involved in biological specificity are mediated by hydrogen bonds. Thus the specificity of DNA base pairing, the formation of protein secondary structures and many of the interactions between enzymes are their substrates depend on hydrogen bonding. Measurements *in vacuo* indicate the intrinsic strength of a OH···O hydrogen bond is around 6 kcal mol^{-1} [6]. However, in aqueous

Figure 3. Hydrogen bonding interactions between residues in the enzyme and enzyme-bound tyrosyl adenylate as determined from the crystal structure of the complex [3].

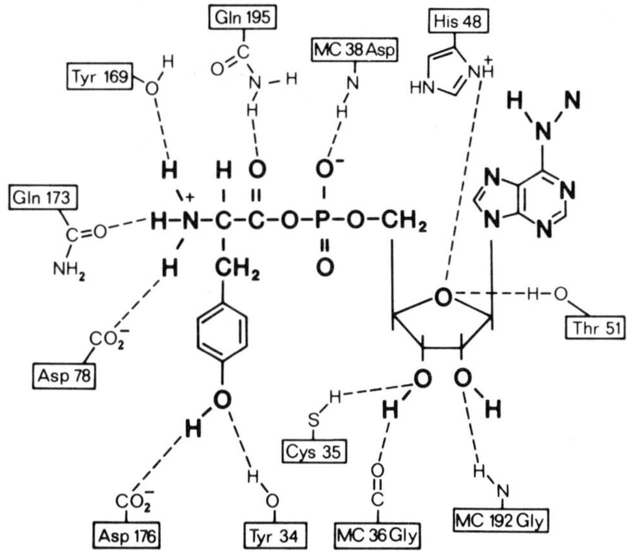

Table I. Apparent hydrogen bond strengths between the enzyme and substrate as determined from the effect of various side chain replacements.

Comparison (Residue and position)		Substrate	ΔG kcal mol^{-1}
Phe34	Tyr34	Tyrosine	0.52
Gly35	Cys35	ATP	1.14
Ala51	Cys51	ATP	0.47
Gly48	Asn48	ATP	0.77
Gly48	His48	ATP	0.96
Ser35	Cys35	ATP	1.18
Phe169	Tyr169	Tyrosine	3.72
Gly195	Gln195	Tyrosine	4.49
Gly35	Ser35	ATP	-0.04
Ala51	Thr51	ATP	-0.44

The two residues compared are in the first two columns. The apparent contributions (ΔG) of the side chains ot the binding energy the the enzyme-transition state complexes were calculated by comparing the ratios of $\underline{k}_{cat}/\underline{K}_M$ for activation by wild type and mutant enzymes, using the equation

$$\Delta G = -RT \ln\{(\underline{k}_{cat}/\underline{K}_M)_{mutant}/(\underline{k}_{cat}/\underline{K}_M)_{wild\text{-}type}\}$$

solution, solvent -OH groups will compete for all hydrogen bonding interactions resulting in a reduced effective hydrogen bond strength [7]. To date, there is little evidence for the actual energetic values of such hydrogen bond strengths in aqueous solution. The X-ray structure of the tyrosyl tRNA synthetase.tyrosyl adenylate complex shows that a large proportion of interactions between the enzyme and this ternary complex are via hydrogen bonds (Fig. 3). Site-directed mutagenesis allows us to measure the contribution of any of these hydrogen bonds by replacing a hydrogen bonding side chain by one which lacks the appropriate donor or acceptor capability. For example, residue Tyr34, which interacts via its phenolic hydroxyl with the -OH of the substrate, was replaced by a Phe, which has a benzenoid rather than a phenol side chain. Comparison of the activity of native and mutated enzyme therefore allows quantitation of the effective bond energy of this interaction.

Several amino acid substitutions were engineered to remove hydrogen bonding interactions between enzyme and substrate (Table I) [8]. The energetic effects of the mutations fall generally into two groups. Removal of hydrogen bonds between uncharged groups leads to a loss of ~0.5 to 1.5 kcal mol^{-1} of interaction energy, whereas removal of a hydrogen bond involving a charged group on the substrate is more deleterious. In this latter case the hydrogen bond provides an effective

Figure 4. The exchange of hydrogen bonds on binding of tyrosine to the tyrosyl-tRNA synthetase. a, Binding of tyrosine by native enzyme involves formation of two hydrogen bonds between its hydroxyl and Asp176 and Tyr34, but two equivalent hydrogen bonds between these residues and water are broken. b, Binding of tyrosine by the Tyr-Phe34 mutant involves just the formation of one hydrogen bond with the enzyme, but only one hydrogen bond is broken when the bound water is displaced. In both cases there is no net change in the number of hydrogen bonds in solution [8].

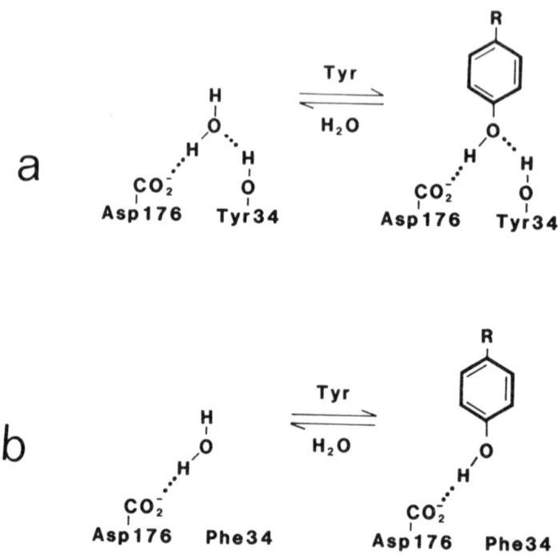

contribution of ~3.5 to 4.5 kcal mol^{-1}. The magnitude of these energies is somewhat less than the *in vacuo* hydrogen bond strengths, and provides a direct measure of the effective hydrogen bond strengths in water [8]. The energetics are best evaluated in terms of net loss or gain of hydrogen bonds. For example, considering the elimination of the hydrogen bond between the -OH of Tyr34 and the substrate hydroxyl. In the absence of substrate the Tyr34 -OH will make a hydrogen bond to water (Fig. 4). When substrate binds this interaction must be displaced, making the net number of interactions gained equal to the number lost, and so leading to a rather small effect on the net energetics. The mutant enzyme will fail to make hydrogen bonds to both substrate and solvent, a situation which again is energetically balanced. The net result is a relatively small energetic effect for this mutation. In the case of hydrogen bonded interactions involving a charged group the net effect of such exchange of hydrogen bonds with water is rather different. This is because of strength of a hydrogen bond involving a charged group is greater than that between two neutral groups. Therefore when an inventory of hydrogen bonds lost and gained on substrate binding is compiled (Fig. 5), it is seen that loss of a strong hydrogen bond between the charged sustrate group and the enzyme in the native enzyme is only compensated by gain of weaker interactions between solvent molecules. This effect, analogous to the burying of an unsolvated charge, is therefore accompanied by a relatively large loss of interaction energy for such mutations.

Figure 5. Schematic illustration of the inventory of hydrogen bonds made and broken when substrate binds either to native enzyme, or to mutant enzyme which lacks the potential to form a hydrogen bond. When the hydrogen bonding groups are both uncharged there is no net alteration in the number of hydrogen bonds formed and broken for both native and mutant enzyme. When one of the hydrogen bonding groups is charged, the loss of the strong hydrogen bond (s) involving this group results in the mutant enzyme having a much weaker interaction with the substrate, since only a weak hydrogen bond (w) between water molecules is gained.

	Inventory
$\mathrm{E{-}H \cdots OH_2 + HOH \cdots B{-}S = [E{-}H \cdots B{-}S] + HOH \cdots OH_2}$	**0**
$\mathrm{E \quad OH_2 + HOH \cdots B{-}S = [E \quad B{-}S] + HOH \cdots OH_2}$	**0**
$\mathrm{E \quad OH_2 + HOH \cdots {}^{-}B{-}S = [E \quad {}^{-}B{-}S] + HOH \cdots OH_2}$	**- s + w**
$\mathrm{E{-}H^{+} \cdots OH_2 + HOH \quad S = [E{-}H^{+} \quad S] + HOH \cdots OH_2}$	**-s + w**

2.2. Use of hydrogen bonds in the catalytic reaction

The reaction between the tyrosine carboxyl and the alpha phosphate of ATP involves the attack of a good nucleophile on a compound with a good leaving group (the pyrophosphate). The mechanism involves an in-line displacement with the alpha phosphorus undergoing inversion, indicating a simple S_N2 process (Fig. 6), which will involve an pentacoordinate intermediate [9]. A likely means for the enzyme to catalyse this reaction is simply by stabilizing the transition state.

Figure 6. The chemical mechanism of formation of tyrosyl adenylate from tyrosine and ATP.

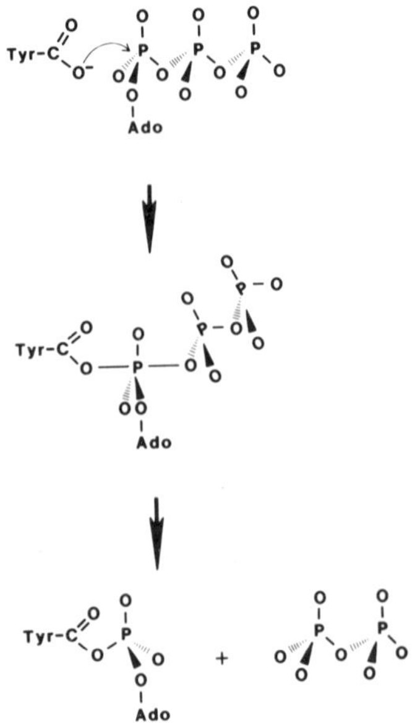

One means of stabilizing the transition state is for the enzyme to make interactions with this state which do not occur in the ground state. The transition state of the reaction will resemble the pentacoordinate intermediate (Fig. 6), and therefore to locate possible groups which could be involved this intermediate was modelled into the enzyme active site [10]. Two potential sites of interaction were found (Fig. 7). The side chains of Thr40 and His45 are located in a position to interact with the gamma phosphate group of the intermediate. To quantitate their role the mutants TyrTS(Thr-Ala40), TyrTS(His-Gly45) and TyrTS(Thr-Ala40;His-Gly45) were constructed in which any hydrogen bonding interactions made by these residues to the transition state would be removed. The kinetics of these mutant proteins show that these changes have large effects on the forward rate constant, but little effect on the dissociation constants for the substates (Table II). These results confirm that the side chains of these residues are involved

Figure 7. Model building of the pentacoordinate transition state of the reaction into the crystallographic structure of the tyrosyl-tRNA synthetase. The transition state structure was extrapolated from the known structure of enzyme-bound tyrosyl adenylate. Interactions are shown between the gamma phosphate and the side chains of Thr40 and His45.

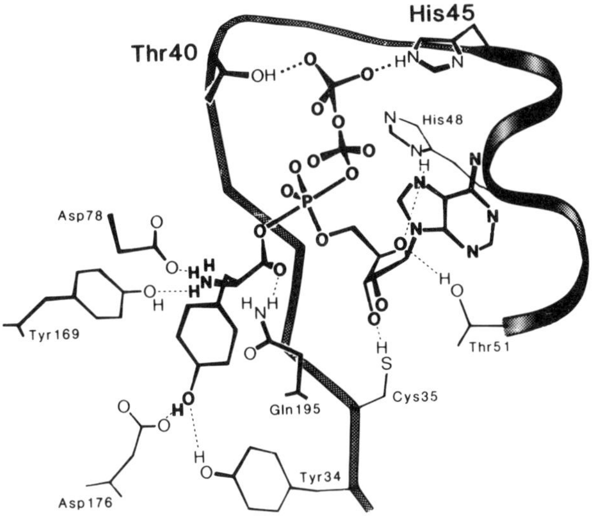

entirely in interactions with the transition state, and do not contribute to the binding of substrate. As shown from the kinetics of the double mutant TyrTS(Thr-Ala40;His-Gly45), in the absence of both these interactions the rate of the reaction is reduced 320,000 fold.

Table II. Pre-steady state kinetic parameters for the formation of tyrosyl adenylate.

Enzyme	k^a /s^{-1}	K_S Tyr /μM	K_S ATP /mM
TyrTS	38	12	4.7
TyrTS(His-Gly45)	0.16	10	1.2
TyrTS(Thr-Ala40)	0.0055	8.0	3.8
TyrTS(Thr-Ala40;His-Gly45)	0.00012	4.5	1.1

k, forward rate constant for the formation of tyrosyl adenylate; K_S, the dissociation constant for the substrate.

[a] Extrapolated to infinite substrate concentrations.

During the reaction, the change in geometry as the alpha phosphorus goes from 4- to 5-coordinate means that there will be a concomitant alteration in the location of the attached pyrophosphate group. It is tempting to suggest that this is sufficient to place the gamma phosphate group of the transition state in position to interact with the side chains of His45 and Thr40 (Fig. 8). The stabilization of the transition state by these residues forms a major contribution to the rate enhancement provided by the enzyme.

Figure 8. Schematic mechanism for the catalysis of tyrosyl adenylate formation by the tyrosyl-tRNA synthetase. Interactions are made by the side chains of Thr40 and His45 to the gamma-phoshate group of the transition state, but not to the enzyme-bound ATP. These interactions stabilize the transition state, and hence lead to an increase in rate.

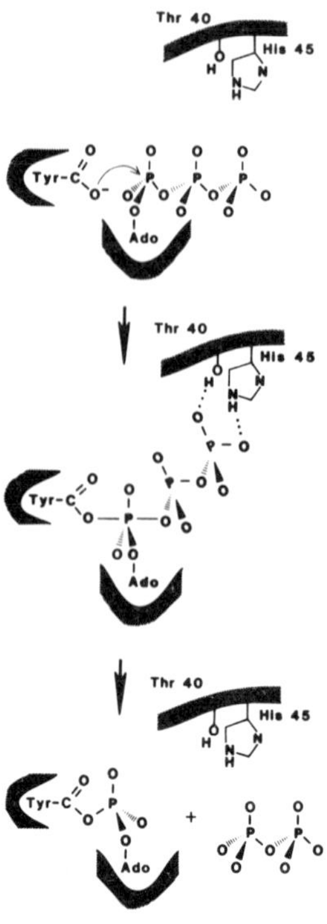

2.3. Fine-tuning of catalysis by hydrogen bonding

To study the contribution of substrate hydrogen bonds to catalysis, a number of mutants with residues involved in binding tyrosine and the ribose moiety of adenosine were produced. The kinetics of these enzymes have been examined in detail [11, 12], and the mutants fall into two broad groups with respect to the stabilization of bound substrate and

Table III. Calculated free energies of stabilization of enzyme-substrate and enzyme-transition state complexes by hydrogen bonded interactions[a].

Mutation	ΔG_b kcal mol^{-1} Side chain binding contribution to enzyme-substrate complex	$\Delta G_t^{\ddagger}$ kcal mol^{-1} Side chain binding contribution to enzyme-transition state complex
(a) Tyrosine binding site mutants:		
Tyr-Phe34	0.48	0.53
Tyr-Phe169	2.93	2.93
(b) ATP binding site mutants:		
Cys-Gly35	-0.08	1.25
Cys-Ser35	-0.22	1.01
His-Gly48	0.82	1.62

[a] The free energy contributions are calculated using the equations:

$$\Delta G_b = RT \ln\{(K_a' K_t)_{mutant} / (K_a' K_t)_{wild\ type}\}$$

$$\Delta G_t^{=} = -RT \ln\{(k_f/K_a' K_t)_{mutant} / (k_f/K_a' K_t)_{wild\ type}\}$$

where k_f is the forward rate constant, K_a' is the dissociation constant for ATP, and K_t is the dissociation constant for tyrosine.

transition state. First, there are those mutant enzymes which have had sidechains within the tyrosine binding pocket removed (Tyr-Phe34 and Tyr-Phe169). Here, there is no significant alteration in the dissociation constant for ATP, or in the forward rate constant, but there is an increase in the dissociation constant for tyrosine. From the enzyme crystal structure, these changes are interpreted [13] as inferring that the hydroxyl groups of Tyr34 and Tyr169 interact equally well with the substrate in both the unreacted form and the transition state of the reaction (Table III). For the forward reaction, therefore, the Tyr34 and Tyr169 hydroxyl interactions with the substrate are an example of uniform stabilization of the bound states.

Second, there are mutant enzymes in which side chains interacting with adenosine have been removed (Cys-Gly35, Cys-Ser35 and His-Gly48). Mutations at residue 35 remove an interaction with the ribose 2'-hydroxyl (Fig. 3), and result in a ten-fold lowering of the forward rate constant with a negligable effect on the dissociation constant for ATP. This demonstrates how the subtle changes in geometry during the reaction can be used by the thiol group of residue 35 to stabilize differentially the transition state. In the His-Gly48 enzyme there is a mixture of effects with a four-fold drop in the rate constant and a two-fold increase in dissociation constant. This implies a slight enzyme-substrate stabilization together with a larger enzyme-transition state stabilization. To summarise, loss of hydrogen bonding interactions in the adenosine binding pocket result in differential effects on the stabilization of the enzyme-substrate and enzyme-transition state complexes (Table III). This implies that these side chains are positioned to optimise hydrogen bonding interactions in the enzyme-transition state complex.

Therefore in all cases the enzyme uses hydrogen bonding groups in the active site region to stabilize the reaction transition state. In some cases this is achieved with out stabilizing the ES complex, preventing the formation of an ES 'thermodynamic pit' which would lower the overall catalytic rate. Individually these transition state stabilizations are rather less than those provided by the side chains of His45 and Thr40 described earlier, however in combination they are significant and provide fine-tuning of the rate of reaction.

2.4. Protein Engineering to Create 'Better' Enzymes

The results given in this paper give some measure of the energetics of individual interactions contributed by amino acid side chains in an enzyme active site. They therefore provide indications as to the extent by which reactions catalysed by other enzymes could be manipulated using suitable amino acid substitutions. The results on the strength of hydrogen bonds between enzyme and substrate show, perhaps disappointingly, that the addition of a single hydrogen bond to an enzyme substrate interaction would only yield an extra 1 kcal mol^{-1} or so of binding energy. However if the interaction was to a charged group on the substrate the effective bond strength would be much greater. Therefore it could be greatly advantageous if such additional interactions could be created. In order to increase rate it is evident that additional interactions made specifically to the transition state may be beneficial. Therefore, using the synthetase as an example, positioning additional hydrogen bonding side chains close to the beta/gamma phosphate groups would be a possible approach to increase rate.

3. REFERENCES

(1) Fersht, A.R., Shi, J-P., Wilkinson, A.J., Blow, D.M., Carter, P., Waye, M.M.Y. & Winter, G. (1984) Ange. Chem. 23, 467-473.

(2) Fersht, A.R. & Jakes, R. (1975) Biochemistry **14**, 3350-3356.

(3) Bhat, T.N., Brick, P. & Blow, D.M. (1985) J. Mol. Biol. (in press).

(4) Zoller, M.J. & Smith, M. (1983) Methods in Enzymology **100**, 468-500.

(5) Winter, G., Fersht, A.R., Wilkinson, A.J., Zoller, M. & Smith, M. (1982) Nature (London) **299**, 756-758.

(6) Weiner, S.J., Kollman, P.A., Case, D.A., Singh, U.C., Ghio, C., Alagona, G., Profeta, S. & Weiner, P. (1984) J. Am. Chem. Soc. **106**, 765-784

(7) Chothia, C. & Janin, J. Nature (London) **256**, 705-708

(8) Fersht, A.R., Shi, J-P., Knill-Jones, J., Lowe, D.M., Wilkinson, A.J., Blow, D.M., Brick, P., Carter, P., Waye, M.M.Y. & Winter, G. (1985) Nature (London) **314**, 235-238.

(9) Lowe, G. & Tansley, G. (1984) Tetrahedron **40**, 113-117.

(10) Leatherbarrow, R.J., Fersht, A.R. & Winter, G. (1985) Proc. Natl. Acad. Sci. U.S.A. (submitted)

(11) Wells, T. & Fersht, A.R. (1985) Nature (London) (in press)

(12) Wells, T. & Fersht, A.R. (1985) (in preparation)

(13) Fersht, A.R. (1985) Enzyme Structure and Mechanism (2nd edn.) W.H. Freeman, pub. Chapter 12.

EVOLUTIONARY GUIDANCE AND THE ENGINEERING OF ENZYMES

Krishnan P. Nambiar, Joseph Stackhouse,
Scott R. Presnell, and Steven A. Benner
Department of Chemistry
Harvard University
12 Oxford Street
Cambridge, Massachusetts 02138 U.S.A.

ABSTRACT. Evolutionary guidance is discussed as a strategy for simplifying the problems associated with the use of recombinant DNA techniques to study structure-function relationships in proteins. Evidence is given that suggests that natural selection "fine tunes" the behavior of proteins. Also progress on the synthesis, cloning, and expression of a gene for pancreatic ribonuclease, designed to permit rapid modification, is reported.

1. INTRODUCTION

Despite efforts in many laboratories to develop enzymes as synthetic reagents, their use in laboratory and industrial organic chemistry remains the exception rather than the rule.

While this may reflect conservatism among organic chemists, it may also simply mean that enzymes are not as useful as traditional organic catalysts in solving synthetic problems. Enzymes are often not easily obtained. Even with recent advances in recombinant DNA technology, expression and purification of proteins are often problematic and expensive. Once in hand, the enzyme may not have the desired substrate or stereospecificity; even rarer are pairs of enzymes with opposite stereospecificities that permit flexibility in synthetic design. Finally, enzymes are often rather unstable, especially in non-native environments.

While enzymes are likely to maintain their position as catalysts for effecting single transformations on biological molecules, will they ever be the catalysts of choice for the broader range of synthetic tasks? Three features of enzymic catalysis make the possibility real:

M. P. Schneider (ed.), Enzymes as Catalysts in Organic Synthesis, 325–340.

1. Enzymes are the products of billions of years of biological evolution. Therefore, enzymes may be inherently better (faster, more specific) than any catalyst that an organic graduate student can construct.

2. Enzymes are "combinable". They can remove a single substrate from a pool containing many compounds, and effect as single transformation on it. Thus, they can participate in designed one-pot multistep pathways that entirely rearrange the carbon skeleton of a precursor, all under isoenergetic conditions, just as in a natural biosynthetic pathway. Therefore, enzymes may form the basis of economical, multistep, one pot synthetic routes to complex chemicals.

3. Enzymes as macromolecules can have subtly different structures and subtly different physical and catalytic properties as a result. A series of proteins displaying a continuum of properties, ranging from rate and selectivity to stability and solubility, are in principle possible given the appropriate variations in amino acid sequence.

While these properties of polypeptide catalysts make them distinct from even the most ideal conventional catalysts, they are not obviously advantages. As products of natural selection, the properties that enzymes have evolved are often undesirable to a bio-technologist. For example, instability is a desirable trait of a protein that functions in an inducible pathway. It is almost never a desirable trait in a protein destined for practical application.

Therefore, in order to take full advantage of the special properties of enzymes to make them the catalysts of choice to solve general synthetic problems, sooner or later we must face the problem of how to engineer the properties of the protein by deliberately altering the amino acid sequence of the protein. If "rules" become available that relate the amino acid sequence of a protein to its physical and catalytic behavior, any undesirable property of a protein could, at least in principle, be eliminated.

Science, not technology, presently limits efforts to engineer proteins. Virtually any protein can now be modified using recombinant DNA technology acting on a gene for the protein that has been cloned and expressed. However, while we are confident theoretically that there exists some set of alterations that will confer almost any desired property on any particular protein, in general we have no idea which alterations to make in the protein's structure to have the desired impact on the protein's behavior.

Nor are recombinant DNA techniques alone sufficient to aevelop an understanding of structure-function relationships in proteins. Naively, one might expect that such understanding would come from changing the structure of a protein and seeing what the effect is. However, even here it is not obvious what changes to make to obtain insights about how the behaviors of proteins (stability, quaternatery structure, substrate specificity) are determined by the amino acid sequence. First, the number of changes possible is astronomically large. For even a small protein like ribonuclease, there are 2356 single variants that are one amino acid different from the native structure, 5,505,972 that are different by two, and nearly 13 billion that are different by three. Introducing mutations via the commonly used strategies is not without effort, requiring several steps, including screening of colonies of bacteria that contain the mutated gene.(1) Even making a variant per week (a generous estimate), it is not likely that a "random walk" through the protein's structure will produce data that are useful for developing a general understanding of protein structure.

It is therefore not surprising that most laboratories exploring structure-function relationships in proteins target residues in the active site that, by inspection, are presumably "important" for catalytic function.(2) These residues are changed, and the altered protein is assayed for activity. If the enzyme is inactive, the residue changes is concluded to be "essential" for catalysis. The "inspect a crystal structure, change an active site residue, and kill the protein" strategy is obviously limited in what it can reveal about changing the structure of protein to improve its physical properties while retaining its catalytic activity. Unfortunately, this information is the kind that the biotechnologist finds most interesting.

The work that I will report today is based on the hypothesis that evolutionary processes have been engineering proteins for billions of years, introducing single amino acid changes into existing proteins to obtain new proteins with desired physical and catalytic properties. If this hypothesis is correct, studying how evolution engineers proteins is a strategy for developing insight that will allow the chemist to do the same. We shall first present evidence supporting this hypothesis, and shall then report our recent progress in synthesizing, cloning, and expressing genes for ribonuclease (RNAse), a protein especially well suited for exploiting "evolutionary guidance" in understanding structure-function relationships in proteins.

2. DOES EVOLUTION ENGINEER THE BEHAVIOR OF PROTEINS?

First, before examining the evolution of proteins to gain insight needed for engineering proteins, we must ask whether there is evidence that the properties of enzymes are indeed engineered by the process of natural selection.

Traits of enzymes are either selected or they are not, depending on whether they influence the survival of the host organism.(3) As natural selection is the only mechanism for obtaining functional behavior in living systems, only selected traits reflect function. Non-selected traits reflect history, including both the traits of ancestral enzymes and any constraints that may prevent those ancestral traits from "drifting."(3) They need not reveal anything fundamental about the workings of biological catalysts.

To use evolutionary history as a guide, we must know whether most of the detailed behavior of proteins reflects function achieved by natural selection, or merely reflects the random accumulation of "neutral" mutations by a process of "drift".(3) For example, if the stability characteristic of pancreatic RNAse is a selected trait, we might expect to see specific amino acids altered in the course of the evolution of the protein to create this stability. If the stability is not a selected trait, its presence in pancreatic RNAse must be an evolutionary accident. The evolution of proteins leading to the stable modern enzyme need not be instructive of how to engineer stability into a protein.

There are many subtle traits of enzymes that bio-organic chemists study where the same question can be asked. For example, dehydrogenases dependent on nicotinamide cofactors transfer only one of the two hydrogens at the 4 position of NADH. About half of the enzymes studied so far transfer the pro-R (or "A") hydrogen, about half transfer the pro-S (or "B") hydrogen.(4) Table 1 shows a list of dehydrogenases listed in the order of the catalog number used by the Enzyme Commission to identify the enzyme.

The impression gained from an inspection of Table I is that the hydrogen transfered by any particular dehydrogenase is "random". Therefore, this trait may not be the product of natural selection, and may not be able to provide fundamental information about protein catalysis. However, in the absence of an alternative testable hypothesis that interprets the same data in terms of function, we cannot be sure whether the appearance of "randomness" reflects actual randomness, or whether it simply reflects our ignorance of the underlying functional significance of the trait.

Table I

E.C.	Reference	Name	Stereochemistry
1.1.1.1	25	Alcohol dehydrogenase (Yeast)	A
1.1.1.3	26	Homoserine dehydrogenase	B
1.1.1.6	27	Glycerol dehydrogenase	A
1.1.1.8	28	Glycerol-3-phosphate dehydrogenase	B
1.1.1.26	29	Glyoxylate reductase	A
1.1.1.27	30	L-Lactate dehydrogenase	A
1.1.1.28	30	D-Lactate dehydrogenase	A
1.1.1.29	31	Glycerate dehydrogenase	A
1.1.1.30	32	3-Hydroxy butyrate dehydrogenase	B
1.1.1.35	33	3-Hydroxy acyl CoA dehydrogenase	B
1.1.1.37	34	Malate dehydrogenase	A
1.1.1.38	34	Malic enzyme	A
1.1.1.40	34	Malic enzyme (NADP)	A
1.1.1.51	35	β-Hydroxysteroid dehydrogenase	B
1.1.1.60	36	Tartronate semialdehyde reductase	A
1.1.1.62	37	Estradiol 17 β-dehydrogenase	B
1.1.1.64	35	Testosterone β-dehydrogenase	B
1.1.1.72	27	Glycerol dehydrogenase (NADP)	A
1.1.1.79	29	Glyoxylate reductase (NADP)	A
1.1.1.81	31	Hydroxypyruvate reductase	A
1.1.1.82	34	Malate dehydrogenase (NADP)	A
1.1.1.91	38	Aryl alcohol dehydrogenase	B
1.1.1.100	39	3-Oxoacyl acyl carrier protein reductase	B
1.1.1.108	40	Carnitine dehydrogenase	B
1.1.1.50	35	3-α-Hydroxysteroid dehydrogenase (P. test.)	B

Stereoselectivity of some dehydrogenases using NADH, listed by Enzyme Commission catalog number. By convention, the "A" and "B" hydrogens are the pro-R and pro-S hydrogens.

Recently, we proposed such a functional theory that suggests that dehydrogenase stereospecificity can be explained as a trait engineered by selective forces to obtain an enzyme with optimal catalytic efficiency. Our model was based on four hypotheses:(5)

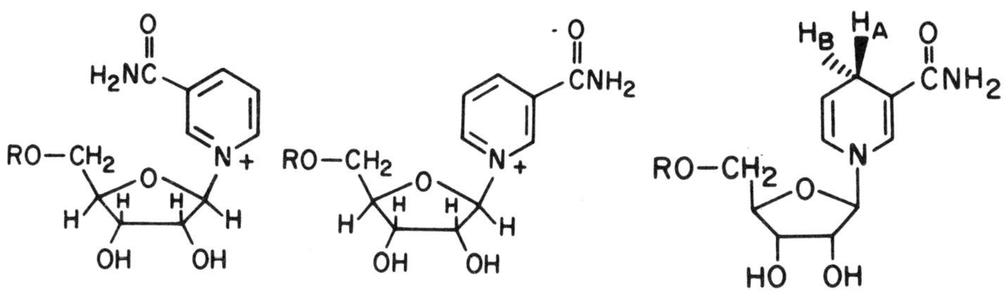

Figure 1: Syn and anti conformations of nicotinamide cofactors, and the two distinct hydrogens (A and B) of NADH.

(a) Dehydrogenases have evolved to reflect stereoelectronic principles. The "A" hydrogen is transfered if the enzyme has evolved to bind the cofactor in the "anti" conformation; the "B" hydrogen is transfered if the enzyme binds the cofactor in the "syn" conformation.(Figure 1).(6)

(b) The "anti" conformation of the cofactor is a weaker reducing agent than the "syn" conformation.

(c) Enzymes are catalytically optimal when they bind cofactor and substrates so that the free energy drop between enzyme-substrate and enzyme-product complexes reflects some of the chemical potential drop of the reaction under physiological conditions.(7)

(d) Dehydrogenases have evolved to be catalytically optimal.

These hypotheses suggest that the stereospecificity of dehydrogenases should correlate with the redox potential of the substrate that the enzyme had evolved to handle. Indeed, upon rearranging the enzymes listed in Table I so as to order them not by catalog number, but rather by the stability of their presumed natural substrate (Table II), a correlation between redox potential and stereospecificity is apparent. Reactive carbonyls are reduced by the pro-R hydrogen, unreactive carbonyls are reduced with the pro-S hydrogen. This correlation is fitted by 125 of the 130 enzymes studied so far.(8)

Table II

enzyme[a]	EC no.	$-\log E_{eq}$[b]	stereo-chem-istry[c]
glyoxylate reductase	1.1.1.26	17.5	*pro-R*
glyoxylate reductase ($NADP^+$)	1.1.1.79	17.5	*pro-R*
tartronate-semialdehyde reductase	1.1.1.60	13.3	*pro-R*
glycerate dehydrogenase[d]	1.1.1.29	13.3	*pro-R*
glycerol 1-dehydrogenase	1.1.1.72	12.8	*pro-R*
hydroxypyruvate reductase ($NADP^+$)[d]	1.1.1.81	12.4	*pro-R*
malate dehydrogenase	1.1.1.37	12.1	*pro-R*
malate dehydrogenase ($NADP^+$)	1.1.1.82	12.1	*pro-R*
malic enzyme	1.1.1.38	12.1	*pro-R*
malic enzyme ($NADP^+$)	1.1.1.40	12.1	*pro-R*
L-lactate dehydrogenase	1.1.1.27	11.6	*pro-R*
D-lactate dehydrogenase	1.1.1.28	11.6	*pro-R*
ethanol dehydrogenase (yeast)	1.1.1.1	11.4	*pro-R*
glycerol 2-dehydrogenase	1.1.1.6	11.3	*pro-R*
glycerol-3-phosphate dehydrogenase	1.1.1.8	11.1	*pro-S*
homoserine dehydrogenase	1.1.1.3	10.9	*pro-S*
carnitine dehydrogenase	1.1.1.108	10.9	*pro-S*
3-hydroxyacyl-CoA dehydrogenase	1.1.1.35	10.5	*pro-S*
3-hydroxybutyrate dehydrogenase	1.1.1.30	8.9	*pro-S*
3β-hydroxysteroid dehydrogenase	1.1.1.50	8.0	*pro-S*
testosterone dehydrogenase	1.1.1.64	7.6	*pro-S*
3-oxoacyl-ACP dehydrogenase	1.1.1.100	7.6	*pro-S*
β-Hydroxysteroid dehydrogenase	1.1.1.51	7.6	*pro-S*
oestradiol 17β-dehydrogenase	1.1.1.62	7.7	*pro-S*

Stereoselectivities of dehydrogenases arranged by the equilibrium constant $[NADH][carbonyl][H^+]/[NAD^+][alcohol]$ for the redox reaction that the enzyme has evolved to catalyze.

The fact that the apparently random data can be organized by a functional model suggests, but does not prove, that evolutionary selection pressures have selected for one stereochemical outcome over the other in different enzymes.

Further evidence that this stereoselectivity reflects selected function rather than random drift can be found by testing predictions that the functional theory makes, especially those that differ from those made by historical models. For example, one generalization regarding dehydrogenase stereoselectivity, known as "Bentley's first rule", is that enzymes from different organisms but acting on the same substrate all have the same stereochemistry.(9)

Historical models treat this as evidence that all enzymes acting on the same substrate are related, with the stereospecificity of the ancestral enzyme being conserved. Functional models such as ours argue that this reflects a selection for a particular stereochemistry for functional reasons related to the redox potential of the natural substrate, the same for all of the enzymes in question. If the functional theory is correct, we would expect that enzymes acting on substrates having redox potential at the break in the correlation would not be under strong selective pressure to favor one stereospecificity over the other. Therefore, we might expect that ethanol dehydrogenases, for example, to come in two stereochemical varieties if the functional model is correct, but not if a historical model is correct.

Therefore, we have examined a number of ethanol dehydrogenases, and find that indeed, some transfer the pro-S hydrogen (the enzyme from Drosophila melanogaster) while others transfer the pro-R hydrogen (the enzyme from yeast). In contrast, all malate dehydrogenases transfer the pro-R hydrogen, while all 3-hydroxyacylcoenzyme A and hydroxybutyrate dehydrogenases transfer the pro-S hydrogen. These data are consistent with our functional model, but not with simple historical models.

Even more subtle details of enzymic catalysis appear to reflect function, and therefore are selected. For example, the free energies of enzyme-substrate and enzyme-product (ES and EP) complexes also appear to reflect selective pressure seeking to evolve enzymes that are catalytically optimal. Based on a simple model (7), enzymes may be catalytically optimized by adjusting the free energies of ES and EP complexes so that they reflect part, but not all, of the chemical potential drop under physiological conditions. This is refered to as a "descending staircase" free energy profile, and is exemplified by the profile displayed by pyruvate kinase.(Figure 2) (7)

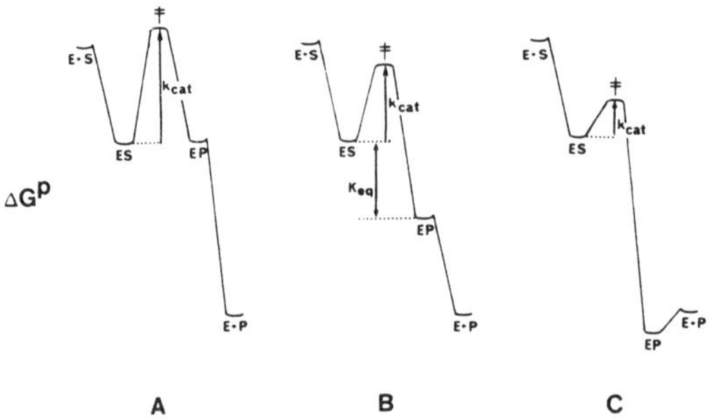

Figure 2: Three possible free energy profiles having different "internal equilibrium constants" between the bound states ES and EP. Profile B is the "descending staircase" profile that gives the fastest rate of catalysis (ref. 7). The ordinate represents physiological free energy, the standard state free energy corrected for physiological concentrations of S and P.

The chemical potential drop under physiological conditions reflects both the intrinsic free energies of the compounds, and their concentrations under physiological conditions. Therefore, the chemical potential drop is different under different conditions, and is reversed when the physiological flux is reversed. Thereofre, the descending staircase model predicts that isozymes whose physiological roles are to catalyze the same reaction in opposite directions should have different internal equilibrium constants. For example, there are two lactate dehydrogenases, one operating under oxidative conditions (from heart), believed to catalyze the conversion of lactate to pyruvate (10), the other under reductive conditions (from muscle), catalyzing the reverse reaction, the conversion of pyruvate to lactate.

The descending staircase model predicts that, for the heart enzyme, the free energy of the Enzyme-pyruvate-NADH complex should be lower than that of the Enzyme-lactate-NAD complex. For the muscle enzyme, the free energies of the two complexes should be reversed. (Figure 2) Estimates of the internal equilibrium constants for both enzymes verify this prediction.(5c,11) The internal equilibrium constant for the heart enzyme appears to be 1.5:1 downhill in the direction of pyruvate;(5c) that for the muscle enzyme appears to be 4:1 downhill in the direction of lactate.(11)

Therefore, enzymes seem to have evolved to adjust the free energies of bound species to reflect both the intrinsic free energies of the compounds that are substrates, but also the nature of the physiological surroundings.

These data do not "prove" functional theories of dehydrogenase stereoselectivity or internal thermodynamics, any more than they "disprove" historical models explaining the same data are incorrect. The statistics associated with stereochemistry requires that many more examples be collected before conclusions can be firmly drawn. Likewise, many more internal equilibrium constants must be measured to provide a large enough number of samples to be confident of our functional interpretation of internal equilibrium constants.

However, the behavior of dehydrogenases appears to conform more closely to functional models than to historical models. It is data such as these that provide the basis of our belief that, in general, enzymes have evolved to be highly refined catalysts optimally suited for effecting specific transformations on specific natural substrates under defined conditions. Thus, within the constraints of Darwinian evolutionary theory, it appears as if natural selection is capable of engineering even subtle details of enzymic behavior without major constraints. This in turn suggests that the process by which evolution engineers proteins is likely to be interesting to those seeking insight into how to engineer proteins in the laboratory.

3. SYNTHESIS, CLONING AND EXPRESSION OF A GENE FOR RNAse

We shall assume that evolutionary processes are constantly searching the "surface" that relates protein structure to protein function, optimizing the protein's structure for optimal survival under the particular conditions it finds itself. Thus, the evolutionary history of a protein that has recently evolved interesting behavior is interesting to study. If we can learn how evolution engineers proteins by changing one amino acid at a time, we may be able to gain insight that will permit the chemist to do the same.

To exploit evolutionary guidance as a rationale for learning about structure-function relationships in proteins, we need to study a protein which has an interesting and reasonably well defined recent evolutionary history, and which can be easily studied structurally and kinetically, and physically. We have chosen ribonuclease (RNAse) as such a system for study, and wish to report our progress in developing the system for structure-function studies. This progress includes synthesis of a gene coding for the protein designed to be easily altered, cloning of that gene into *E. coli*, and expressing the protein as a fusion protein with beta-galactosidase.

RNAse is a small protein with approximately 124 amino acids. It is found widely in the digestive tract of ruminants (artiodactyls), where it has undergone a rapid recent evolutionary divergence as ruminants themselves have diverged to occupy a wide range of environmental niches.(12) Thanks in large part to the extensive work of Beintema and coworkers,(13) the amino acid sequences of many of these modern RNAses are known. From there modern enzymes, a tree can be constructed (Figure 3) which includes hypothetical amino acid sequences of the ancestral RNAses from the (now extinct) ancestors of the modern ruminants. This tree contains a number of evolutionary paths connecting proteins which have altered properties. These include:

1. Primary Structure: Some 70% of the residues in RNAse have been varied by evolutionary processes within the sequences that have been examined from higher organisms.(12)

2. Stability: Certain RNAses are exceptionally stable, others are less stable, all within the same evolutionary series.

3. Quaternary structure: Certain of the related RNAses are monomers, others are dimers covalently linked by disulfide bonds, while still others may transiently form dimers.(14)

4. Anti-tumor activity: Certain RNAses have potent anti-tumor activity, others have no anti-tumor activity at all.(15)

Of course, we would be unable to predict with any particular RNAse what residues to change to alter the proteins stability or quaternary structure simply by inspection of a crystal structure. Further, given the fact that two thirds of the structure is a candidate for alteration, a random walk is unthinkable. Even making good guesses based on fundamental organic principles is unlikely to lead to a dimeric pancreatic RNAse from a monomer it likely to produce countless of inactive failures before the goal is achieved.

However, within the tree in Figure 3, paths connect proteins with such different properties. Proteins that are stable are connected with those that are unstable. Proteins that are monomers are connect by paths involving single amino acid changes with those that are dimers. The paths proceed by single amino acid substitutions, and the intermediate proteins can be made with molecular biological techniques and studied, providing a new rationale for studying the relationships between protein structure and function.

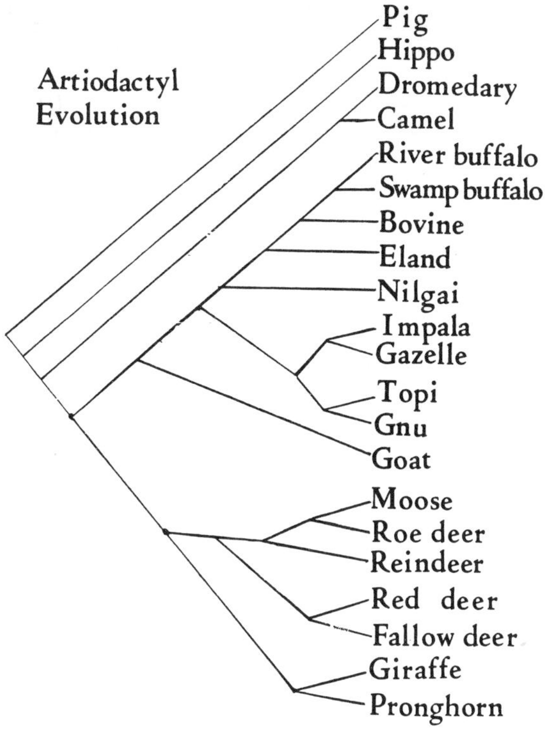

Figure 3. A hypothetical evolutionary tree connecting species whose RNAses have been sequenced. Branch points represent ancestral RNAses with hypothetical sequences that may be deduced from the sequences of the modern enzymes.

In addition to a new rationale for studing structure-function relationships in proteins, we are implementing a new approach to permit rapid alteration of RNAse. This approach relies on a totally synthetic gene.(16) The redundancy of the genetic code is exploited in the design of the synthetic gene to introduce unique restriction sites at regular intervals. These restriction sites facilitate modification of the gene. To introduce a mutation, the gene is cut by digestion with the restriction endonucleases whose sites flank the region to be replaced. The gene is then religated with a segment of synthetic duplex DNA that codes for the desired mutation.

The following criteria were used to construct a sequence best designed to facilitate rapid and controlled alteration:

1. Sites for commercially available restriction endonucleases should be distributed at intervals in the sequence.

2. These restriction sites must be unique; no restriction site should appear more than once in the gene.

3. Repetitions of sequences (or complementary sequences) longer than four base pairs should be eliminated.

4. Provisions should be made for the cloning of the gene into a cloning vector between EcoR1 and BamH1 sites, and into expression vectors between EcoR1 and HindIII sites.

The sequence of the gene for RNAse is in Figure 4.

```
                                 1
      5'    Ile Glu Gly Arg  Lys Glu Thr Ala Ala Ala Lys Phe Glu Arg      10
      AATTC-ATT-GAG-GGT-CGT==AAA-GAA-ACT-GCG-GCC-GCA-AAA-TTT-GAA-CGT-
          G-TAA-CTC-CCA-GCA==TTT-CTT-TGA-CGC-CGG-CGT-TTT-AAA-CTT-GCA-
        EcoRI      MnlI                          NotI,     DraI
                                                 XmaIII

      Gln His Met Asp Ser Ser Thr Ser Ala Ala Ser Ser Ser Asn Tyr Cys     26
 48   CAG-CAT-ATG-GAC-AGT-TCC-ACG-TCC-GCC-GCT-TCT-TCT-TCG-AAT-TAT-TGT-
      GTC-GTA-TAC-CTG-TCA-AGG-TGC-AGG-CGG-CGA-AGA-AGA-AGC-TTA-ATA-ACA-
          NdeI                                   MboII   TaqI

      Asn Gln Met Met Lys Ser Arg Asn Leu Thr Lys Asp Arg Cys Lys Pro     42
 93   AAT-CAA-ATG-ATG-AAG-TCT-AGA-AAC-CTC-ACC-AAG-GAC-CGT-TGC-AAG-CCC-
      TTA-GTT-TAC-TAC-TTC-AGA-TCT-TTG-GAG-TGG-TTC-CTG-GCA-ACG-TTC-GGG-
                              Xba I                Sau96 I

      Val Asn Thr Phe Val His Glu Ser Leu Ala Asp Val Gln Ala Val Cys     58
138   GTT-AAC-ACT-TTT-GTG-CAC-GAA-TCC-TTA-GCG-GAT-GTG-CAA-GCC-GTT-TGC-
      CAA-TTG-TGA-AAA-CAC-GTG-CTT-AGG-AAT-CGC-CTA-CAC-GTT-CGG-CAA-ACG-
       Hpa I              HgiA I  Hinf I Dde I   Fok I

      Ser Gln Lys Asn Val Ala Cys Lys Asn Gly Gln Thr Asn Cys Tyr Gln     74
183   AGC-CAA-AAA-AAC-GTT-GCA-TGC-AAG-AAT-GGC-CAA-ACA-AAC-TGT-TAC-CAA-
      TCG-GTT-TTT-TTG-CAA-CGT-ACG-TTC-TTA-CCG-GTT-TGT-TTG-ACA-ATG-GTT-
      Bbv I                 Sph I           Bal I  Tth 111II
      Fnu 4HI

      Ser Tyr Ser Thr Met Ser Ile Thr Asp Cys Arg Glu Thr Gly Ser Ser     90
228   TCG-TAC-TCA-ACT-ATG-TCG-ATC-ACA-GAC-TGC-AGG-GAG-ACT-GGA-AGC-TCA-
      AGC-ATG-AGT-TGA-TAC-AGC-TAG-TGT-CTG-ACG-TCC-CTC-TGA-CCT-TCG-AGT-
        Rsa I         Taq I  Mbo I       Pst I              Alu I

      Lys Tyr Pro Asn Cys Ala Tyr Lys Thr Thr Gln Ala Asn Lys His Ile    106
273   AAA-TAT-CCA-AAC-TGC-GCA-TAT-AAA-ACT-ACC-CAG-GCA-AAC-AAA-CAC-ATC-
      TTT-ATA-GGT-TTG-ACG-CGT-ATA-TTT-TGA-TGG-GTC-CGT-TTG-TTT-GTG-TAG-
                        Hha I               BstN I

      Ile Val Ala Cys Glu Gly Asn Pro Tyr Val Pro Val His Phe Asp Ala    122
318   ATC-GTC-GCG-TGT-GAA-GGT-AAC-CCC-TAT-GTC-CCG-GTT-CAC-TTT-GAC-GCA-
      TAG-CAG-CGC-ACA-CTT-CCA-TTG-GGG-ATA-CAG-GGC-CAA-GTG-AAA-CTG-CGT-
              FnuD II     Bst EII            Hpa II       Hga I SfaN I

                              3'
      TCT-GTC-TAA-TAA-G____5'
363   AGA-CAC-ATT-ATT-CCTAG
                            BamH I
```

Figure 4: Sequence of one synthetic gene coding for RNAse

Oligonucleotides were synthesized using phosphoramidite reagents to elongate nucleotide chains attached to solid supports.(17) While most of the synthetic work was done by hand, more recent synthesis has been performed on an Applied Biosystems automated DNA synthesizer. Synthetic oligonucleotides were then released from the supports and purified by thin layer chromatography. Purified fragments were sequenced by the Maxam-Gilbert method (18) prior to subsequent ligation. Oligomers were then enzymatically ligated to give the assembled gene.

The synthetic gene was incorporated into a plasmid, and the plasmid was introduced into _E. coli_ via the "calcium shock" method. We adapted a procedure of Hanahan's (19), in which the dimethylsulfoxide was replaced by sulfolane. Transformants were selected for ampicillin resistance, and the plasmids within them were identified by restriction digestion and sequencing.

We have constructed several genes using this general method. The first coded for the ribonuclease S-protein, the proteolytic fragment of RNAse that consists of residues 21-124.(15) Subsequent constructions code for the complete RNAse protein.

As often the case with the production of eukaroytic proteins cloned in _E. coli_, one challenge is to express the cloned gene. RNAse presents some special problems in this regard, as an active ribonuclease may digest the messenger RNA that codes for its production. While a complete description of the work that we have done cannot be presented here, I would like to show one example of how the protein may be expressed.

One of the constructions prepared (shown in Figure 4) was designed to create a "fusion protein" in which RNAse is covalently fused to beta-galactosidase, a protein found naturally in the expression host, _E. coli_. The two proteins are joined by a short peptide that is the recognition sequence for Factor Xa, a protease that cleaves after the sequence Ile-Glu-Gly-Arg. Nagai and Thogersen (20) have suggested this protease as generally useful for cleaving fusion proteins.

An SDS electrophoresis gel of the RNAse fusion protein expressed in _E. coli_ shows a band at the appropriate molecular weight. The fusion protein is expressed in substantial quantities, and is found primarily as inclusion bodies within the _E. coli_ cell. The band is identified both by molecular weight, and by "Western blotting" with radioactive antibodies against RNAse.

4. CONCLUSION

We have presented evidence that natural selection has finely tuned the properties of enzymes so as to engineer in those proteins the behaviors that contribute most to the survival of the host organism. Using this as a hypothesis, we have introduced a new strategy for the use of recombinant DNA methods to study structure-function relationships in proteins. Using both synthetic and enzymatic methods, we have synthesized a gene coding for RNAse, a gene that is specially designed to permit rapid alteration. This gene has been cloned and expressed in *E. coli*.

We are now prepared to retrace the recent evolution of RNAse, using the hypothetical evolutionary tree deduced from the amino acid sequences of the RNAses that are the products of the evolutionary divergence. We hope that by retracing this tree, we will be able to gain insight into the process by which nature has engineered the properties of RNAse. We are hopeful that our next report will contain evidence that this approach has been fruitful in allowing us to mimic the engineering done by Nature in our chemical laboratory.

ACKNOWLEDGEMENT

This work would have been impossible without the support of the National Science Foundation (CHE-8213267 and PCM-8111659), the National Institutes of Health (GM-30110), and many private organizations, including the Searle Foundation, Stuart Pharmaceuticals, Alfred Bader, the Dreyfus Foundation, the Petroleum Research Fund, the Research Corporation, the Clay Science Fund, the Anna Fuller Fund, the Milton Fund, Owens Corning Fiberglass, the Xerox Corporation, the Harvard Society of Fellows, and a gift in memory of Baron Langer von Langensdorf. S. B. is a Fellow of the Alfred Sloan Foundation.

REFERENCES

(1)(a) Gillam, S.; Waterman, K.; Smith, M. *Nucl. Acids Res.*, 1975, *2*, 625. (b) Dalbadie-McFarland, G.; Cohen, L. W.; Riggs, A. D.; Morin, C.; Itakura, K.; Richards, J. H. *Proc. Nat. Acad. Sci.*, 1982, *79*, 6409-6413.

(2)(a) Wallace, R. B.; Schold, M.; Johnson, M. J.; Dembek, P.; Itakura, K. *Nucl. Acid Res.*, 1981, *9*, 3647. (b) Inouye, S.; Soberon, X.; Franceschini, T.; Nakamura, K.; Itakura, K.; Inouye, M. *Proc. Nat. Acad. Sci.*, 1982, *79*, 3438. (c) Sigal, I. S.; Harwood, B. G.; Arentzen, R. *Proc. Nat. Acad.*

Sci., 1982, 79, 7159. (d) Winter, G.; Fersht, A. R.; Wilkinson, A. J.; Zoller, M.; Smith, M. Nature, 1982, 299, 756. (e) Villafranca, J. E.; Howell, E. E.; Voet, D. H.; Strobel, M. S.; Ogden, R. C.; Abelson, J. N.; Kraut, J. Science, 1983, 222 782-788.

(3)(a) Kimura, M. Nature, 1968, 217, 624. (b) Jukes, T. H. Science, 1980, 210, 973. (c) Kimura, M. The Neutral Theory of Molecular Evolution, N.Y., 1983, Cambridge University Press.

(4)(a) You, K. Meth. Enzymol., 1982, 87, 101-126. (b) You, K.; Arnold, L. J.; Allison, W. S.; Kaplan, N. O. Trends in Biochem. Sci., 1978, 3, 365-368.

(5)(a) Benner, S. A. " Experientia, 1982, 38, 633-37. (b) Benner, S. A.; Stackhouse, J. Chemical Approaches to Understanding Enzymic Catalysis, Proceedings of the 26th OHOLO Conference, Zichron Yaacov, Israel, B. S. Green, Y. Ashani, and D. Chipman, editors, 1982, Elsevier, 32-43. (c) Nambiar, K. P.; Stauffer, D. M.; Kolodziej, P. A.; Benner, S. A. J. Am. Chem. Soc., 1983, 105, 5886-5890.

(6) Garavito, R. M.; Rossmann, M. G.; Argos, P.; Eventoff, W. Biochem., 1977, 16, 5065-5071.

(7) Stackhouse, J.; Nambiar, K. P.; Burbaum, J.; Stauffer, D. M.; Benner, S. A. J. Am. Chem. Soc., 1985, 107, 2757-2763.

(8) Benner, S. A.; Nambiar, K. P.; Chambers, G. K. J. Am. Chem. Soc., 1985, in press.

(9) Bentley, R. Molecular Asymmetry in Biology, 1970, 2, 1-89.

(10) Cahn, R. D.; Kaplan, N. O.; Levine, L.; Zwilling, E. Science, 1961, 136, 962-969.

(11) Gutfreund, H. Prog. Biophys. Mol. Biol., 1975, 29, 161-195.

(12) Barnard, E. A. Nature, 1969, 221, 340.

(13) J. J. Beintema and J. A. Lenstra, in Macromolecular Sequences in Systematics and Evolutionary Biology, M. Goodman, Ed., (Plenum, N. Y., 1981).

(14) DiDonato, A.; D'Alessio, G. Biochem. Biophys. Res. Comm., 1973, 55, 919-928.

(15) Vescia, S.; Tramontano D. Augusti-Tocco, G.; D'Alessio, G. Cancer Res., 1980, 40, 3740-3744.

(16) Nambiar, K. P.; Stackhouse, J.; Stauffer, D. M.; Kennedy, W. G.; Eldredge, J. K.; Benner, S. A. Science, 1984, 222, 1299-1301.

(17) Beaucage, S. L.; Caruthers, M. H. Tetrahedron Lett., 1981, 22, 1859-1862.

(18) Maxam A. M.; Gilbert, W. Meth. Enzymol, 1980, 65, 499.

(19) Hanahan, D. J. Mol. Biol., 1983, 166, 557-580.

(20) Nagai, K.; Thogersen, H. C. Nature, 1984, 309, 810-812.

Posters

BAKER'S YEAST MEDIATED SYNTHESIS OF PROTECTED α-HYDROXYALDEHYDES

G. Guanti,* L. Banfi, E. Narisano, and A. Guaragna
Institute of Organic Chemistry, University of Genova
Centro C.N.R.
Palazzo delle Scienze, Corso Europa
16132 Genova
Italy

ABSTRACT. Some β-substituted protected α-hydroxyaldehydes with high optical purity have been readily prepared by acylation of the lithium salt of di-*p*-tolylthiomethane and successive baker's yeast reduction of the corresponding ketones.

α-Hydroxyaldehydes are compounds of primary importance for the synthesis of many biologically active derivatives.[1] Unfortunately, although the (*R*)-glyceraldehyde and (*S*)-lactaldehyde are readily available substrates, more difficulty exists in other cases. More recently direct access to (*S*)-glyceraldehyde,[2] (*R*)-lactaldehyde,[3] and other α-hydroxyaldehydes[4] has been reported. However, no simple methods are till now available for obtaining optically pure β-substituted-α-hydroxyaldehydes wich represent a very useful chiral building block.

As an extention of our previous research on α-hydroxyaldehydes[5] we wish to report here on a simple and straightforward synthesis of some β-substituted protected α-hydroxyaldehydes. The strategy is based on the following disconnection.

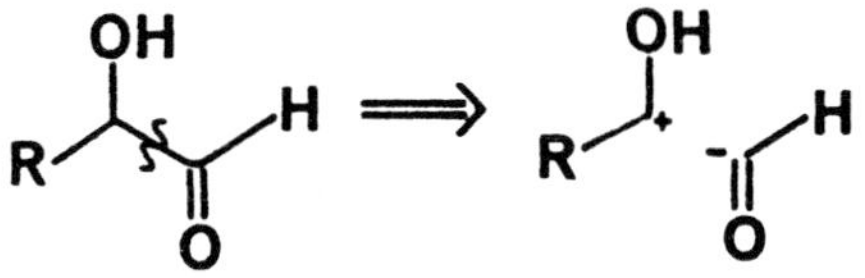

The procedure involves the use of α-substituted ethyl esters as starting materials for acylating a synthon of a formyl anion (in our case we used the lithium salt of di-*p*-tolylthiomethane) (Table 1). Then the corresponding ketones were reduced with baker's yeast to the corresponding alcohols and the asymmetric induction was determined by NMR (Table 2).

M. P. Schneider (ed.), Enzymes as Catalysts in Organic Synthesis, 343–345.

TABLE 1. Synthesis of 1,1-di-*p*-tolylthio-2-ketones

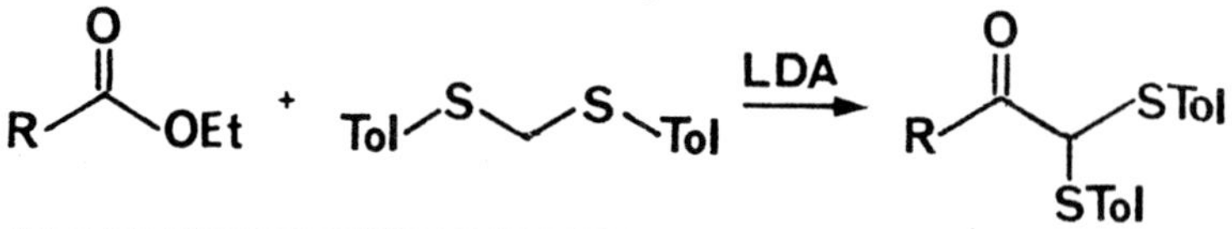

R	CH_3	CH_3CH_2	CF_3	$CH_3OCH_2OCH_2$	CH_2F	CH_2Cl
Yield	70%	68%	60%	85%	60%	70%

TABLE 2. Enantioselective reduction of 1,1-di-*p*-tolylthio-ketones with baker's yeast

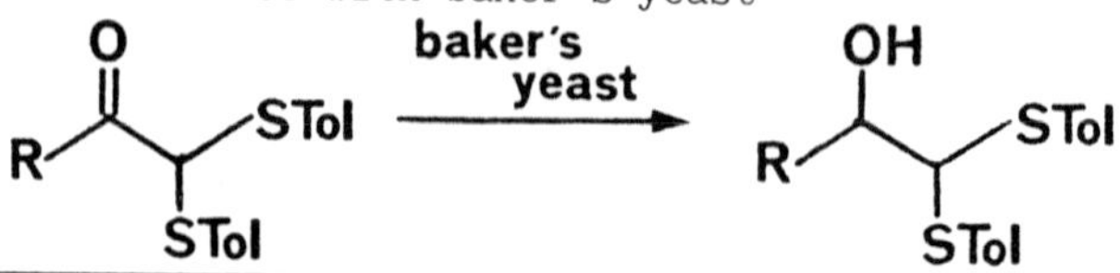

R	Time (days)	Temp.	Yield[a]	e.e.
CH_3	3.5	27°C	50% (85%)	>95%[b]
CH_3CH_2	7	27°C	<10%	-
CF_3	3	27°C	25%	>95%[c]
CH_2F	4	30°C	42% (70%)	>95%[c]
$CH_3OCH_2OCH_2$	6	30°C	50% (85%)	>95%[b]
CH_2Cl	7	30°C	<5%	-

[a]Yield from non-recovered starting material in brackets. [b]Determined by ^1H NMR in the presence of $Eu(hfc)_3$. [c]Determined by ^1H NMR of Mosher's esters.

The results till now collected indicate that the rate of reduction is strongly dependent on the length of the hydroxycarbon chain and on the type of substituent in the chain. The reduction yields are satisfactory for R = CH_3, CH_2F, CF_3, $CH_2OCH_2OCH_3$ but fall when R changes to CH_2CH_3, CH_2Cl. However a very striking result is the large enantiomeric excess observed in the cases when reduction occurs with the formation of practically only one enantiomer.

In the light of this encouraging result we are now working in order to determine the absolute configuration of the compounds and their transformation into many other derivatives to be used as simple building block in the synthesis of natural compounds.

REFERENCES

1. G. Fronza, C. Fuganti, P. Grasselli, and G. Pedrocchi-Fantoni, *Carbohydrate Research*, **136**, 115 (1985) and previous papers.
2. M. E. Jung and T. J. Shaw, *J. Am. Chem. Soc.*, **102**, 6304 (1980).
3. L. E. Overman, K. L. Bell, and F. Ito, *J. Am. Chem. Soc.*, **106**, 4192 (1984); M. Braun and W. Hild, *Angew. Chem. Int. Ed. Engl.*, **23**, 723 (1984) and ref. 8 herein.
4. Y. Le Merrer, A. Duréault, C. Gravier, D. Languin, and J. C. Depezay, *Tetrahedron Lett.*, **26**, 319 (1985).
5. G. Guanti, E. Narisano, F. Pero, L. Banfi, and C. Scolastico, *J. Chem. Soc., Perkin 1*, 189 (1984).

ENANTIOSELECTIVE PLE-CATALYZED HYDROLYSIS OF MESO-DIMETHYL TETRAHYDROPHTHALATE ON A 100 MOLE SCALE - PROTECTION OF THE ENZYME BY ADDITION OF BOVINE SERUM ALBUMIN

B. F. Riefling and W. K. Brümmer
E. MERCK DARMSTADT, Frankfurter Str. 250, D-6100 Darmstadt
H. J. Gais
Institut für Organische Chemie der Technischen Hochschule, Petersenstr. 22, D-6100 Darmstadt
W. Germany

Enantioselective saponification of meso-dimethyl tetrahydrophthalate (1) with commercial pig liver esterase (PLE) results in practically exclusive attack at the (1R)-ester group to afford the enantiomerically pure (1S, 2R)-monoester 2 in 93 % yield (e. e. > 96 %).

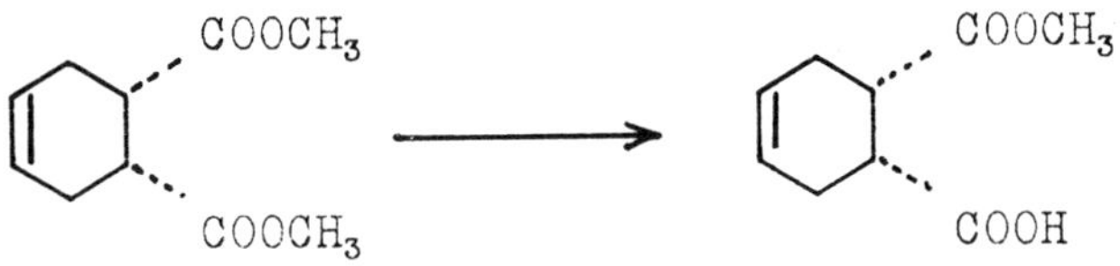

To prevent inactivation of the enzyme, bovine serum albumin (BSA) was added. The hydrolysis of 20 kg diester 1 in the presence of 3 g PLE (390000 U) + 100 g BSA in pH 7.5 - (phosphate) buffer, titrated with 50 l 2n NaOH, thus took place within 11 days until completion.
When BSA was omitted (on a smaller scale) the saponification reaction came almost to a halt after 4 days. However, consecutive addition of BSA caused an immediate and full reconstitution of the enzyme.
The enzyme could be recycled without major loss of activitiy by ultrafiltration of the reaction mixture and precipitation with ammonium sulphate.

References:

H. J. Gais, Habilitation Thesis Darmstadt 1981
H. J. Gais and K. L. Lukas, Angew.Chem.Int.Ed.Engl. 23 (1984) 142

M. P. Schneider (ed.), Enzymes as Catalysts in Organic Synthesis, 347.

ENANTIO- AND DIASTEREOSELECTIVITY OF MICROSOMAL EPOXIDE HYDROLASE: POTENTIAL APPLICATIONS TO THE PREPARATION OF NON-RACEMIC EPOXIDES AND DIOLS

G. Berti
Institute of Organic Chemistry, Faculty of Pharmacy
University of Pisa
Via Bonanno, 6
56100 Pisa, Italy

ABSTRACT. The conversion of epoxides into *trans*-diols, catalyzed by microsomal epoxide hydrolase (MEH) is reviewed from the point of view of the relations between substrate structure and the regio-and stereoselectivity of the enzyme. Several examples of reactions exhibiting a high enantioselectivity are given and the possibility of using the method for the purpose of preparing chiral epoxides and diols of high optical purity is discussed.

Enzymatic routes to epoxides and diols. Epoxides are very useful and ex-extensively employed intermediates for the synthesis of biologically important compounds and their availability in optically pure form greatly increases their usefulness as synthons. Several 'chemical' methods are known for preparing them from resolved precursors, but, with the notable exception of the excellent Sharpless asymmetric epoxidation (1) that is limited to allylic alcohols, no efficient resolutions or asymmetric syntheses are as yet known for the preparation of other types of optically active epoxides, as pointed out by the most recent review of the field (2). Microbiological epoxidations on the other hand can provide an easy access to simple epoxides of high optical purity, as shown by several reported cases (3), but the preparative scope and the limitations of these techniques have so far been studied only in a very limited way and would certainly deserve a more thorough investigation.

An enzymatic method alternative to the microbiological one could be found in the use of the important metabolizing enzymes cytochrome P-450 dependent monooxygenases, that convert alkenes and arenes into epoxides, and epoxide hydrolases that further transform epoxides into *trans*-diols. These enzymes are widely present in the microsomal fractions of animal tissues, particularly in the liver, and are often highly regio-, diastereo- and enantioselective in their reactions. Unfortunately, monooxygenases, that exhibit significant, but still mostly unexplored stereoselectivities, appear to have very little, if any, preparative value, owing to the fact that the hydrolytic step is usually much faster than the oxidative one, so that the epoxide cannot accumulate in the incubation medium (a disadvantage in comparison with bacterial epoxida-

M. P. Schneider (ed.), Enzymes as Catalysts in Organic Synthesis, 349–354.

tion, since epoxide hydrolase activity is usually absent in microorganisms). Furthermore, microsomal epoxidation requires expensive NADPH as a co-factor. Epoxide hydrolase promoted hydrolysis, on the other hand, could offer an inetersting novel approach to the preparation of some epoxides and diols of high optical purity. It is the purpose of this paper to shortly review some aspects of the stereoselectivity of microsomal epoxide hydrolase (MEH), of its reaction mechanism, and to examine its potential use for the preparation of chiral compounds in high enantiomeric excess.

Epoxide hydrolases. At least three types of epoxide hydrolases have been described, differing in their substrate specificities: the microsomal and the cytosolic hydrolases and a specific enzyme that hydrolyzes 5,6-epoxysteroids. MEH is by far the best characterized one (4). It is a single chain polypeptide of about 50,000 molecular weight, not requiring co-factors, and is deeply buried in the membrane of the endoplasmic reticulum, from which it can be freed by detergents and purified through several procedures, even if at the price of a substantial loss of activity. However, a crude microsomal preparation, obtained by ultracentrifugation, or even the 9000xg centrifugation surnatant of homogenized liver (S-9 fraction) can be used for preparative purposes, with little difference from the use of purified enzyme. The pH optimum of the isolated enzyme and of the enzymic activity in microsomes is between 8.9 and 9.4.

Mechanism of hydrolysis. It is by now firmly established that MEH promoted hydrolysis proceeds through a general base catalyzed mechanism (5), involving a histidine unit present in the active site, the imidazole ring of which provides nucleophilic assistance to the attack by a water molecule on one of the oxirane ring carbon atoms and possibly transfers a proton to the anionic oxygen originating in the ring opening process.

Chemoselectivity. Owing to the role in detoxifying foreign compounds, epoxide hydrolases act on a wide range of substrates, but with several limitations. MEH hydrolyzes mono- and cis-disubstituted epoxides, whereas trans-disubstituted, tri- and tetrasubstituted ones are mostly unreactive. Cytosolic epoxide hydrolase, on the other hand, hydrolyzes much better trans- than cis-stilbene oxide (6). The presence of a lipophilic substituent near the oxirane ring enhances the MEH hydrolysis rate, whereas polar groups, such as hydroxyl as found in allylic and homoallylic epoxyalcohols and in epoxy diols, strongly inhibit the enzymatic reaction. This and other evidence point to an essentially hydrophobic nature of the MEH active site, in accordance with the fact that most of its xenobiotic substrates are apolar compounds that need to be converted into more hydrosoluble metabolites in order to facilitate elimination.

Regio- and enantioselectivity. Monosubstituted epoxides are hydrolyzed by MEH exclusively through attack by water on the primary oxirane carbon, in accordance with a base promoted S_N2-type mechanism (7).A significant preference for attack on the (R)-enantiomer has been observed in racemic styrene oxides (8) and prelimary results show a similar trend in aliphatic terminal epoxides (9). This allows to isolate (R)-1,2-diols of fairly high optical purity during the first stage of the reaction and (S)-epoxide with an enantiomeric excess (e.e.) increasing with increasing conversion, a typical case of kinetic resolution.

Also the high preference often exhibited by MEH for attack at one of the two oxirane carbons of opposite chirality in *cis*-disubstituted epoxides could be very useful for asymmetric synthesis purposes. In the large majority of cases a preference for attack at the (S) carbon to give the (R,R)-diol is found, particularly when the oxirane ring is fused to a six-membered ring. Thus, cyclohexene oxide, a *meso* compound, is opened by MEH to give the (R,R)-and (S,S)-*trans*-cyclohexane-1,2-diols in a ratio of 85:15 (10), a very simple way to prepare this compound with an e.e. of 70%. This has been rationalized by assuming that the epoxycyclohexane molecule fits into the MEH active site preferentially in its conformation in which the six-membered ring has the 3,4-M helicity and undergoes diaxial opening by attack of a water molecule on the (S) carbon, this orientation being determined by the relative position of the catalyzing histidine unit with respect to the complexed epoxide molecule (11). This hypothesis is confirmed by the results obtained with substituted epoxycyclohexanes and with 3,4-epoxytetrahydropyran and its derivatives, for which enantiomers having the 3,4-M helicity (or the equivalent 1,2-M one for tetrahydropyrans) are usually hydrolyzed much faster than their antipodes. This is illustrated,for instance, by the kinetic resolution of racemic *trans*-4-*tert*-butyl-1,2-epoxycyclohexane, in which the (+)(1S,2R,4S) enantiomer having the 3,4-M helicity is hydrolyzed at a rate about 40 times as fast as its antipode to give near 50% conversion the (-)(1R,2R,4S)-diol of about 90% e.e., the (-)(1R,2S,4R)-epoxide being recovered unchanged (11). In the corresponding *cis* racemate, the MEH promoted hydrolysis is much slower and (+)(1S,2S,4R)-diol of 70% e.e. is obtained. This is an exception to the rule of preferential formation of the (R,R)-diol and reveals a second important general feature of MEH substrate specificity, overwhelming the effect of ring helicity: a bulky apolar substituent preferentially fits the active site when situated to the right back side of the oxirane ring, viewed with oxygen on the top side, a fact that implies a deep hydrophobic pocket

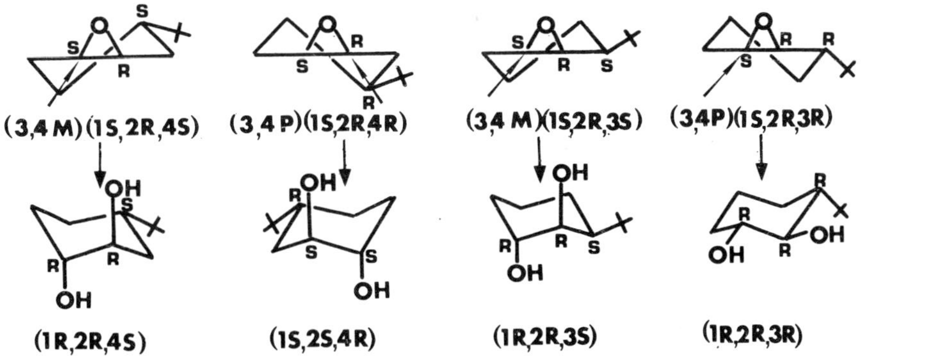

in the corresponding region of the active site. This is confirmed by the hydrolysis of 3-*tert*-butyl-1,2-epoxycyclohexane, in which the (+) (1R,2R,3S) diaxial diol of 96% e.e. and the optically pure (+)(1R,2R, 3R) diequatorial diols were obtained, respectively from the racemic *cis*- and *trans*-epoxides, at about 50% conversion (12). The latter case is an unusual example of diequatorial opening of an epoxycyclohexane. Other results agreeing with our hypotheses were obtained, for instance, in the kinetic resolutions of racemic *cis*-3-bromo-1,2-epoxycyclohexane (9) and *trans*-3,4-dimethyl-1,2-epoxycyclohexane (13).

3,4-Epoxytetrahydropyran exhibits a particularly interesting behaviour (14). Its two enantiomers are good substrates for MEH and are hydrolyzed by it in an entirely regioconvergent fashion, attack by water occurring at similar rates on both at the (S) carbon, that is exclusively at C-4 in the (3R,4S) enantiomer and at C-3 in the (3S,4R) one, so that the (-)(3R,4R) -diol is the sole product at any stage of conversion, a most convenient way to obtain this compound in optically pure form from the easily available racemic epoxide. Here again the preference for diaxial opening on the 1,2-M conformer is of paramount importance.

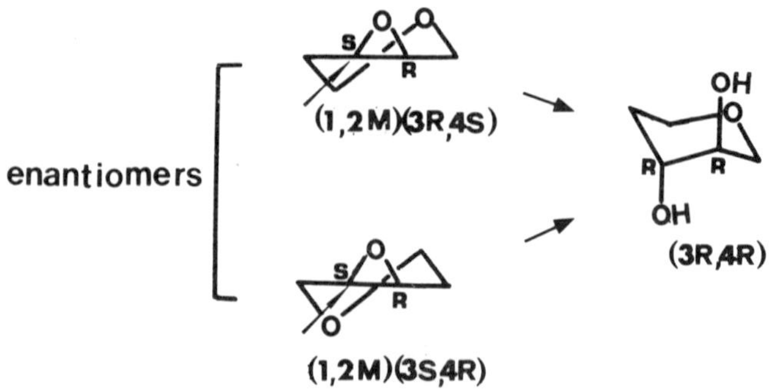

A high enantiodifferentiation has also been observed for *trans*-3,4-epoxy-2-methyltetrahydropyran from which the (-)(2R,3R,4R)-diol of 98% e.e. was obtained after about 40% conversion (15). Preliminary work has also shown that enzymatic hydrolysis could be of some interest for the synthe-

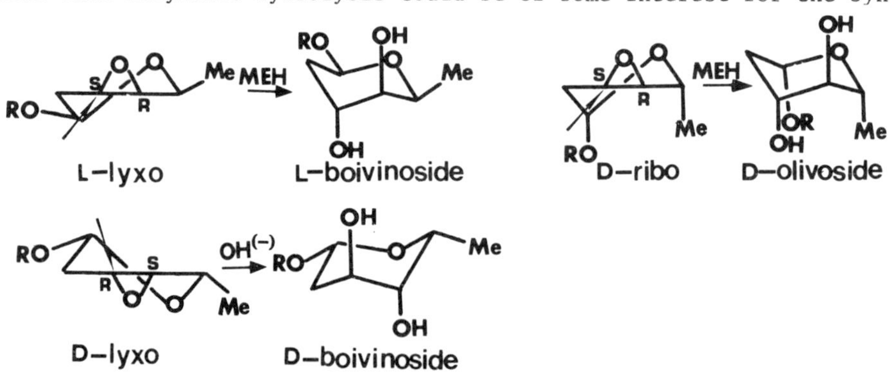

sis of anomalous sugars. For instance the 2,6-dideoxyhexopyranosides L-boivinosides and D-olivosides can be obtained in fair optical yields, respectively from the easily available racemic lyxo- and ribo-2,6-dideoxy-3,4-anhydrohexopyranosides, through partial hydrolysis with MEH. The D-boivinoside is obtained by alkaline hydrolysis of the recovered D-lyxo epoxide (9).

The reaction of MEH with polycyclic arene oxides shows variable enantioselectivities. It is low with naphthalene-1,2-oxide, but the hydrolysis of racemic benzo(a)pyrene-4,5-oxide is highly enantioconvergent yielding the (-)(R,R)-dihydrodiol of 78% e.e. (16).. The anti-epilectic drug carbamazepine is metabolyzed in humans with formation of the (S,S)-trans-dihydrodiol, via the intermediate meso-9,10-epoxide (9). This is one further case of exception to the rule of attack at the (S) carbon, together with that of the structurally similar phenanthrene-9,10-oxide that is converted into the (S,S)-diol, even if in an optical yield of only 25%.

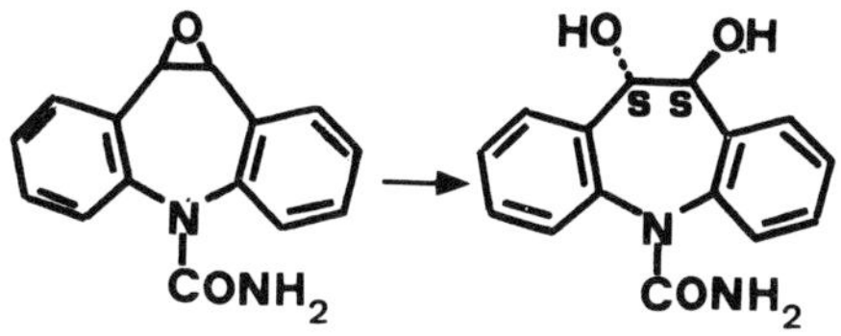

Perspectives of preparative uses of MEH. The data presented above and other literature reports clearly show that MEH promoted reactions are often highly stereoselective and in several cases produce trans-1,2-diols and epoxides of high optical purity. Most of them, however, have so far been obtained at the millimolar scale, without preparative purposes. The present report wants to suggest that at least in some cases there should not be particular difficulties in scaling up incubations to provide access to larger, even if probably still limited amounts of products by a simple method that could be competitive with more traditional approaches. This technique could have several advantages.

1) The enzyme source, animal liver, is easily available anywhere at negligible cost, and its processing does not require lengthy operations, nor special equipment, except a homogenizing device and a refrigerated centrifuge. Crude 9000xg surnatant can often be used, even if in some cases purified microsomes, requiring ultracentrifugation, could be preferable.

2) The enzyme is remarkably stable and crude as well as purified preparations can be stored for a long time under refrigeration. Less than 10% loss of activity was found after storage for 2 months at 0°C in a phosphate buffer (4). Incubations have been carried out for up to 24 hr at 36°C with limited decrease in activity.

3) MEH does not require co-factors, a great advantage for an enzymatic preparative procedure.

4) Microsomal preparations can be immobilized on several types of car-

riers; this allows reuse of the enzymatic preparation.
5) The method can be used both for the near quantitative conversion of a *meso* epoxide into a large excess of one enantiomer of the glycol, or for the kinetic resolutions of racemic epoxides with recovery of resolved diol and of one enantiomer of the epoxide. Their separation is very easy owing to the large difference in the hydrophilicity of epoxides and glycols. Several methods are available for the reconversion of the resolved diol into the antipode of the recovered epoxide.
6) In contrast with acid or base catalyzed hydrolyses of epoxides that often produce rearrangement products, such as aldehydes, ketones, allyl alcohols, side products are usually formed in negligible amounts in MEH catalyzed reactions. Only in particular cases other functional groups present in the substrate may undergo reactions catalyzed by other enzymes, particularly when working with crude S 9 fraction. For instance, ester groups are hydrolyzed by esterases concomitantly with the oxirane ring hydrolysis (9). However, many of the microsomal enzymes, such as the monooxygenases, requiring NADPH or other co-factors, are not active in their absence.

In conclusion, many aspects have still to be investigated before a practical use of epoxide hydrolases in synthesis becomes a reality. Their use will probably be limited to special cases, but studies on the application of this technique may well be rewarding.

References

(1) T. Katsuki, K.B. Sharpless, *J. Am.Chem.Soc.*, **102**, 5974 (1980).
(2) J.W.Scott in J.D.Morrison,J.W.Scott, eds. "Asymmetric Synthesis", Academic Press, Orlando, Vol. 4, p.5.
(3) A.Q.H. Habets-Crützen, S.J.N. Carlier, J.A.M. deBont, D.Witsuba, V. Schurig, S.Hartmans, J.Tramper, *Enzyme Microbiol. Technol.*, **7**, 17 (1985) and references cited therein.
(4) J.Seidegard, J.W.Pierre, *Biochem. Biophys. Acta*, **695**, 251 (1983)
(5) G. Bellucci, G. Berti, M. Ferretti, F. Marioni, F. Re, *Biochem. Biophys. Res. Commun.*, **102**,838(1981) and references cited therein.
(6) S.M.Mumby, B.D.Hammock, *Pestic.Biochem.Physiol.*, **11**, 275 (1979)
(7) J.G.Buchanan, H.Sable, in B.S. Thyagarajan "Selective Organic Transformations", Wiley, New York, 1972, vol.2, p.1.
(8) T. Watabe, N.Ozawa, A.Hiratsuka, *Biochem. Pharmacol.*, **32**,777 (1983).
(9) Unpublished work from this laboratory
(10) D.M.Jerina, H.Ziffer, J.W.Daly, *J.Am.Chem.Soc.*, **92**, 1056 (1970).
(11) G.Bellucci, G. Berti, G.Ingrosso, E.Mastrorilli, *J.Org.Chem.*, **45**,299 (1982)
(12) G.Bellucci, G. Berti, R.Bianchini, P. Cetera, E.Mastrorilli, *J. Org. Chem.*, **47**, 3105 (1982).
(13) G.Bellucci, G.Berti, M.Ferretti, E. Mastrorilli, L.Silvestri, *J. Org. Chem.*, **50**, 1471 (1985).
(14) G.Bellucci, G. Berti, G.Catelani, E.Mastrorilli, *J.Org.Chem.*, **46**, 5148 (1981).
(15) G.Catelani, E.Mastrorilli, *J. Chem. Soc., Perkin 1*, 2717 (1983).
(16) D.R.Thakker, H. Yagi, W.Levin, A.Y.H.Lu, A.H.Conney, D.M.Jerina, *J.Biol. Chem.*, **252**, 6328 (1977).

EFFICIENT ENZYMIC PRODUCTION OF ENANTIOMERICALLY PURE AMINO ACIDS

W.H.J. Boesten, B.H.N. Dassen, P.L. Kerkhoffs, M.J.A. Roberts, M.J.H. Cals, P.J.H. Peters, J.A.M. van Balken, E.M. Meijer and H.E. Schoemaker
DSM, Research and Patents
P.O. Box 18
6160 MD Geleen
The Netherlands

ABSTRACT. A generally applicable process for the production of enantiomerically pure amino acids is described. The method is based upon an enzymic resolution of DL-amino acid amides. As a biocatalyst aminopeptidase from Pseudomonas putida is used, which will stereospecifically hydrolyze the L-from of the amino acid amide. Efficient methods for the synthesis of substrates, the separation of L-amino acid from D-amino acid amide and the racemization and recycling of the unwanted isomer have been developed.

1. INTRODUCTION

Enantiomerically pure amino acids are becoming increasingly important as intermediates for the production of pharmaceuticals, food- and agrochemicals. Production methods include fermentation, asymmetric synthesis and resolution processes [1].

As opposed to fermentation, resolution methods are especially suited for the production of unnatural amino acids and for amino acids possessing the unnatural configuration. Although on a laboratory scale asymmetric synthesis has already reached a high level of sophistication, uptill now practical application of this type of process in the manufacture of amino acids has been limited to a few isolated cases.

In general, asymmetric synthesis has some advantage over resolution methods in that there is no need for racemization and recycling of the unwanted isomer. In the specific case of amino acids, the highly competitive position of resolution procedures may in part be due to the relative ease of racemization of amino acid derivatives.

Next to efficient recycling procedures, good accessibility of the substrates and a high optical purity of the products are essential features of an industrially applicable resolution process. In this context enzymic methods deserve special attention. In contrast to resolving agents used in industrial chemical resolution procedures,

M. P. Schneider (ed.), Enzymes as Catalysts in Organic Synthesis, 355–360.

which often are very specific and therefore have a rather narrow applicability, the hydrolytic enzymes used in resolution processes often couple a high degree of stereoselectivity with a broad substrate specificity, i.e. one enzyme can be used in the resolution of a number of amino acids.

At DSM an enzymic process has been developed based upon the stereospecific hydrolysis of amino acid amides by aminopeptidases.

Both D- and L-amino acids with a high optical purity can be produced under mild reaction conditions and starting from simple raw materials (aldehydes and ketones).

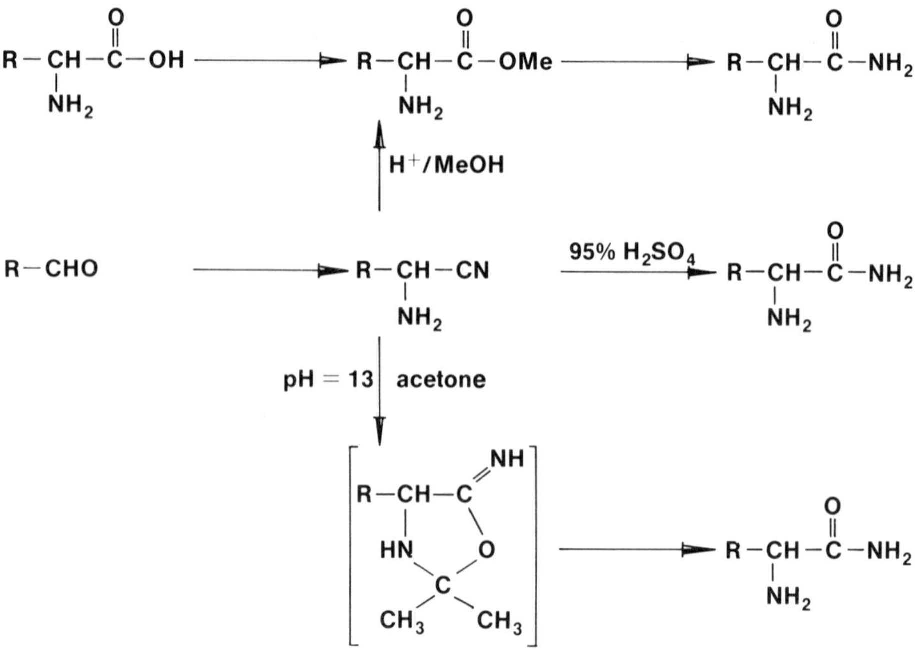

Fig. 1. Several methods for the preparation of α-amino acid amides.

2. PREPARATION OF α-AMINO ACID AMIDES

The amino acid amides used as substrates for the enzymic reaction can be prepared from the parent amino acids via initial conversion to the corresponding amino acid esters. However, from a practical viewpoint, it is more attractive to start from the corresponding α-amino-nitriles, intermediates in amino acid synthesis, obtained via Strecker reaction (HCN, NH_3) of aldehydes.

Several routes can be envisaged for the conversion of amino-nitriles in amino acid amides. The reactions depicted in fig. 1 are generally applicable and proceed in high yield.

A very efficient procedure has been discovered in our laboratory [2], comprising the one step conversion of the amino-nitrile, via the intermediate oxazolidine, to the amino amide under alkaline conditions in the presence of catalytic amounts of acetone. High yields (≥ 90 %) can be obtained in this manner.

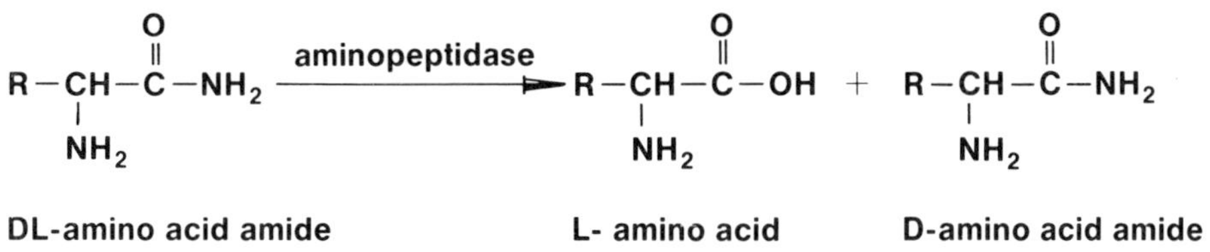

Fig. 2. General principle of the enzymic resolution.

3. ENZYMIC HYDROLYSIS OF α-AMINO ACID AMIDES

The general principle of the enzymic resolution process [3] is depicted in fig. 2. L-amino acid amides are stereospecifically hydrolyzed by the aminopeptidase from Pseudomonas putida ATCC 12633. The crude enzyme preparation is used in the soluble form in a batchwise process. In this manner also poorly soluble amino acids can be resolved without technical difficulties. If desired, the enzyme can also be used in immobilized form, but in general there is no economical incentive for re-use of the enzyme preparation.

Although other micro-organisms (e.g. Aspergillus species) have been known to produce aminopeptidases, the applicability of these enzyme preparations was not feasible due to low activities, narrow substrate specificity and undesired enzymic side effects. In our laboratories, however, it was discovered [4] that crude enzyme preparations of the aminopeptidase from Pseudomonas putida showed superior characteristics. A nearly 100 % stereoselectivity in hydrolyzing only the L-amino acid amide is combined with a very broad substrate specificity. Relative activities are ranging from 20 % (L-Phenylalanine amide) to 100 % (L-Phenylglycine amide).

Moreover, no enzymic side effects are observed in using the crude enzyme preparation (permeabilized cells).

Substrate concentrations up to 20 % by weight can be used without affecting the enzyme activity. Product inhibition largely depends on the ratio of substrate- and product-concentration. From a practical viewpoint the broad pH-range in which the enzyme is active should also be considered as a major advantage.

A high yielding fermentation process for the biocatalyst has been developed by NOVO Industri A/S. The cells are harvested during the exponential growth phase. A preparation with high aminopeptidase activity can be obtained by precipitating the cells using a flocculating agent. To render the enzyme more accessible the cell walls are modified, e.g. by heat treatment.

The α-hydrogen atom and an unsubstituted amino-group are essential structural features for enzyme activity. Therefore, in order to have access to the commercially interesting class of α-C-disubstituted amino acids, a new aminopeptidase biocatalyst from a Mycobacterium species has been developed in our laboratories, capable of stereoselectively hydrolyzing a range of α-C-disubstituted amino acid amides (patent pending).

Finally, a third biocatalyst has been developed, capable of hydrolyzing a broad range of D- and/or L-amino acid amides. Application of this novel amidase from cell preparations of Rhodococcus erythropolis (or high yielding mutants) is especially advantageous in the production of highly soluble D-amino acids (patent pending). Due to the lack of stereoselectivity of this enzyme preparation, D-amino acid amides, obtained after separation of the corresponding L-amino acids, will be hydrolyzed to the D-amino acids. Isolation yields are very high compared to the conventional chemical hydrolysis-neutralization procedure (no removal of salts).

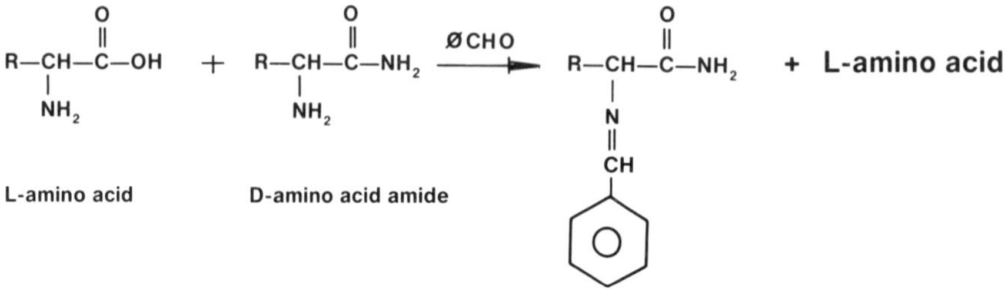

Fig. 3. Separation of the D-amino acid amide by formation of the insoluble schiff base with benzaldehyde (D-N-benzylidene amino acid amide).

4. PRODUCT RECOVERY AND RECYCLING OF THE UNWANTED ISOMER

Following enzymic hydrolysis, the L-amino acid formed has to be separated from the unchanged D-amino acid amide.

Conventional techniques like ion-exchange and extraction can be used.

In specific cases product separation is easily accomplished by filtration of the insoluble product.

A very convenient method has been discovered at DSM [5]. It was found that the Schiff base of benzaldehyde and D-phenylglycine amide is almost insoluble in water and therefore can easily be separated.

Surprisingly it was found that this formation of an insoluble Schiff base derivative was a general phenomenon and applicable for a broad range of D-amino acid amides (fig. 3). Subsequent acid hydrolysis at temperatures below 100 °C will yield the D-amino acid without racemization. The benzaldehyde is recycled. In this manner both D- and L-amino acids can be obtained with a very high optical purity (⩾ 99 %).

Depending on the specific example a number of racemization and recycling procedures have been developed. Recently an especially attractive procedure for the racemization of N-benzylidene amino acid amides under very mild conditions has been developed (patent pending).

5. EXAMPLES

The broad substrate specificity of the aminopeptidase from *Pseudomonas putida* is reflected in the synthesis of a wide variety of enantiomerically pure amino acids, both with the D- and the L-configuration.

The added advantage that reactions can be performed with solubilized enzyme preparations allows for the practical production of poorly soluble amino acids like e.g. L-homophenylalanine, an attractive intermediate for the synthesis of Angiotensin-Converting Enzyme inhibitors. Other examples include D-phenylglycine and D-p-hydroxyphenylglycine, intermediates for broad-spectrum antibiotics. If desired optically pure L-phenylglycine and L-p-hydroxyphenylglycine can be produced along with the corresponding D-amino acids, illustrating the versatility of the enzymic method compared to classical chemical resolution processes. In addition, both L- and D-forms of other substituted phenylglycines can be produced with high optical purity (e.g. D-o-chlorophenylglycine $[\alpha]^{25}_{578} = -118.1$, c = 1; 1 N HCl; lit. $[\alpha]^{25}_{578} = -89.1$, c = 1; 1 N HCl [6]).

Also other commercially attractive intermediates like D-Valine, starting material for fluvalinate, a pyrethroid-like insecticide and L-Phenylalanine, building block for the novel dipeptide sweetener aspartame can be produced.

In addition to these examples the method can also be used for the resolution of hetero-atom containing amino acids, like methionine, lysine, tryptophan, thienylglycine and derivatives.

This list is by no means exhaustive. The scope and limitations of the enzymic resolution process based upon the stereospecific hydrolysis of L-α-amino acid amides have yet to be determined.

6. CONCLUSION

A multi-purpose method for the production of enantiomerically pure D- and L-amino acids (both natural and unnatural) has been developed. The method is characterized by:

- good accessibility of the starting materials and substrates
- high optical purity of the final products
- mild reaction conditions both in substrate preparation and in the racemization and recycling procedure.

7. REFERENCES

1. See for reviews:
   a. G. Schmidt-Kastner and P. Egerer, in H.J. Rehm and G. Reed (Eds). Biotechnology Vol 6a, Verlag Chemie, Florida-Basel, p. 387-421 (1984)
   b. E.M. Meijer, W.H.J. Boesten, H.E. Schoemaker and J.A.M. van Balken.
      Symposium Biocatalysts in organic syntheses, Noordwijkerhout, The Netherlands, Elsevier Science Publishers, in press (1985)

2. W.H.J. Boesten (DSM/Stamicarbon) British Patent, 1,548,032 (1976)

3. W.H.J. Boesten (DSM/Stamicarbon) U.S. Patent, 3,971,700 (1976)

4. W.H.J. Boesten and L.R.M. Meijer-Hoffman (NOVO-DSM), U.S. Patent, 4,080,259 (1978)

5. W.H.J. Boesten (DSM/Stamicarbon) U.S. Patent, 4,172,846 (1979)

6. H. Schutt, G. Schmidt-Kastner, A. Arens and M. Preiss Biotechnology and Bioengineering, Vol 27, 420-433 (1985)

BIOHYDROXYLATION OF NON ACTIVATED CARBON ATOMS.
A MODEL FOR THE HYDROXYLATION SITE OF THE FUNGUS BEAUVERIA SULFURESCENS.

R. Furstoss * and A. Archelas, J.D. Fourneron , B. Vigne.
Laboratory of Organic and Bioorganic Chemistry,
Faculté des Sciences de Luminy, Case 901,
70 Route Léon Lachamp, 13288 MARSEILLE CEDEX 9, FRANCE.

ABSTRACT. The biohydroxylation of a number of bridged bicyclic or tricyclic amide-type molecules by the fungus Beauveria sulfurescens (ATCC 7159) has been studied. These hydroxylations lead to regio, stereo and sometimes enantioselective functionalization of non activated carbon atoms. The results have been analyzed in order to get informations concerning the topology of the enzyme active site. A model , based on computerized structure calculations, is proposed which allows to enhance the predictability of these processes.

1. INTRODUCTION

Because of the ease and the high selectivity shown by microbial enzymes which are able to perform hydroxylations of non activated carbon atoms, biohydroxylations constitue a highly powerfull tool for the organic chemist. However, one important drawback to the use of these techniques lies in the fact that it is usually very difficult to make any previsions concerning the specificities of such hydroxylation on a given substrate. One way to approach this problem is to study the bioconversion of different structurally related molecules, which can then be considered as being some kind of molecular probes, and may therefore allow the chemist to get informations concerning the influence of different geometric or steric parameters involved in these transformations. With these results in hand, one can hope to rationalize the metabolic pathways and, as a consequence, try to reach a high degree of predictability for these enzymatic processes (see for instance ref. 1-3).

Bearing this point in mind, we decided to study the biohydroxylation of various polycyclic amides or lactams by the fungus Beauveria sulfurescens (ATCC 7159). Indeed, this fungus has been described by a group directed by Fonken to achieve regiospecific hydroxylation of a number of essentially linear or monocyclic amides (4,5). On the basis of

M. P. Schneider (ed.), Enzymes as Catalysts in Organic Synthesis, 361–370.

the thermodynamically more stable conformations of these compounds, these authors have proposed a model (6) of the hydroxylating enzyme active site, which implies :

- first, that the molecule is fixed to this site by an interaction between the electron rich oxygen atom of the amide function and the enzyme,
- secondly that, as a subsequent step, hydroxylation occurs on a carbon atom located at a distance of about 5,5 Å of this oxygen atom (Fig.1).

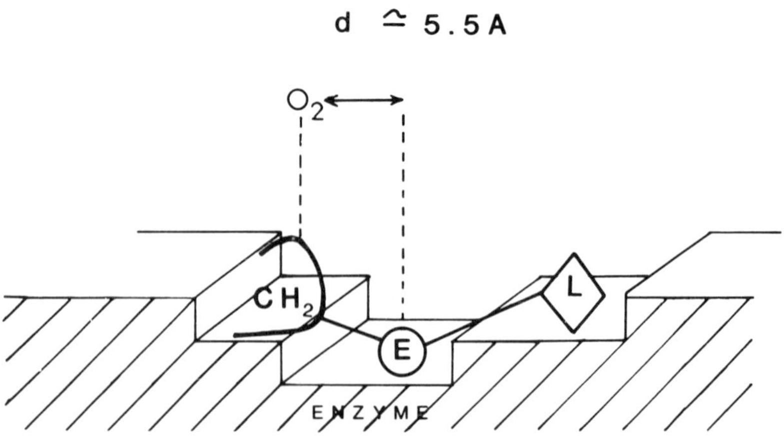

Figure 1 : *Proposed topology for the hydroxylation site of Beauveria sulfurescens.*

These results led us to raise several questions, in particular :

- would it be possible to influence the specificity of the hydroxylation by varying some structural factors of the starting compound as for instance the stereochemistry of the amide function ?
- would it be possible to use these biohydroxylations in order to achieve resolution of the substrate and, thus, to prepare optically active compounds from racemates ? Indeed, whereas numerous very interesting studies have shown that hydrolases and oxydoreductases are able to achieve enantioselective reactions (See for instance ref. 7-12), this aspect has not , to our best knowledge, been studied up to now for monooxygenase catalyzed hydroxylations (13).
- what would be the influence of the intrinsic nature of the amide group (amide versus lactam), the change from one compound to the other implying restricted rotation as well as a different location of the anchoring oxygen atom ?
- finally, would it be possible to rationalize all the obtained informations into a model based on the various substrate structures, in order to be able to predict the selectivities of these hydroxylations ?

The following results do shed some light on these four questions.

2. RESULTS

2.1. Influence of the amide group stereochemistry.

As can be seen on figure 2, hydroxylation of the two epimeric amides derived from norbornene do lead to one single product, formed by exclusive exo hydroxylation on C-5. This does indicate that the amide group conformation does not influence the stereochemistry of the introduced hydroxyle group. The same result is observed starting from the epimeric N-methylated amides having a camphor skeleton. Here again, the reaction is highly regio and stereoselective and no change is observed dependent on the stereochemistry of the amide.

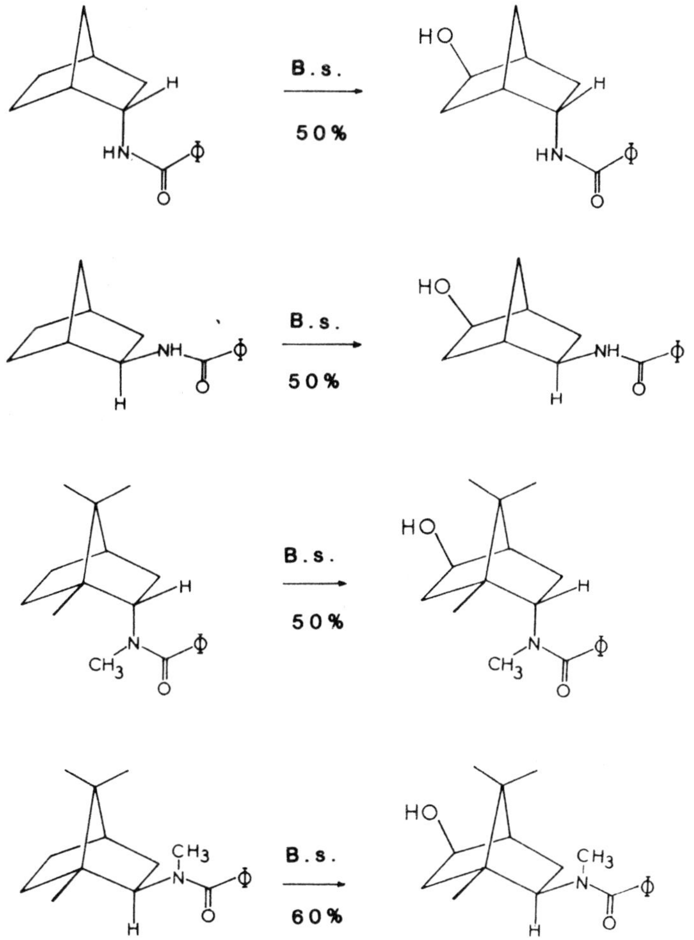

Figure 2 : *Hydroxylation of norbornene and camphor amide derivatives.*

The same behaviour has been observed for the different amides prepared from α-pinene. These compounds, which differ from one another by the stereochemistry of either the amide or the C-2 methyl group, should also be different as far as their conformations are concerned. In spite of these facts, biohydroxylation of these molecules always occurs at the C-8 methyl group.

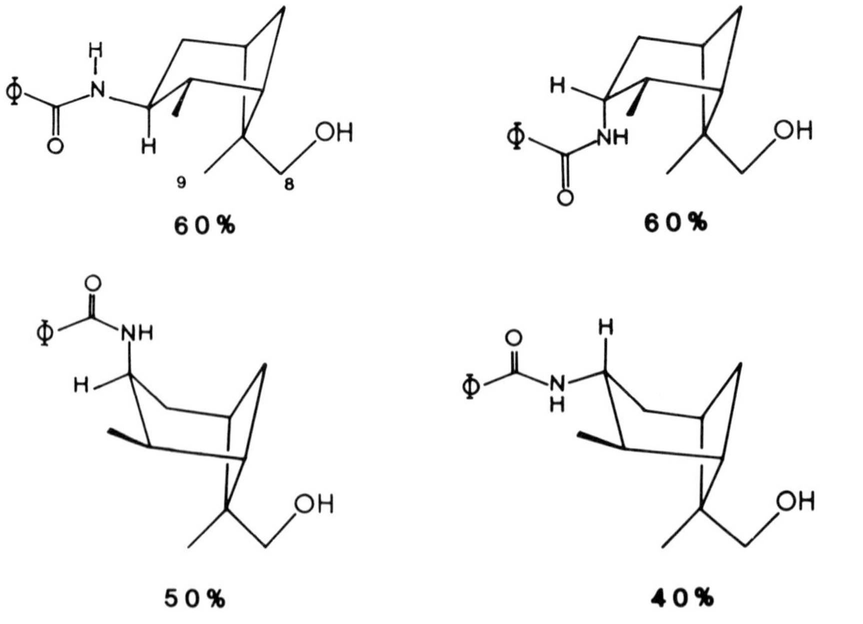

Figure 3 : *Hydroxylation of pinane amides.*

Interestingly enough, one must emphasize that, in all these cases, none of the obtained alcohols are optically active, which means the hydroxylations are not enantioselective.

2.2. Optical resolution

This is , however, no longer true for the isomeric secondary amides , derived from camphor where we observe a largely different behaviour between the two epimers. In these cases, the racemic exo isomer still leads to a single racemic alcohol, whereas the endo isomer affords three optically active compounds (Fig.3). This means that, for this compound, the regioselectivity of the hydroxylation varies from one enantiomer to the other. Thus, the 1R isomer is preferentially hydroxylated at the C-5 position , whereas the 1S isomer is essentially attacked on the C-8 methyl group. Each of the starting compounds can of course be prepared optically pure, starting from the pure enantiomers of camphor, which allows direct determination of the optical purity for each alcohol (fig.4).

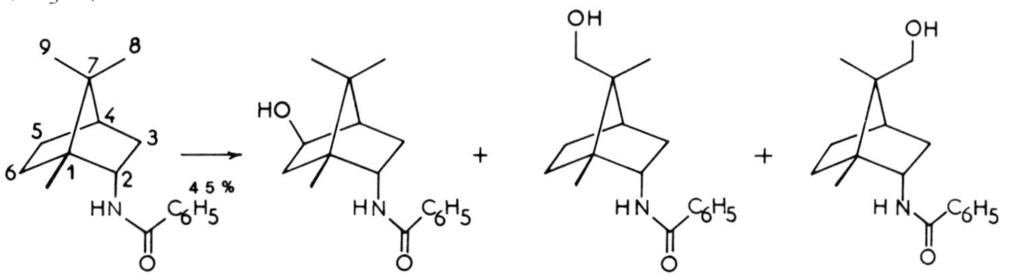

PROPORTIONS

1 RS	**36 (o.p. 25%)**	**20 (o.p. 25%)**	**44 (o.p. 65%)**
1 R	**48**	**22**	**30**
1 S	**11**	**15**	**74**

Figure 4 : *Hydroxylation of the secondary amide of camphor.*

Even much higher optical purities have been obtained in the case of the following compound, where the amide function has been blocked into a tricyclic carbon framework.

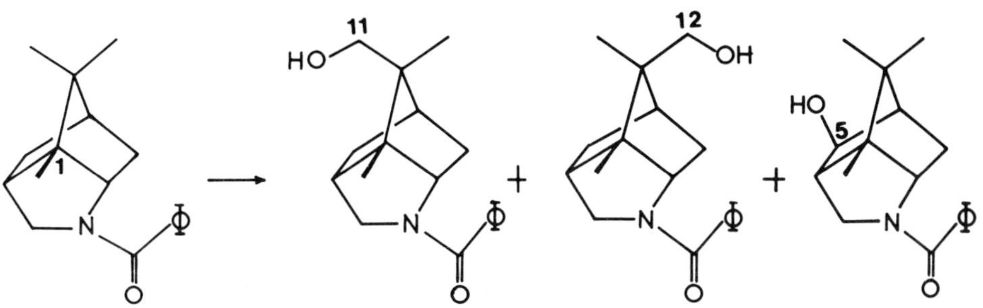

1RS	30% (o.p. 85%)	30% (o.p. 95%)	5% (o.p. 46%)
1R	10%	75%	5%
1S	30%		

Figure 5 : *Optical resolution by enantioselective hydroxylation.*

In this case, it appears that the regioselectivity of the hydroxylation is completely different from one enantiomer to the other, leading to high optical purity for each one of the alcohols formed by hydroxylation of a geminal methyl groups.

2.3. Comparison of the hydroxylation of amides and lactams.

Because of the previously proposed anchoring role of the oxygen atom, it also was of interest to study the behaviour of various structurally related substrates, for which the only difference lies into the location of the carbonyl function. Moreover, whereas the amides are known to exist in two different cisoïd and transoïd conformations, this possibility is suppressed in the corresponding lactam type structure. As a consequence, the different location of the carbonyl oxygen atom on the carbon skeleton should lead to different regioselectivities for the hydroxylation.

As shown on figure 6, this appears not to be the case, since the conversion of each of these compounds do lead to a single alcohol, resulting from stereospecific hydroxylation on the same carbon atom . This quite surprising result is also observed in the case of the bridged tricyclic amides and lactams . Here again, hydroxylation occurs on the same carbon atom, independently of the localization of the carbonyl function (14).
These observations do of coarse raise the question, whether the oxygen atom does play any role at all in the interaction between the enzyme and the substrate. The answer is still yes : indeed, whereas the alcohols obtained by hydroxylation of the racemic amides are optically inactive, those formed from the racemic lactams are optically active. As the only difference between the two types of models is precisely the localization of the oxygen atom on the carbon skeleton , we must conclude that this atom is somehow involved in the process.

These results do however emphasize the fact that it is not possible to make any prevision on the basis of the previously proposed 5.5 Å distance between the carbonyl oxygen atom and the hydroxylated carbon atom. Indeed, it appears that this distance does vary, for the different substrates we have studied, from 3.3 to 6.2 Å, making this criteria unusable as a predictive tool.

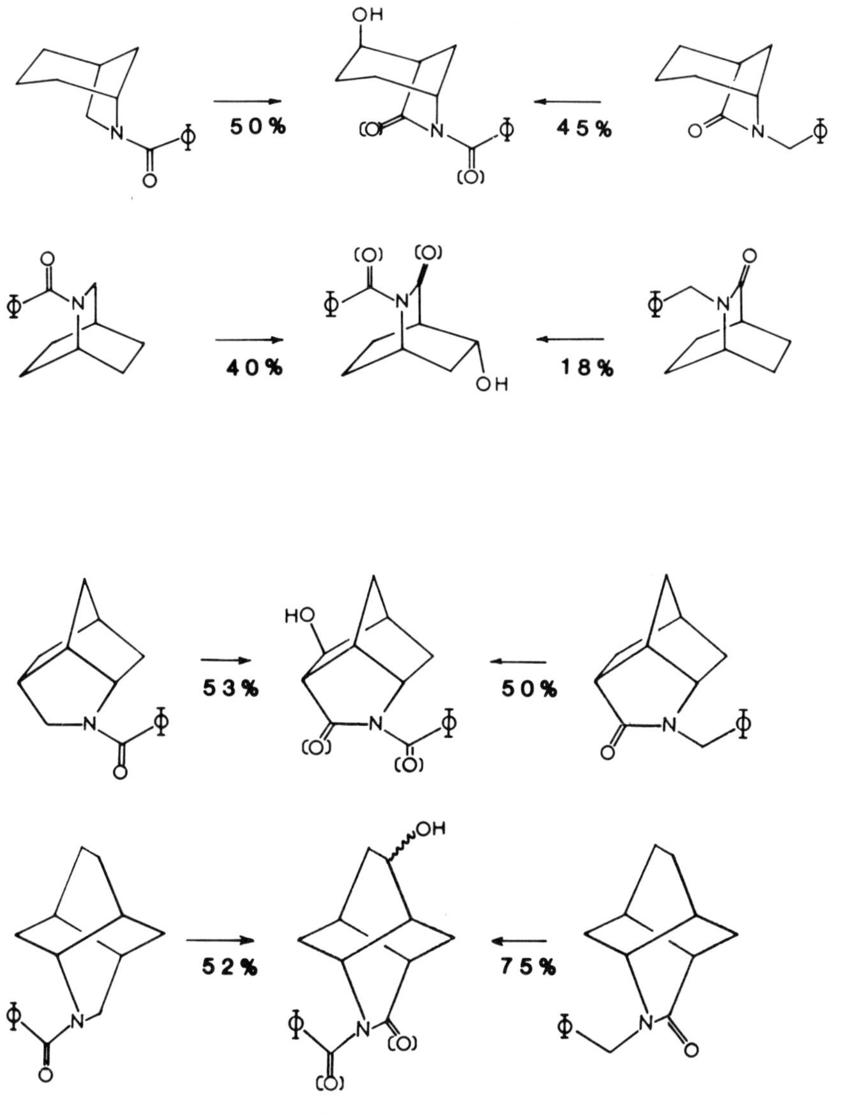

Figure 6 : *Comparative hydroxylation of amides and lactams.*

2.4. Proposed predictive model for the hydroxylation site.

However, the surprisingly high selectivities observed for these various hydroxylations still do suggest that it should be possible to rationalize these apparently puzzling results into a logical model. This would be particularly attractive since, on the basis of this model, one should subsequently be able to make previsions concerning the hydroxylation site on a given substrate.

Therefore, we decided to elaborate a model where, after positioning the different substrates in an unambiguous manner, one would locate precisely the hydroxylation site in the space. Also, we had to take into account the various results we had obtained i.e. - lack of influence of the nature of the amide moiety (amide versus lactam) - formation of optically active compounds from racemates - extensive variation of the distance between the carbonyl oxygen atom and the hydroxylation regioselectivity in structurally related models.

Examination of all these factors led us to conclude that everything seems to happen as if a two steps process would occur, - the first one being positioning of the substrate on the enzymatic site by interaction with the amide moiety and, eventually, the aromatic ring - the second step being hydroxylation, by way of an oxygen activating site located on a relatively flexible arm , as represented on figure 7.

This model does account for two important features - the very low substrate specificity of the fungus, which is able to hydroxylate numerous structurally different compounds -also, the very high regio and stereoselectivity observed for these hydroxylations, which would occur because of a relatively precise trajectory of the oxygen activating locus. We have determined this trajectory on the basis of the three dimentional structures of our different substrates , obtained either from previously described X-Ray structures, or by computerized calculations using appropriate programs. This trajectory is represented on figure 8.

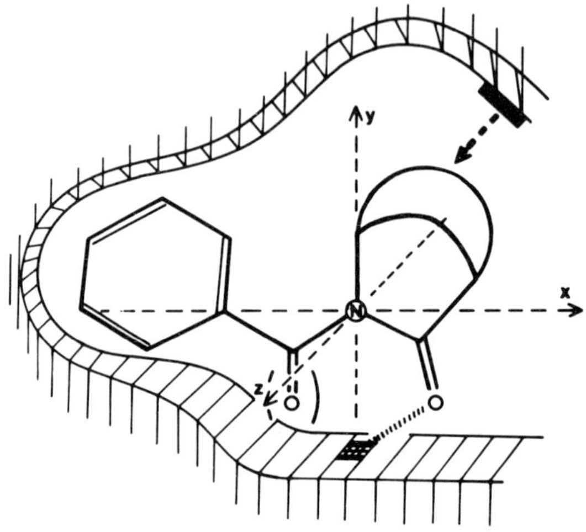

Figure 7 : *Positioning of the substrates in the model proposed for the hydroxylating site of Beauveria sulfurescens.*

It shows clearly that the various hydroxylation points are indeed distributed in the space in a relatively precise manner and that, therefore, the hydroxylation points may be located differently relative to the amide function, dependent on the intrinsic structure of the molecules. This corroborates the previously proposed flexibility of the enzymatic "arm" bearing the oxygen activating locus. On the basis of this model, it should now be possible , after positioning a given substrate as previously proposed, to predict the regio, the stereo and even the enantioselectivity for its hydroxylation. Work is in progress in our laboratory in order to ascertain this approach.

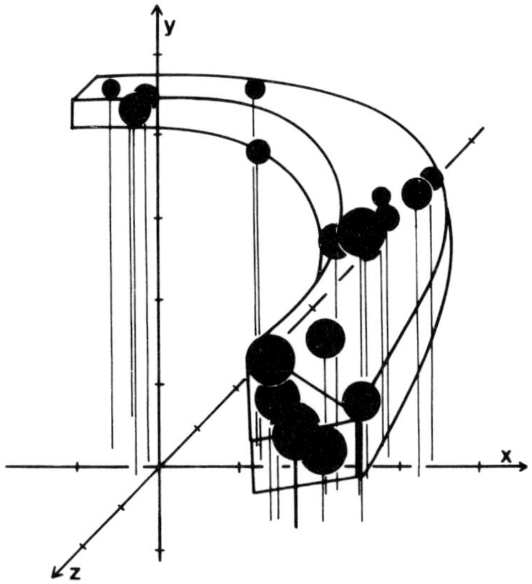

Figure 8 : *Trajectory of the hydroxylating site.*

3. CONCLUSION

As a conclusion, we would like to emphasize the fact that biohydroxylation may be a very powerfull tool in the hands of the organic chemist who wants to achieve functionalization of non activated carbon atoms. Indeed, these transformations do perform highly regio and stereoselective hydroxylations and even sometimes enantioselective ones, allowing synthesis of optically active compounds from racemic material. Also, we have achieved in this work the rationalization of the various results obtained on several substrates, leading to a model which should allow to use hydroxylations by Beauveria sulfurescens in a predictive manner.

REFERENCES

1. V. Prelog, Pure Appl.Chem., **9**, 119 (1964).
2. M. Nakasaki, J. Org. Chem., **45**, 4432 (1980).
3. J.B. Jones and I.J. Jakovac, Can. J. Chem., **60**, 19 (1982).
4. R.A. Johnson, Oxidation in Organic Chemistry, Part C, Academic Press , p. 131, (1978).
5. G.S. Fonken and R.A. Johnson, Chemical Oxidation with microorganisms, Dekker, New York (1972).
6. R.A. Johnson, M.E. Herr, H.C. Murray and G.S. Fonken, J. Org. Chem., **33**, 3217 (1968).
7. B. Cambou and A.M. Klibanov. Biotech. and Bioeng., **26**, 1449 (1984).
8. A.S. Gopalan and C.J. Sih ,Tetrahedron Lett., **25** , 5235 (1984).
9. K. Laumen and M. Schneider, Tetrahedron Lett., **26**, 2073 (1985).
10. C.J. Sih and C.S. Chen, Angew. Chem. Int. Ed., **3**, 570 (1984).
11. G.S.Y. Ng, L.C. Yuan, I.D. Jakovac and J.B. Jones, Tetrahedron, **40**, 1235 (1984).
12. K. Nakamura, K. Ushio, S. Oka and Ohno, Tetrahedron Lett., **25**, 3979 (1984).

"STEROLS BIOCONVERSIONS IN SYSTEMS WITH AN ORGANIC PHASE"

JOSE M. C. DUARTE
DEPARTMENT OF CHEMICAL INDUSTRIES
LNETI
2745 QUELUZ/PORTUGAL

ABSTRACT. The oxidation of cholesterol to cholestenone, the first step in cholesterol utilization by microorganisms, was catalyzed by cholesterol oxidase of *Nocardia rhodochrous* NCIB 10554. The use of an immiscible organic solvent phase allowed the use of higher cholesterol concentrations with higher rates of conversion observed. Several apolar solvents were tested and it was concluded that for this reaction, properties such as interfacial tensions, the molecule chain-length and presence of double bonds may influence the outcome of the enzyme activity. Entrapment of the *Nocardia* cells in polymers of different hydrophilicity resulted in an increase of the enzyme stability. However entrapment of the cells in the polymers seem to further enhance the mass-transfer control of the reaction. The usefulness of these systems for carrying these type of reactions was established and criteria for the use of enzymatic conversions on systems with an organic phase could be derived.

I. INTRODUCTION

Substrate concentration in most sterol and steroid conversions rarely exceeds one gram per litre. Low substrate concentration is one of the reasons for the lack of competitiveness of microbiological methods as against chemical transformations. Several techniques have been used to increase the steroid substrate concentration : Weaver *et al* (1960) were able to use finely ground progesterone in concentrations as high as 20 to 50 g/l with very good yields ; advantages of the higher concentrations were : dihydroxylation was almost non-existent and extraction of the steroid was easier because 95% of the total steroid was associated with the micellium. This same technique was used by Constantinides (1980) with freeze-dried *C. simplex* cells immobilized in collagen for the dehydrogenation of cortisol to prednisolone. Optimal substrate concentration was found to be 15 mg/ml with 80% of conversion in 48 hours. Bhasin *et al.* (1976) used a silicone polymer as a steroid reservoir : the bacterial hydroxysteroid dehydrogenase seems to be more stable in the presence of the polymer

M. P. Schneider (ed.), Enzymes as Catalysts in Organic Synthesis, 371–382.

than in the presence of either ethyl or butyl acetate. However , the rate of reaction seems to be limited by the mass transfer of steroid from the hydrophobic to the aqueous phase ; another disadvantage is that the polymer has a poor reservoir capacity for steroids compared with the organic solvents.

Organic solvents are commonly used to solubilize steroid substrate . Water miscible organic solvents such as methanol (Jones and Gordon, 1973) are normally employed ; their concentration is maintained as low as possible : a compromise between inhibition and steroid reservoir capacity must be made to ensure the high productivity of the reaction. Organic solvents with very low solubilities in water may be preferable due to the fact that they can form a heterogeneous phase and therefore contact with enzymes may be minimized ; they also normally have higher steroid solubilities, so that they can achieve a high reactor load. This was firstly shown by the works of Cremonesi and co-workers (Lugaro *et al*. 1973) ; Cremonesi *et al* . , 1974 , 1975 , 1977) on the enzymatic preparation of steroids in emulsions of water and organic solvents , mainly ethyl and butyl acetate where they were able to recycle the cofactors and enzymes due to the fact they are not in the same phase of the steroid. Enzymatic reduction within these biphasic system gave better results than chemical reaction (Cremonesi *et al*., 1975).

Immobilization of the enzymes by covalent linkage to CNBR-activated sepharose led to catalysts that were highly stable in the presence of organic solvents : β -hydroxy-steroid dehydrogenase maintained 60% of its original activity two months after continuous use in a water--ethyl acetate system ; its value for the reduction of cortisone by 20 β -HSDH-sepharose , a high stereospecific enzymatic reaction , on a preparative scale has been demonstrated (Carrea , 1979). Buckland *et al* . (1975) used whole cells of *Nocardia rhodochrous* , with cholesterol oxidase activity, in the presence of a high proportion of several organic solvents. They were able to obtain conversion rates of 7 g/hr of cholesterol to cholestenone using a 300 ml reaction volume. This rate shows how far one can go by using appropriate techniques for steroid conversions and the potential for the use of immiscible organic solvents. Yamane *et al*. (1979) extended this thecnique by using entrapped cells in either hydrophobic or hydrophilic gels. A benzene : n-heptane (1:1 by volume) mixture was used with *Nocardia rhodocrous* previously induced to synthesize steroid 1,2-dehydrogenase. AD was converted into ADD in the presence of an electron acceptor (Phenazine methosulphate). The lipophylic gel exhibited a slightly higher conversion rate, and mass transfer was a limiting factor in both types of gel. Entrapped cells had a considerably higher stability compared with free cells, but even so their life-times were relatively short (after 6 runs , equivalent to 6 hours of operation at 20º C , the reaction rate fell to half for both gels) for consideration as worthy of commercial application. Sonomoto *et al*.(1979) also used immobilized cell preparations in gels with different percentages of hydrophylic prepolymer (ENT-4000) with hydrophobic prepolymer (ENTP-2000) to study the influence of polymer hydrophobicity on the specific activity of the entrapped cells. *Arthrobacter simplex* activity to convert hydrocortisone to prednisolone increased about 5 times when the ENTP-2000 content was increased from 0 to 30% in the prepolymer mixture. Steroids were dissolved in methanol (10% of the reaction volume). A. simplex. cells immobilized in another polymer (PBM-2000 prepolymer) were able to effect the same conversion in a completely aqueous medium, but steroid transformation in organic solvents or organic/water mixtures did not succeed(Omata et al. , 1979 a). The influence of the hydrophobicity of gels in the transformation of 3β-hydroxy-5-en-steroids to corresponding 3-Keto-4-en-steroids by entrapped whole cells of Nocardia rhodocrous was also studied by Omata et al. (1979 b) : in the cases of sterols (cholesterol, sitosterol and stigmasterol) only hydrophobic gel entrapped cells showed activity. Hydrophobicity of the substrate was also important to the activity of the preparations. Nocardia erythropolis cells were also used to convert cholesterol to cholestenone (Atrat et al., 1980) in the presence of toluene or carbon tetrachloride ; these cells were immobilized in different supports ; adsorption on DEAE-cellulose was the most effective method.

These works document well the potential for the use of organic solvents in heterogeneous systems in conjunction with enzymes or micro organisms. However , not many studies have been devoted to elucidating the influences that the complexity of such systems may have on enzyme kinetics and reaction. To elucidate some of these we used Nocardia rhodochrous to convert cholest 5-en-3 β -ol (cholesterol) to cholest--4-en-3-one in the presence of water-immiscible organic solvents (Buckland et al., 1975); the reaction is taken to be first step in cholesterol degradation by microorganisms and no external cofactors are required. The enzyme that catalyzes the reaction is located on the outside of the plasma membrane (Richmond , 1973) and possesses two catalytic activities : 3 β -hydroxysteroid oxidase (HSDH) and 4,5--isomerase (OSI) ; these enzymatic activities probably are associated with the same active site (P. Cheetham, 1979) with the 3 β -hydroxysteroid oxidase activity limiting the overall reaction :

$$\text{cholest-5-en-3}\beta\text{-ol} + O_2 \xrightarrow{\text{HSDH}} \text{cholest-5-en-3-one} + H_2O_2$$

$$\text{cholest-5-en-3-one} \xrightarrow{\text{OSI}} \text{cholest-4-en-3-one}$$

$$H_2O_2 \xrightarrow{\text{Cat}} H_2O + 1/2\ O_2$$

Enough catalase (Cat) seems to be present in the Nocardia cell to transform hydrogen peroxide quickly into water and oxygen (Richmond, 1973). When mixed with water-immiscible organic solvents, and in particular with carbon tetrachloride the Nocardia , maintains its cholesterol oxidase activity , yielding higher conversion rates than in the completely aqueous system;no degradation of cholestenone was observed with these cells (Buckland et al. 1975).

A diagramatic representation of the system used is shown in Figure 1. This system is composed of an organic immiscible phase with high substrate solubility and of an aqueous phase where the Nocardia cells are located , it is not easy to say a"priori"whether the reaction will be either controlled by the enzyme kinetics or by mass-transfer of any

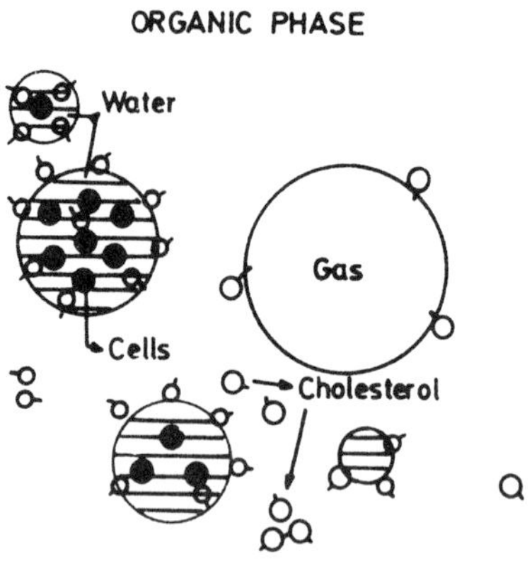

Figure 1. Diagramatic representation of an emulsion of Nocardia paste in a water immiscible solvent containing cholesterol. The organic solvent constitutes the continuous phase.

of the reactants : one of the reactants, cholesterol, is supplied from the organic phase, whereas the other , oxygen , must be supplied from air, which introduces a new phase to the system.

In the work reported here it is attempted to further elucidate the mechanisms by which organic solvents influence the conversion rates of this system , in particular we studied the influence of the nature of the organic solvent on the conversion rates and how immobilization techniques afford protection against the denaturing effects of organic solvents.

2. MATERIALS AND METHODS

Growing the cells. The microorganism used was Nocardia rhodochrous NCIB 10554 previously grown in a 1500 litre volume fermenter on a glycerol and yeast extract medium. At the end of growth the fermenter content was harvested with a Westphalia KO 8006 multichamber centrifuge. The wet cells recovered were packed and immediately placed in the freezer at -20º C until required.

Measuring cells activity. The cell slurry direct from the fermenter was diluted with 50 mM potassium phosphate buffer, pH 7.0, and 2 ml were placed in 20 ml screw-top bottles and brought to 30º C . At zero-time, 100 ul of 1% cholesterol in ethanol were added to form a colloidal suspension of the substrate. The bottles were then flushed with oxygen, stoppered and incubated with shaking (3.8 cm stroke, 150 strokes/min) for 10 minutes. The reaction was stopped by the addition of 5 ml ethanolic potassium hydroxide and the mixture was extracted using 6 ml of spectrograde cyclohexane and shaken vigorously for 30 seconds. The amount of cholestenone formed was determined by measuring the absorbances of the cyclohexane extracts at 237 nm against the appropriate blanks.

Cells in the organic solvent reactors - Frozen Nocardia paste was mixed with carbon tetrachloride or other used solvents. Samples of this emulsion were taken at regular intervals centrifuged and analysed. Experiments were conducted in shake flaks mounted in a reciprocal shaker (225 strokes/min) . Experiments were perfomed at a temperature of 22º C ± 1. Oxygen concentration was monitored with Beckman 39553 O_2 electrodes, using plain Teflon membranes. Cholesterol oxidase activity could be calculated from the time progress curve of the reaction.

Entrapment of whole cells . Nocardia grown as described was entrapped in gels to test its ability to support the cholesterol oxidase activity of the cells. Alginate gels were prepared by mixing a sodium alginate solution with the Nocardia cells and adding it dropwise to a 0.2 M solution of Ca Cl_2; spherical pellets of calcium alginate were instantaneously formed at the contact of the two solutions. Polyacrylamide gel entrapped cells were prepared in the following way : 2.5 g of the acrylamide monomer and 0.13 g of N, N'-methylene-bis-acrylamide were dissolved in Tris buffer 0.05 M, pH 7.0 ; wet cells were then added and volume made up to 24 ml ; to this solution 0.5 ml of the catalyst N, N, N', N', -tetramethylene diamine (10% w/v, aqueous solution) and 0.5 ml of ammonium persulphate (5% , w/v , aqueous solution) were quickly added and mixed. Polymerization occurred in a few seconds. The solidified gel was cut into small cubes and then sieved through a mesh size.

Steroid Analysis . Cholesterol was analysed by the method of Parekh and Jung; cholest-4-en-3-one (cholestenone) dissolved in cyclohexane was determined at 233 nm in a Unicam SP 1700 spectrophotometer. TLC chromatograms of samples (silicagel 60 F_{254} plates) were made and developed in tanks containing chloroform : diethyl ether (10 + 1) to confirm that cholestenone was the only steroid being formed.

Wet cell paste analysis - The dry weights of cell paste used in the experiments were routinely measured ; dried samples were extracted with solvents and their steroids content determined. Nitrogen analysis of the cell paste was done to give a direct indication of the biomass ; a kjedahl type digestion method was used to liberate the nitrogen from biomass as NH_4^+ which was assayed by the use of the Nessler reagent method.

3. RESULTS

Reactor runs. Consisted basically of mixing a definite weight of the harvested cell paste with the organic solvent. The main components of this paste were the wet Nocardia cells mixed with cholesterol which has been used as inducer and some cholestenone not degraded during the fermentation. It was particularly important to ensure that the amount of biomass per unit weight of cell paste was constant in each run for each series of experiments. Analysis of the biological nitrogen of samples showed that the deviations were not significant. The ratio between wet weight and dry weight of the cell paste was about 4.2 ;the harvested cell paste composition was :

water content 76.0% , cholesterol 9.2%, cholestenone 0.8%

biomass 14.0%

Comparison of aqueous and organic solvent reactors. Two reactor runs were conducted for comparison of the use of a completely aqueous medium and of another where only organic solvent was used (Figure 2). In the

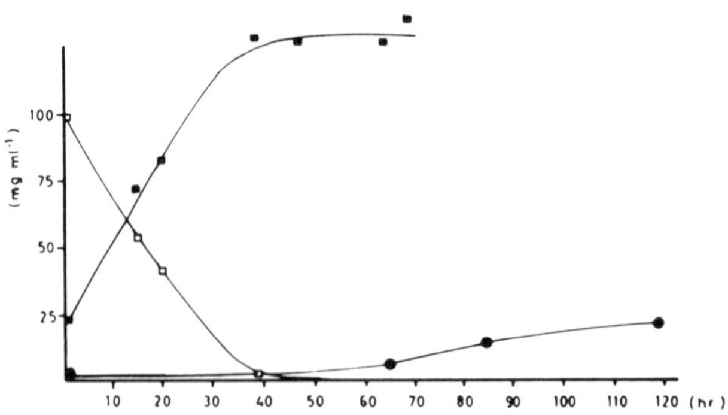

Figure 2. Comparison of the time course for cholesterol (□) oxidation to cholestenone (symbols in full) in a completely aqueous reactor (●) with another where the chlolesterol in dissolved in carbon tetrachloride (■) . The solvent constitutes the larger phase on the biphasic system.

organic solvent reactor more than 100 g cholesterol were converted to cholestenone-4 in about 40 hours; when using the completely aqueous medium to which the cholesterol was added as a fine powdered solid ,on a larges excess, pratically no conversion was observed in the first 40 hours ; later on cholestenone-4 began accumulating in the medium at a rate of about 0.25 $mg.ml^{-1}$. This value is about 24 times smaller than for the comparable run in carbon tetrachloride ; the observed lag could be dependent on the shearing of the cholesterol particles until a critical area of contact was obtained

Factors affecting the expression of cell paste cholesterol oxidase activity . The influence of increasing cholesterol and oxygen, concentrations in the cell paste-organic solvent system was tested in shake flasks experiments. To increase the oxygen concentration the flask contents were fluxed with pure oxygen for 1 minute: the flasks were stoppered and sealed . Results are shown on Table I.

Increasing the cholesterol concentration in the organic solvent,used in this experiment further increases the rate of cholestenone production.

TABLE I. Cholesterol Oxidase Activity of *Nocardia* Cells

Experiment	Activity **
Standard *	16
Oxygen flux	51
water added at 0.03 ml/ml	20
cholesterol at 50 gl^{-1}	28

* A wet cell paste of 50 mg per ml of carbon tetrachloride was used. Cholesterol concentration in the organic solvent was 15 gl^{-1}.

** As mg of cholestenone$\cdot$ g^{-1}. hr^{-1}

Fluxing with oxygen is very effective in increasing the rate of reaction : a more than threefold increase was observed. Water, although added in a very small proportion, also seems to have some effect : an almost 30% increase was observed.

Reactor stability of the cholesterol oxidase in the presence of organic solvents. To study how the re-use of cells affected the cholesterol oxidase activity, the cells were left in a reactor for 22 hours until all the cholesterol was transformed. Cells were taken out of this reactor and its re-use activity tested by using them in a fresh batch of carbon tetrachloride. This re-use activity of the cells is compared (Figure 3) with the activity which they maintenained if cells were not taken out of the original batch; in this case the activity was measured by adding fresh substrate to the emulsion.

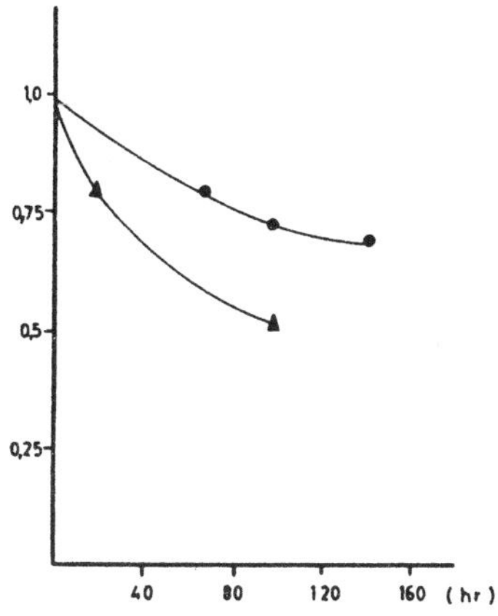

Figure 3. Cholesterol oxidase activity of reused *Nocardia* cells as a function of the time of permanence in the two-phase reactor system. Both activities of cells reused without changing the solvent (●) and by using "fresh" solvent (▲) are compared. V^o is the initial activity of cells and V^r the activity after a time of operation.

It is evident that when cells were removed from the original emulsion their loss of activity was greater than when they were re-used in the original emulsion ; about 20% more activity was conserved after 100 hours in the original than in a new batch.

The effect of the organic solvent. The experiments described above were done with carbon tetrachloride as the solvent for cholesterol. It has been reported (Cremonesi *et al.*, 1977) that different organic solvents affect the apparent K_M value of enzymes, the K_M increasing with increasing concentration of the solvent but with a constant Vmax, suggesting a competitive type of inhibition. However, Buckland *et al.* (1975) found different maximal velocities for the range of solvents used. No simple correlation could be observed. We tested the effects of two series of solvents on the cholesterol oxidase activity. The results for a serie of chloro derivaties of methane and ethane are shown in Table II and for several ethers in Table III.

TABLE II. Effect of Chloro derivatives of Methane and Ethane on the Rate of Cholesterol Oxidation by Non-growing *Nocardia* Cells

Solvent Molecular Formula	Solubility of Solvent in Water (% w/w)	Comparative Rates
$CHCl_2$	2.00	0.65
CHI_3	0.8	0.39
CCl_4	0.08	1.00
$ClCH_2CH_2Cl$	0.8	1.11
CH_3CCl_3	-	1.31

TABLE III. Effect of different Ethers on the Rate of Cholesterol Oxidation by Non-growing Nocardia Cells

Solvent Molecular Formula	Solubility of Solvent in Water (%w/w)	Comparatives Rates (1,1,1 Trichloroethane = 1.0)
$(C_2H_5)_2O$	6.9	0.55
$((CH_3)_2CH)_2O$	0.9	1.10
$(CH_3(CH_2)_3]_2O$	0.3	0.93
$(C_5H_{11})_2O$	-	0.85
$CH_3(CH_2)_3OCH=CH_2$	-	1.31

The reactions were all done with excess cholesterol and in the presence of air.

These tables show that although solubility of the solvent in water is probably an important factor it does not entirely explain all the differences in the observed reaction rates and other properties of the solvent (e.g. molecular weight,density, interfacial tension, oxygen and water solubility) may be important.

Use of gel-entrapped cells. From the previous experiments it was concluded that the organic solvent was affecting the enzyme activity which was most probably due to damage of the cell membrane where the enzyme is located. Entrapment of the cells in a hydrophilic gel should help in increasing the enzyme stability by decreasing the probability of contact between the cells and the organic solvent.1,1,1 - Trichloroethane was the solvent used in these experiments since it displayed the largest rate of a number of chloro derivatives of methane and ethane.

TABLE IV. Comparison of Cholesterol Oxidase Activities and Stability of Entrapped Nocardia

Entrapment method	Activity (mg/g/hr) Initial	Reuse *
Polyacrylamide (0.5 mm diam.)	7.6	3.8
Alginate (2.0 mm diam.)	1.9	1.3

* After 300 hours operation.

The results are shown in Table IV it shows that the half-life activity of cells in polyacrylamide extended to 300 hours compared with about 100 hours for free cells ; with alginate gels which are more hydrophilic and had larger pellet diameter the stability was even higher;however the specific rate of reaction was reduced about four times, which is about the same as the ratio of the diameters of the pellets. Thus the entrapped preparations have better stability and may be suitable for repeated batch or continuous operation.

4. DISCUSSION

The advantages of the use of organic solvents in the oxidation of cholesterol to cholestenone by cholesterol oxidase result from the much higher conversion rates obtained (Figure 2) . The mechanism of "activation" by the organic solvents is not completely clear and is certainly related with the apolar character of the organic solvent. Apolar surfaces were recognized to be an essential feature for the activation of lipase(Brockman et al. 1973) a surface acting enzyme ; these enzymes are activated when the substrate is presented in a micellar form. The observed effect of an aqueous buffer on the rate of reaction may be related with the maximization of the available surface area for substrate transfer by influencing the emulsion droplet diameter. However the use of organic solvents affects the enzyme stability which decays in an exponential manner with the time of exposure of cells to the organic solvent (Fig.3). An interesting feature is that cells taken out of the reactor for re-use lose activity more quickly. This fact may be due to the loss of some lipids that are essential to enzyme activity and/or substrate micellization ; progressive delipidation by means of phospholipase A2 normally lowers both the Vmax and K_M of menbrane-bound enzymes. (Lenaz et al. 1978).

Tables II and III show that the maximum velocity of the reaction (at air saturation) is strongly affected by the organic solvent used, even for closely related compounds. Ethyl ether, with one of the highest solubilities for both oxygen and cholesterol(almost twice those of Carbon Tetrachloride), showed only about half the activity ; solubility of the solvent in water, being a good indication of its polarity, correlates better with activity. Solvents may affect the reaction rate depending on their solvation power and the nature of the reaction ; there is some evidence for a radical mechanism in the enzymatic cholesterol-cholestenone reaction by cholesterol oxidase of *Nocardia* (McGuiness *et al.*, 1978) and by soybean lipooxygenase (Teng and Smith, 1974) ; this may be a common feature of the mechanism of flavoprotein oxidases (Walsh, 1979). Solvents with lower polarities and capacity for interfering with hydrogen bonds must be preferred ; these solvents will also normally have lower solubility in water, what can increase the enzyme stability; more polar compounds also have a higher capacity for interfering with lipids of the cell membrane, mainly the more polar ones such as phospholipids, which are important for their ability to create a spatial structure (and therefore influence the enzyme conformation) ; there is also the possibility of binding to the enzyme resulting in some type of inhibition ; a more conclusive answer is difficult because the changes in the K_M(app.) could not be evaluated from the time-progression curves of the reaction. It is therefore not surprising that, chloroform and methylene chloride, with the higher polarity parameters, (Snyder, 1978) show less activity in the chloride derivatives series, the same happening with the ethyl and benzyl ethers. The highest levels of activity were obtained with the chloroderivatives of ethane and ethers with 6 carbon atoms (possibly with a length similar to or a little greater than the cholesterol side chain) ; it may be speculated that smaller molecules interfere with the binding of cholesterol, that is, at least for the isomerization reaction (Smith and Brooks , 1977) through the cholesterol side-chain, and as such probably in a highly hydrophobic zone of the active site; on the other hand this could be related with the fact that these solvents solvate the cholesterol molecule in the right way for the presentation of the cholesterol molecule, e.g., leaving the OH group more accessible to the reaction. It must be said that whether the influence of the nature of the organic solvents on the reaction rates is predominantly an effect on the mass--transfer characteristics of the system (Duarte, 1982) or on the reaction mechanism and/or enzyme turnover power cannot be decided without more fundamental studies.

Cell entrapment increased the stability of the *Nocardia* cholesterol oxidase to the presence of the organic solvents (Table IV). It looks that the more hydrophilic the polymer the more protection it can afford. This contrasts with the fact that the use of highly hydrophobic polymers (Omata *et al.* 1979) did not extended the enzyme half-life. As might be expected the specific rates also decreased when cells were entrapped ; this is probably due to the presence of intraparticle substrate diffusion-limitation which is a function of the substrate diffusivity in the support, of the particle radius and of the cell content; even though the activity of the entrapped cells in an organic milieu is high when compared with the activity of free cells in aqueous buffer solution. Thus it can be concluded that the transport properties of the medium in relation to the substrates (cholesterol and oxygen) may be a decisive factor for expression of the activity of the enzyme (Duarte, 1982). The advantages of the use of apolar organic solvents in these sterols transformations are summarized in Table V.

TABLE V. Comparison of Enzymic Aqueous and Water-immiscible Organic Solvents Reactors for use with Low Water Solubility Substrates

	Aqueous	Organic
Conversion Rates	small	large
Enzyme Deactivation	small	large
Substrate/Product Inhibition	problem	may be avoided
Oxygen Sink capacity	weak	good
Product Recovery	difficult	easy
Reuse of Cells/Enzymes	difficult	easy
Reactor Volume	large	small
Foaming	large	non-existant
Energy consuming	high	may be low
Limitations	substrate solubility	surface area
Catalyst Immobilization	not convenient	convenient

Among others operational advantages are reinforced by the entrapment of the cells in mainly hydrophilic polymers which makes easier the re-use of cells and their use in continuous reactors.

It is reasonable to believe that development of the use of organic solvents along these lines will enlarge the scope of enzymatic reactions for use in organic synthesis and in industrial applications, and will enlarge the scope for the use of organic solvents in enzymatic catalysis. A few enzymatic processes using water immiscible organic solvents were already found to be suitable for industrial application. Among these are fat interesterification (Coleman and Macrae, 1976) and preparation of side - chain amino acids for antibiotic synthesis (Bayer AG, 1979). The extension of the use of these systems to more complex transformations (multi-reaction, co-factor regeneration) looks promising as demonstrated recently by Duarte (1983) and Duarte, Lilly (1984) . All these features are an indication that the use of these systems in organic synthesis and industrial applications will go to increase in the future.

5. REFERENCES

Atrat, P., E. Huller and C. Horold : Zeitschrift Allg.Mikrob., 20, 79 (1980).

Bayer AG: Euro Patent 22492 (07.07.79)

Bhasin, D., C. Gryte and J. Studebaker : Biotechnol.Bioeng. 18 , 1777 (1976)

Brockman, H., J. Law and F. Kezdy : J.Biol.Chem. 248 , 4965 (1973)

Buckland, B.P. Dunnill and M. Lilly : Biotechnol. Bioeng. 17, 815.(1975)

Carrea, G : Biotechnol Bioeng ., 21 , 39 (1979)

Cheetham, P. : J.Appl Biochem. , 1, 51 (1979)

Coleman, M. and A. Macrae. U.K. Pat. 1577933 (1976)

Constantinides, A.: Biotechnol.Bioeng , 22 , 119 (1980)

Cremonesi. P., G. Carrea, L.Ferrara and E. Antonini : Eur. J.Biochem., 44 , 401 (1974).
Cremonesi, P.G.Carrea, L.Ferrara and E. Antonini .: Biotechnol.Bioeng., 17, 1101 (1975).
Cremonesi, P., G. Mazzola and L. Cremonesi: Annali di Chimica , 67 , 415 (1977).
Duarte J.C. : Enzyme Engineering, 6 , 157 (1982)
Duarte J.C. Ann.N.Y.Acad.Sci , 413 "Biochemical Engineering III " 548 (1983)
Duarte, J.C.and M. Lilly : Ann. N.Y.Acad. Sci. 434 "Enzyme Engineering 7" 573 (1984)
Jones , J. and K.Gordon : Biochemistry, 12 , 71 (1973)
Lenaz, G., G. Curatola, L. Mazzanti and G. Parenti-Castelli : Molec.Cell. Biochem., 22, 3. (1978)
Lugaro, G.,G. Carrea, P.Cremonesi, M. Casellato and E.Antonini : Arch. Biochem. Biophys ., 159 , 1 (1973).
McGuiness, T., H. Brown, S. Chattopadhyay and F. Chen: Biochim.Biophys. Acta, 530 , 247 (1978).
Omata, T. , A. Tanaka , T. Yamane and S. Fukui : Eur. J.Appl.Microbiol. Biotechnol ., 6 , 207. (1979-a)
Omata, T., T. Iida, A. Tanaka and S. Fukui : Eur. J. Appl.Microbiol. Biotechnol. 8 , 143 (1979-b).
Parekh, A. and D. Jung : Analytical Chemistry, 42 , 12 (1976)
Richmond, W. : Clin.Chem., 19 , 1350 (1973)
Smith, A. and C. Brooks : Biochem. Soc. Trans. (569 th Meeting), 5 , 1088 (1977).
Snyder , L. : J. Chromatogr. Sci. 16 , 223 (1978)
Sonomoto, K., A. Tanaka, T. Omata, T.Yamare and S. Fukui: Eur.J.Appl. Microbiol. Biotechnol , 6 , 325 (1979)
Teng, J., and L. Smith : J. Steroid Biochem . 7 , 577 (1974).
Walsh, C. : Enzymatic Reaction Mechanisms. W.H. Freeman & Co (1979).
Weaver , E. , H. Kenney and M. Wall : Appl. Microbiol., 8 , 345 (1960)
Yamane, T., H. Nakatani, E.Sada, T. Omata, A.Tanaka and S. Fukui : Biotechnol. Bioeng., 21 , 2133 (1979).

Acknowledgements. I wish to thank Professor Malcolm Lilly for his advice during this work done on the Chemical and Biochemical Engineering Department at University College London. I am also thankfull to Dr.Schneider and the NATO Scientific Committee for inviting me to this workshop.

SECOND-GENERATION BIOCATALYSIS

J. Tramper[1], C. Laane[2] and J.A.M. de Bont[3]
[1]Department of Food Science, Food and Bioengineering Group,
[2]Department of Biochemistry and [3]Department of Microbiology,
Agricultural University, Wageningen, The Netherlands.

ABSTRACT. Applied biocatalysis is a rapidly expanding area, but there are still several bottlenecks which restrict the number and scope of applications. Especially availability of efficient biocatalysts, water as essential reaction medium, toxicity of substrates and products, cofactor regeneration and product recovery are limiting factors. At the Agricultural University Wageningen, within the framework of the Working Party Biocatalysts, these problems are attacked in a multidisciplinary fashion. A short account of the joint efforts is given in this paper. Approaches to solve these problems include screening for suitable microorganisms; optimization of desired microbial activity by cultivation in chemostats, by classical genetics and by recombinant-DNA work; integration of biocatalysis and electrochemistry; introduction of water-immiscible organic solvents; omission of free water; and immobilization of biocatalysts. These approaches are illustrated using the examples of stereospecific epoxidation of alkenes, regiospecific chlorination of barbituric acid and transformations of steroids.

1. INTRODUCTION

Seldom missing in argumentations advocating the application of biocatalysts for syntheses is a statement in the style of: Biocatalysis is non-polluting due to the necessary aqueous reaction medium. This ostensible advantage, however, also brings about several fundamental restrictions which limit the range of applications. Especially instability and low solubility of substrate(s) and/or product(s) in water exclude many interesting biocatalytic reactions from economic and convenient practice. Other fundamental problems inherent to many bioprocesses are substrate and/or product inhibition, and difficult product/biocatalyst separation. These problems more generally make dilute solutions unavoidable and laborious work-up procedures rather a rule than an exception. When in addition it is considered that biocatalysts usually have a low efficiency and may require cofactor regeneration, it is obvious why the number and scope of applications of biocatalysts is still limited.

M. P. Schneider (ed.), Enzymes as Catalysts in Organic Synthesis, 383–390.

Our approach to solve the problems associated with water as essential medium for the biocatalysts, is the introduction of a water-immiscible organic solvent, which is used as substrate and/or product reservoir. The biocatalyst however still remains in an aqueous environment, either by its immobilization in a water-containing solid carrier, or by confining it to the water pool of a reversed micelle. In case of volatile or gaseous substrates these water-related problems are attacked by employing a gas/solid bioreactor, consisting of the solid immobilized biocatalyst and a continuous gas phase containing the substrate(s) and product. In other words, free water is omitted as much as possible.

Problems associated with low overall productivity of the biocatalyst may be solved by screening for more efficient biocatalysts [1] and by immobilization [2]. For more effective cofactor regeneration the approach may either be using whole cells, multi-enzyme catalysis or integration of biocatalysis and electrochemistry. In this paper various aspects of these approaches are illustrated by discussing some representative systems under investigation.

2. STEREOSPECIFIC EPOXIDATIONS OF SHORT CHAIN ALKENES

Fundamental and practical knowledge of the behaviour of immobilized biocatalysts in multi-liquid phase and gas/solid bioreactors hardly exists, both from a microbial and engineering point of view. Therefore we have initiated a study with the aim to build up fundamental knowledge of these second-generation bioreactors, using the microbial epoxidation of short-chain alkenes as the model reaction.

$$R\text{-}CH{=}CH_2 + O_2 + XH_2 \longrightarrow R\text{-}\underset{\diagdown O \diagup}{CH\text{-}CH_2} + H_2O + X$$

Immobilized bacteria producing epoxyethane from ethene and oxygen are used in the investigations on the gas/solid bioreactor [3]; the epoxidation of higher alkenes as for instance propene and butene is the model reaction in the two-liquid-phase bioreactor. These epoxidation reactions were chosen as model because of our experience [4] with relevant bacteria and because the epoxides produced have pertinent properties like inhibitory effects [5] and low solubility and stability in water, while the substrates (alkenes and oxygen) also are poorly soluble in water.
The most important criteria in the selection of a suitable strain are site-specific and possibly stereospecific catalysis, specific activity of the enzyme involved, absence of product-degrading enzymes, range of possible co-substrates, and sensitivity of the microorganisms used for the product formed. An additional interesting aspect of the epoxida-

tion reactions is the involvement of mono-oxygenases. These enzymes require a reduced cofactor for catalysis and therefore cofactor regeneration and cosubstrate utilization by whole cells may be studied as well. Finally, the easy and accurate gaschromatographic analysis of substrates and products make the chosen system very convenient and attractive [6].

From a synthetic viewpoint stereospecificity is by far the most important factor, as chiral epoxides are very interesting synthons, which are difficult to synthesize by chemical means. We have analysed the enantiomers by various strains and Tables I and II show some of the results which are promising with respect to this synthetic aspect [7].

Table I Enantiomeric purity of 1,2-epoxypropane and 1,2-epoxybutane produced by various alkene and alkane-grown strains

Strain	Growth subtrate	R-1,2-Epoxy-propane (%)	S-1,2-Epoxy-propane (%)	R-1,2-Epoxy-butane (%)	S-1,2-Epoxy-butane (%)
E3	Ethene	93 ± 1	7 ± 1	85 ± 1	15 ± 1
E4	Ethene	93 ± 1	7 ± 1	84 ± 1	16 ± 1
E44	Ethene	93 ± 1	7 ± 1	83 ± 1	17 ± 1
Eul	Ethene	93 ± 1	7 ± 1	84 ± 1	16 ± 1
2W	Ethene	93 ± 1	7 ± 1	83 ± 1	17 ± 1
Tul	Ethene	91 ± 1	9 ± 1	83 ± 1	17 ± 1
E20	Ethene	88 ± 1[a]	12 ± 1[a]	81 ± 1[b]	19 ± 1[b]
E20	Ethane	47 ± 1	53 ± 1	41 ± 1	59 ± 1
B	1,3-Butadiene	93 ± 1	7 ± 1	89 ± 1	11 ± 1
Py1	Propene	96 ± 1[a]	4 ± 1[a]	n.d.	n.d.
Py2	Propene	97 ± 1[a]	3 ± 1[a]	94 ± 1[b]	6 ± 1[b]
Py10	Propene	n.d.	n.d.	95 ± 1[b]	5 ± 1[b]

[a]Inhibited with 10 mM 1,2-epoxybutane; [b]Inhibited with 10 mM 1,2-epoxypropane; n.d., not determined.

Table II Enantiomeric purity of 1-chloro-2,3-epoxypropane produced by various alkene-grown strains

Strain	Growth substrate	R-1-Chloro-2,3-epoxypropane (%)	S-1-Chloro-2,3-epoxypropane (%)
E3	Ethene	4 ± 1	96 ± 1
E4	Ethene	3 ± 1	97 ± 1
2W	Ethene	1 ± 1	98 ± 1
Py2	Propene	20 ± 1	80 ± 1

For applications in the two-liquid phase bioreactor sensitivity of the organism used to organic solvents is obviously a key parameter. In order to find a correlation between type of organic solvent used and retention of activity, free and immobilized cells were confronted with about 40 common solvents [8,9]. In case of free cells the presence of a second liquid phase caused often immediate aggregation of the cells at the interface with complete loss of activity. Although immobilization of the cells in gels like alginate and carrageenan prevents this phenomenon, no additional protection is afforded by immobilization per sé. The results given in Figure 1 make clear that low molecular weight solvents with a high polarity are most detrimental to the activity of the cells.

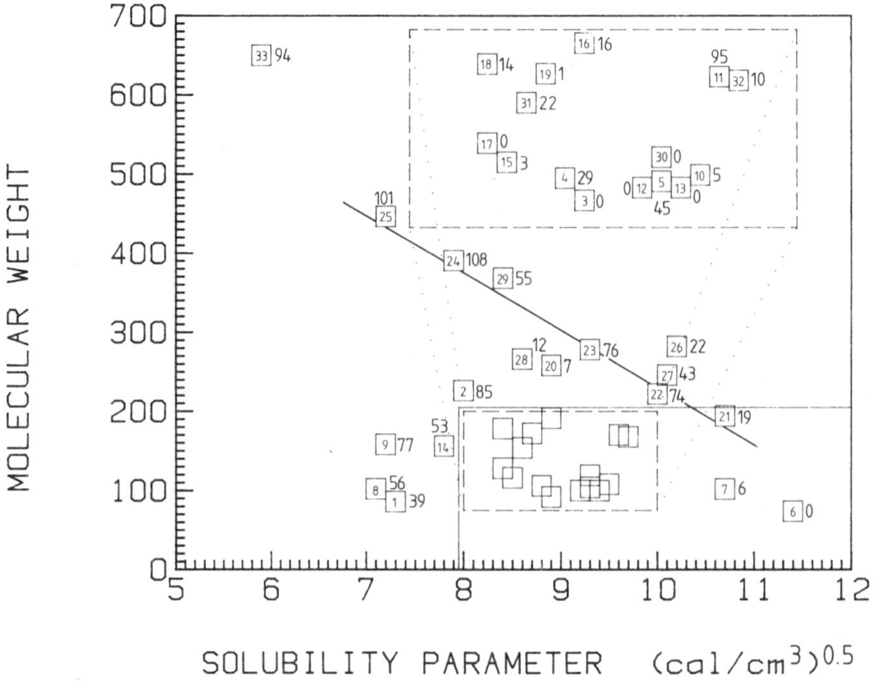

Figure 1. Measured activity retentions (%) of immobilized cells exposed to organic solvents as a function of solvent molecular weight and Hildebrand solubility parameter; number in block refers to particular organic solvent (identification in ref. 8) e.g. 33 = perfluoro chemical and 21, 22, 23, 24 and 25 di-n-alkylphthalates.

Work presently in progress is stabilization and kinetic description of the biocatalyst.

3. BIOELECTROCHEMICAL CHLORINATION OF BARBITURIC ACID

To date, the main objective of bioelectrochemical devices is the generation of electricity, or the development of sensitive detectors for (bio)chemicals. In addition to these objectives we have demonstrated recently that these systems can also be applied for the synthesis of various (bio)chemicals [10-12].
One of these bioelectrochemical systems currently under investigation converts barbituric acid (I) into monochlorobarbituric acid (II) using chloroperoxidase.

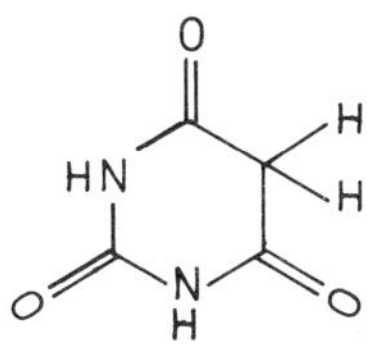

I

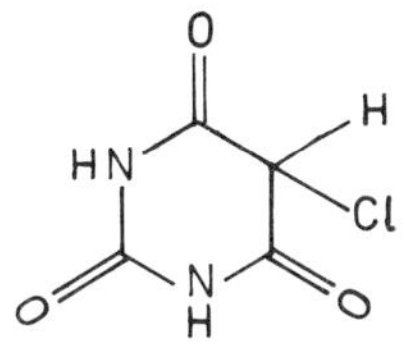

II

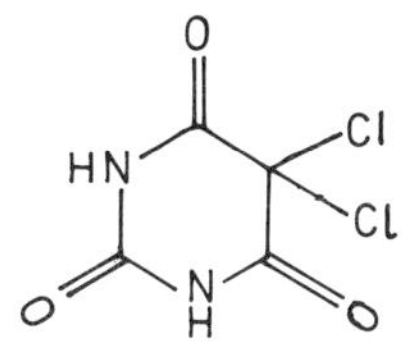

III

The system is developed in collaboration with the department of Organic Chemistry and consists of three interconnected units, i) an electrolytic cell that oxidizes H_2O at the anode to O_2 and generates the required H_2O_2 for chloroperoxidase action at the cathode using electricity as energy source, ii) a hollow-fiber membrane reactor in which the electrochemically produced H_2O_2 is used by chloroperoxidase for the halogenation of barbituric acid; and iii) an anionexchanger that scavenges the monochlorobarbituric acid produced in the membrane reactor (Figure 2).
In this system only monochlorobarbituric acid is formed and not the usual mixture of mono- and dichlorobarbituric acid (III), since in the electrolytic cell dichlorobarbituric acid is converted electrochemically into the monohalogenated compound according to the reactions shown in figure 3.

Another advantage of this electrochemical system is that the operational lifetime of the enzyme is longer than usual, since in an electrochemical cell the production rate of H_2O_2 can be tuned easily to the reaction rate of chloroperoxidase, thus avoiding toxic concentrations of H_2O_2. The results so far, although preliminary, are encouraging. In our model system about 10 mg of pure monochlorobarbituric acid can be produced per hour per unit enzyme.

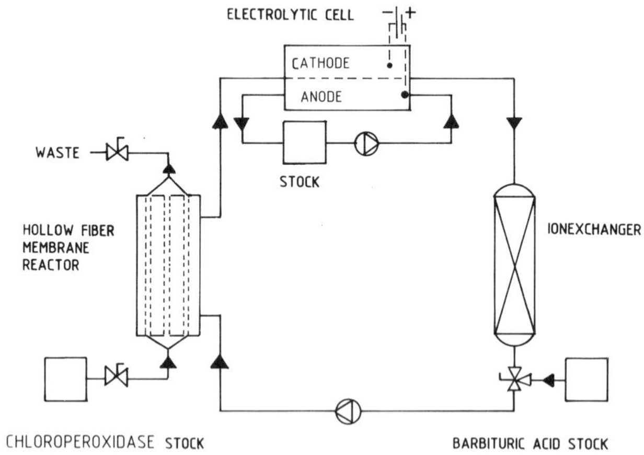

Figure 2. Scheme for the bioelectrolytic production of monochlorobarbituric acid.

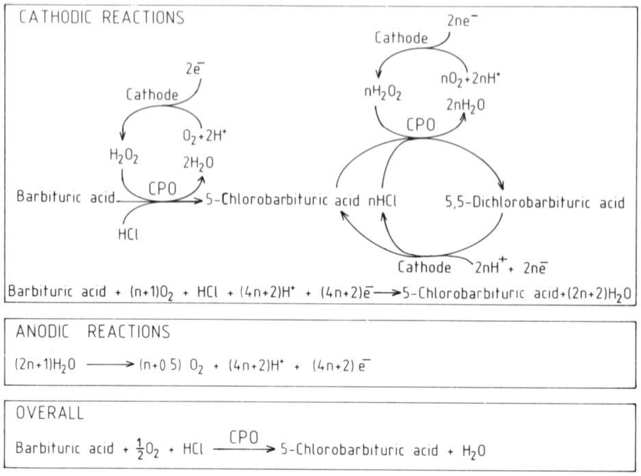

Figure 3. Reactions taking place in an electrolytic cell that converts barbituric acid into monochlorobarbituric acid using chloroperoxidase (CPO).

4. REGIOSPECIFIC TRANSFORMATIONS OF STEROIDS

In addition to the organic solvent systems described in chapter 2, biocatalysis in micellar media is also investigated.
These systems contain the following components; i) a water immiscible organic solvent, ii) a surfactant, and iii) an aqueous phase. Enzymatic reactions are performed preferentially in media containing relatively small amounts of water (water-in-oil microemulsions), while bacterial transformations are carried out in media containing relatively large amounts of water (oil-in-water microemulsions). Figure 4 shows a schematic representation of these two media.

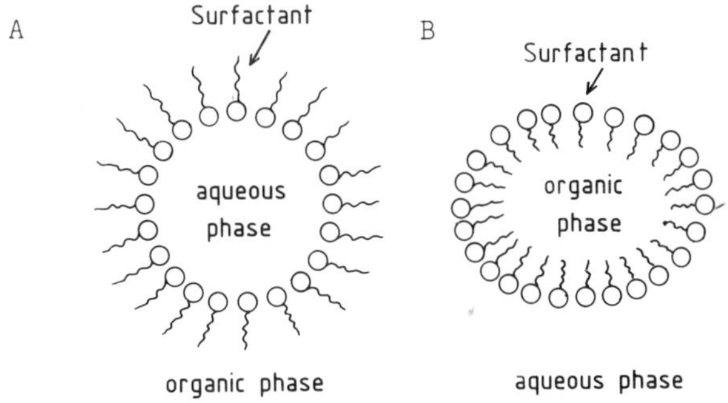

Figure 4. Multiphase systems containing the components surfactant, water, and an organic solvent. (A) w/o-microemulsion, or reversed micellar medium; (B) o/w-microemulsion.

At present most of our research is focused on the enzymatic conversion of apolar compounds in w/o-microemulsions, also known as reversed micellar media. We have shown that many redox enzymes such as cholesterol oxidase [13] and 20β- hydroxysteroid dehydrogenase [14-16] are not only very active in reversed micelles but remain it for many days.
Another project is the transformation of androstenolene acetate via dehydro-epiandrosterone into 4-androsterone-3,17-dione by *Flavobacterium dehydrogenans*. Usually these reactions are carried out in aqueous media even though the reactants are hardly soluble in water. Consequently the steroids are present in the solid state, and hence the reaction rate is hindered by the limited availability of the substrate. Furthermore, the recovery of the product, and its yield is restricted through entrapment of substrate molecules in the crystalizing product. In industry surfactants such as Tween and Span are used to increase the substrate concentration in the medium. Although this procedure increases the rate of steroid conversion it does not prevent the cocrystalization process. To solve this problem the product should be kept in a soluble form till the conversion is complete. This can be achieved by adding an organic solvent to the medium that does not inhibit the desired catalytic activity of the microorganism.

Such a solvent appeared to be octane. Just by adding octane to the culture medium reaction rates could be increased 2-7 times compared with the rate obtained in Tween-media. Highest conversion rates were achieved when the bacterium was in the end of its logarithmic growth phase. Under these conditions flavobacterium excretes an emulsifying agent which probably facilitates the overall reaction by enhancing the mass transfer of substrate in the medium.
Work is in progress to explain in more detail the beneficial effect of octane in combination with this emulsifying agent on the conversion of androstenolone acetate and related apolar steroids.

5. REFERENCES

1. Habets-Crützen, A.Q.H., Brink, L.E.S., Ginkel, C.G. van, Bont, J.A.M. de, and Tramper, J. (1984) Appl.Microbiol.Biotechnol. 20, 245-250.
2. Tramper, J. (1985) Trends in Biotechnology 3, 45-50.
3. Bont, J.A.M. de, Ginkel, C.G. van, Tramper, J. and Luyben, K.Ch.A.M. (1983) Enzyme Microb.Technol. 5, 55-59.
4. Bont, J.A.M. de, Attwood, M.M., Primrose, S.B. and Harder, W. (1979) FEMS Microbiol. Lett. 6, 183-188.
5. Habets-Crützen, A.Q.H. and Bont, J.A.M. de (1985) Appl.Microbiol.Biotechnol., in press.
6. Brink, L.E.S., Tramper, J., Riet, K. van 't, and Luyben, K.Ch.A.M. (1984) Analytica Chimica Acta 163, 207-217.
7. Habets-Crützen, A.Q.H., Carlier, S.J.N., Bont, J.A.M. de, Wistuba, D., Schurig, V., Hartmans, S. and Tramper, J. (1985) Enzyme Microb.Technol. 7, 17-21.
8. Brink. L.E.S. and Tramper, J. (1985) Biotechnol.Bioeng., in press.
9. Tramper, J., Brink, L.E.S., Hamstra, R.S., Bont, J.A.M. de, Habets-Crützen, A.Q.H. and Ginkel, C.G. van (1984) Third European Congress on Biotechnology, Volume II, Verlag Chemie, Weinheim, pp. 269-276.
10. Laane, C., Pronk, W., Franssen, M. and Veeger, C. (1984) Enzyme Microb.Technol. 6, 165-168.
11. Laane, C., Hilhorst, R., Spruyt, R., Dekker, K. and Veeger, C. in: Flavins and Flavoproteins (Bray, R.C., Engel, P.C. and Mayhew, S.G., eds.) Walter de Gruyter & Co., New York, 1984, pp. 879-892.
12. laane, C., Dekker, K. and Veeger, C. (1984) Proc. Third Eur.Congress on Biotechnology, Verlag Chemie, Weinheim, Vol. 1, pp. 395-399.
13. Laane, C. and Spruyt, R. (1985) Proc. NATO-ASI Meeting "Recent Developments in Biotechnology" Tróia, Portugal, 17-29 March, in press.
14. Hilhorst, R. (1984) Thesis, Agricultural University, Wageningen, The Netherlands.
15. Hilhorst, R., Laane, C. and Veeger, C. (1983) FEBS Lett. 159, 225-228.
16. Hilhorst, R., Spruyt, R., Laane, C. and Veeger, C. (1984) Eur.J.Biochem. 144, 459-466.

ROUND-TABLE DISCUSSIONS

In a series of round-table conferences special topics within this rapidly expanding field were discussed. The major conclusions and statements are summarized below. We are grateful to Prof. Dr. H. Simon, München, for assistance in the preparation of this summary.

Preliminary remarks: It was generally agreed upon that enzymatic reactions will become increasingly important in organic synthesis under a variety of different practical aspects:

1. They are already highly attractive and widely used in the selective preparation of optically active compounds on a laboratory scale and in all areas where small quantities of compounds must be made available rapidly, e. g. for testing purposes. Here, a straightforward synthesis, using known enzymatic steps may be advantageously employed without much optimisation.
   Costs of products are irrelevant if time and labour is saved instead. Synthetic organic chemists can already achieve a lot without the assistance of specialists.

2. In cases where
   a) one or more enzymatic reaction steps have to be employed frequently in a research laboratory within a larger programme; or
   b) a steady supply of a commercially unavailable research chemical has be ensured

   an increased cooperation of the synthetic organic chemists with specialists will be of great advantage. The application also of unorthodox protocols, microorganisms and enzymes should be considered.

3. A compound with an established market prize or a new product has to be prepared in a competitive way using enzymes. On an industrial scale prizes and added values of products will determine the feasibility of an enzymatic process. Enzymatic transformations are particularly attractive if there is no good chemical alternative or chemical processes are difficult to carry out due to practical or environmental reasons.
   Not surprisingly some large scale chemical processes are about to be terminated in favour of enzymatic methods.
   It is most likely that this trend in industry will continue.

These different points of view, expressed by academic and industrial participants, although not always mentioned, were kept in mind during the round-table discussions.

M. P. Schneider (ed.), Enzymes as Catalysts in Organic Synthesis, 391–401.

I. NICOTINAMIDE RECYCLING

Oxidoreductases are catalyzing a multitude of highly enantioselective oxidations and reductions of suitable substrates, thereby leading to valuable optically active products.
Although other cofactors (e. g. PQQ for oxidations with methanol dehydrogenases) are known or will probably become known in the future, the overwhelming portion of presently applied oxidoreductases is pyridine nucleotide-dependent. Due to the high costs of these cofactors they cannot be used preparatively in equimolar amounts. In order to reduce their quantities to catalytic amounts efficient methods for their recycling had to be developped. In order to achieve high numbers of cycles (50000 - 100000) enzymatic methods are clearly preferred. They must, however, fulfill the following requirements:

- a thermodynamically favored reaction and a stable enzyme must be employed;
- neither substrate nor product in the enzymatic regeneration should interfere with the rest of the system;
- no interference with the product isolation should occur.

For the regeneration of NADH these problems are basically solved by the application of:

(a) NAD^+ dependent formate dehydrogenase from _Candida boidinii_ catalyzing the oxidation of formate ions to carbon-dioxide (CO_2):

$$HCO_2^- + NAD^+ \longrightarrow NADH + CO_2$$

The method ideally fulfills the above requirements, formates are cheap and the product (CO_2) is readily removed from the reaction mixture. The main disadvantage of this system is the high prize of the enzyme and the fact that it accepts NAD only.

(b) Glucose dehydrogenase from _Bacillus cereus_ (C.-H. Wong, D.G. Drueckhammer, H.M. Sweers, _J. Am. Chem. Soc._ **107**, (1985) 4028-31). Turnover number of up to 40000 cycles are reported. The enzyme accepts both NAD and NADP which is important for the application of oxidoreductases from thermophilic microorganisms, many of which are NADPH-dependent. It uses glucose as substrate and produces equimolar amounts of D-gluconate which have to separated from the desired product.

(c) Indirect electrochemical or enzymatic regeneration of NAD(P)H using viologen dependent pyridine nucleodide reductases (see contributions in these proceedings by: H. Simon, H. Günther, p. 32; S.Y.R. Pugh, p. 211).

Although in dehydrogenase reactions the reoxidation of NAD(P)H can be achieved on a small scale with the existing techniques (G. Lemière, p. 16 of this book; J.B. Jones, K.E. Taylor, Can. J. Chem. **54** (1976) 2969; Review: G.M. Whitesides, C.-H. Wong, Angew. Chem. Int. Ed. Engl. **24** (1985) 617), these methods must be improved considerably in order to become really useful tools in organic synthesis. *

Another problem is the stability of the pyridine nucleotides. They are basically labile compounds which are subject to a number of chemical and enzymatic degradations. Their stability depends very much on the conditions such as pH, presence of oxygen, kind of buffer (e.g. instability in phosphate buffer even at pH 7 due to general acid catalysis), effects well known in biochemistry. Spontaneous redox exchange producing the 6-dihydropyridine (an inhibitor), is observed; another reason to keep cofactor concentrations low.
Usually most cells contain enzymes (amidases, glycosidases, phosphatases, NAD pyrophosphatases) which degrade pyridine nucleotides. Therefore the application of resting cells or crude enzyme preparations may be limited due to enzymatic NAD(P) degradation.

Despite all these difficulties the regeneration especially of NADH is a solved problem. For continous homogeneous catalysis in an enzyme membrane reactor the coenzyme has been covalently linked to polyethylene glycol. The material is already being prepared in kg quantities and cycle numbers of up to 90000 have been obtained by the groups of M.R. Kula and C.H. Wandrey (p. 257 of this book). Wong et al. (see above) reported cycle numbers of 40000 with non-modified NAD during a period of 6 days. With such cycle numbers it becomes clear that the prize for the coenzyme no longer dominates the cost of a synthetic process. The use of this method on an industrial scale shows great promise and the scale-up of NADH-dependent processes for L-aminoacid production in a pilot plant is under investigation.
The molecular weight enlarged NAD often reveals kinetic data not very different from those of the unmodified material. However, there are exceptions known where it proved to be less effective.(N. Katayama, I. Urabe, H. Okada, Eur. J. Biochem. **132** (1983) 403-409.)
The material has, however, the advantage of higher stability, it can be easily recovered and used repeatedly. Because of the great advantages in practical applications it should be made commercially available and its application to other systems than L-aminoacid production should be studied.

* For a very important step in this direction see: L.G. Lee, G.M. Whitesides, J. Am. Chem. Soc., **107** (1985) 6999-7008 (added after completion of the manuscript).

The use of whole cell systems as an alternative to cofactor recycling is not always attractive. As already mentioned above, resting cells are frequently degrading the cofactors. Growing cultures have the general disadvantage of creating a lot of undesired biomass. Next to the large volumina produced, exocellular side products or impurities by secondary reactions are creating work-up and separation problems. Wherever the isolated enzyme, catalyzing the desired reaction step, can be made available its application together with cofactor regeneration should be seriously considered.

Suggestions and requests of interested users, also regarding the alcohol dehydrogenases, will be found in the general discussion (IV. "Where do we go?" Conclusions and Suggestions).

II. ENZYMATIC HYDROLYSIS OF NITRILES, AMIDES, URETHANES AND CARBONATES.

Hydrolases are highly attractive enzymes in organic synthesis and many useful applications of enzymatic ester hydrolysis are shown in this book. Particularly important arguments for the introduction of hydrolases with new activities are the very mild conditions under which they can hydrolize functional groups which usually need rather vigorous conditions for the corresponding chemical hydrolysis, e. g. in nitriles and amides.

Nitriles are synthetic intermediates which are readily accessible and which are frequently also used for the introduction of nitrogen into organic molecules. Their mild hydrolysis in presence of nitrilases may be particularly useful, if these enzymes display broad substrate tolerances and interesting regio- and stereoselectivities (eq. 1).

$$R - CN \longrightarrow R\text{-}CONH_2 \longrightarrow RCO_2H \qquad (1)$$

Very little seems to be known so far about the enantioselectivities of these reactions. In cases, where two enzymes are involved (Yamada, Kyoto) in the hydrolysis of nitriles (a nitrile hydratase and an amidase) only the second step seems to be stereospecific (eq. 2).

$$R\text{-}CN \xrightarrow{\text{nitrile hydratase}} R\text{-}CONH_2 \xrightarrow{\text{amidase}} R\text{-}CO_2H \qquad (2)$$

In dinitriles the first function is hydrolized much more rapidly than the second group (eq. 3).

$$NC\text{-}(CH_2)_n\text{-}CN \xrightarrow{\text{fast}} NC\text{-}(CH_2)_n\text{-}CONH_2 \xrightarrow{\text{slow}} H_2NCO\text{-}(CH_2)_n - CONH_2 \qquad (3)$$

The reasons for this behaviour, similar to observation with diesters, became not clear even after a long discussion. Among other arguments, a good reason could be the fact that the produced amide group is much better solvated than the nitrile, resulting in a drastic change of the binding constant to the enzyme. Not many kinetic data seem to be known as yet. However, this selectivity could be used for the synthesis of multifunctional compounds and perhaps even chiral building blocks. In this respect the problems and likely approaches to their solution are similiar to the ones with esterases.

Organic chemists indicated that hydrolases capable of converting cyclic urethanes and carbonates would be of great synthetic interest. It was also mentioned that hydrolases of epoxides may be of great value.

It became obvious that, before practical synthetic methods will emerge, a lot of work still has to be invested in order to find more well defined hydrolases with useful and defined specifities. In this context it should be mentioned that even widely used enzymes like porcine liver esterase preparations are ill defined and may be mixtures of enzymes with different specifications.

III: ENZYMES IN ORGANIC SOLVENTS. ENZYMATIC ESTERIFICATION.

The application of enzymatic reactions in organic solvents is of increasing importance and in recent years a rapid development in this area has taken place (Review: S. Fukui, A. Tanaka, Endeavour, New Series **9** (1985) 10 - 17).
Enzymes in organic solvents may be of interest with respect to:

- better solubility of hydrophobic substrates and products;
- shifts of equilibria, e. g. from hydrolysis to synthesis;
- enzymes may exhibit unusual catalytic properties when exposed to unusual environmental conditions.

With regard to the reaction mechanisms of enzymes one of the best known examples is the formation of an acylated enzyme intermediate during the hydrolysis of esters, peptides or amides. In the first step of such a hydrolysis the enzyme acts as a nucleophile leading to the acylated enzyme RCO-E. The most common nucleophilic groups in such enzymes are the serine hydroxyl (serine proteases, cholinesterases, esterases, lipases, acid and alkline phosphatases) and the thiol group of cysteine in thiol proteases (papain, ficin, bromelain etc.).

Two step mechanisms of many hydrolases:

$$\text{R-COX} + \text{E} \xrightarrow{\text{1. step}} \text{R-CO-E} + \text{HX}$$

$$\text{R-CO-E} + \text{Nu} \xrightarrow{\text{2. step}} \text{R-CO-Nu} + \text{E}$$

X = OR, NHR etc.

In an aqueous environment the second step consists in an attack of the acylated enzyme by the nucleophile water. If the concentration of water is diminished, (e. g. by carrying out the reaction in an organic solvent) and that of other nucleophiles (amines, aminoacids, alcohols etc.) is increased there is a good chance that the acyl residue is transferred to such a nucleophile Nu. In such a way peptide, amide or ester bonds may be formed. Recent examples clearly demonstrate that these reaction are often highly enantioselective and can be used for the preparation of optically active esters from racemic alcohols in presence of lipases. (G. Kirchner, M.P. Scollar, A.M. Klibanov, J. Am. Chem. Soc., **107** (1985) 7072-7076; S. Koshiro, K. Sonomoto, A. Tanaka, S. Fukui, J. Biotechnol. 2 (1985) 47 - 57; G. Langrand, M. Secchi, G. Buono, J. Baratti, C. Triantaphylides, Tetrahedron Lett. **1985**, 1857) Proteases have been successfully employed in insulin semisynthesis and in the preparation of enkephalin. Shifts of equilibria can be achieved either by mass action (see above) or, alternatively by removal of a product e. g. due to insolubility.

An interesting application of this principle is the stereospecific enzymatic synthesis of L-α-aspartame from benzyloxycarbonyl-D,L-aspartic acid and D,L -phenylalanine methylester. In constitutes both an enantio- and regioselective biocatalytical transformation.

The use of organic solvents as co-solvent or as a second liquid phase has been employed for:

- conversions in the steroid field
- transesterifications with lipases, e. g. in glycerides
- stereoselective hydrolyses of racemic menthylesters for the production of l- und d-menthol
- enzymatic resolution of racemic aminoacids, e. g. production of D-phenylglycine

and many others.

IV. GENERAL DISCUSSION "WHERE DO WE GO?" CONCLUSIONS AND SUGGESTIONS

All participants agreed that the application of enzymes in organic syntheses will rapidly grow in the near future. Enzymatic reactions are efficient and highly selective, their use is frequently advantageous from an energetic or environmental point of view. They can be used for the preparation of enantiomerically pure compounds and also to achieve simple chemical transformations under very mild conditions.
In the very near future chemoenzymatic approaches to target molecules, in which the enzymatic reaction is only one step in a sequence, will probably have the best chances to be successful.
From an industrial point of view the feasibility of an enzymatic process will be determined by product prizes ("added value") and the efficiency of existing chemical methods. If the compounds in question (substances with biological activity, pharmaceuticals, agrochemicals, unnatural aminoacids, enzyme inhibitors, biochemical research tools, fine chemicals etc.) have the required added value and cannot be prepared chemically in an easy way, enzymes catalyzing the desired transformations can usually be found and applied.

It was generally accepted that the use of biocatalysts in organic synthesis is not so much limited by the lack of suitable enzymes and methods. It seems to be limited more by a lack of knowledge regarding the needs and possibilities.

- Most of the enzymes used more generally in organic synthesis have not been selected for this special purpose.

- They have been detected in biochemical or medical studies using natural substrates.
  Only slowly their synthetic usefulness is becoming obvious.

- Clearly, for many conversions more suitable enzymes may exist or even be known without realizing their range of possible applications.

- It is therefore recommended that biochemists increasingly include substrates of synthetic interest into their studies in order to test substrate tolerances and reaction selectivities of their known or newly discovered biocatalysts.

- From the viewpoint of stability, enzymes from thermophilic organisms can be superior.

The regeneration of most coenzymes is solved in such a way that it can be exercised in most organic laboratories.
Already available are also generally applicable protocols for the immobilization of biocatalysts even by non-specialists.

With these points in mind the following topics were briefly discussed:

- Chiral building blocks
- Role of enzymes in protection and deprotection
- Regioselective reactions using enzymes
- New areas for enzyme applications
- Artificial coenzymes, genetic- and proteinengineering

Chiral building blocks.

The enzymatic preparation of these, usually small and well defined, molecules is a very successful area of research which is likely to expand further in the future. Similiarly, isotopically labelled compounds could be made availabe increasingly by these methods.

Oxidoreductases. The use of commercially available alcohol dehydrogenases on a preparative or industrial scale still seems to be somewhat problematic. Yeast alcohol dehydrogenase (YADH) is rather instable. The mammalian enzyme horse liver alcohol dehydrogenase (HLADH) displays high selectivities and sufficient stability. Product inhibition, however, largely prevents the necessary scale-up. Nevertheless, the methodology for biocatalytic, enantioselective, reductions of unsaturated compounds has undergone considerable development over the past few years.
Alcohol dehydrogenases from thermophilic microorganisms (e. g. Thermoanaerobium brockii) have now become commercially available. They are highly attractive catalysts due to their high stabilities and selectivities.*
Clearly, thermophilic and anaerobic microorganisms should be screened increasingly for ADH activities. It is likely that further preparatively or even industrially useful oxidoreductases will be discovered, since an intensive search for such enzymes under the aspect of synthetic usefulness only began a few years ago.

Hydrolases. These coenzyme-independent enzymes are already used extensively for the preparation of chiral building blocks on a laboratory and industrial scale. Several examples for their application can be found in this book.
It is to be recommended that the possibilities of introducing heteroatoms (e. g. S, N etc.) by transfer of acyl-enzyme complexes onto other nucleophiles than H_2O or ROH should be studied intensively. Hydrolases should be used increasingly also for purely "chemical" transformations like saponifications under mild and neutral conditions.

* For a very recent example see: E. Keinan, E.K. Hafeli, K.K. Seth, R. Lamed, J. Am. Chem. Soc., **108** (1986), 162 - 169.

In this connection screening for hydrolases transforming other functional groups (nitriles, urethanes, carbonated and substituted ureas) should be carried out.

Role of enzymes in protection and deprotection.

The selective enzymatic deprotection of acylated carbohydrates and nucleosides would be of great interest and should be studied intensively. Which esterase deacetylates pentaacetylglucose selectively? Although some work in this area has been done already, the methods are far from being clean and practical.

Glycoside transferases for the synthesis of oligosaccharides with unprotected monosaccharides would be a highly attractive field of research. Unfortunately most of the involved enzymes are membrane bound, instable, and therefore difficult to isolate. Nevertheless, the latest enzyme list contains 153 hexosyl- and 33 pentosyl transferases, most of which have hardly been tested under preparative aspects. Transferases for the selective fixation of glycosidic residues to peptides etc. as posttranslational modifications would also be desirable.

The synthesis of oligopeptides with unprotected aminoacids is of great preparative interest. A japanese enzyme producer is now offering eight different t-RNA-synthetases, which, together with an ATP-regeneration system, can be used effectively for that purpose.

Regioselective reactions using enzymes.

Next to already known applications of oxidoreductases and hydrolases, the regiospecific epoxidation of carbon-carbon-double bonds in olefines would be of great preparative interest. Of equal value is the regio- and enantioselective opening of epoxides to the corresponding 1,2-diols by epoxide hydrolases (e. g. from fresh liver preparations, G. Berti, p. 338). Particularly attractive would also be selective hydroxylations of non activated C-H-bonds (compare whole cell hydroxylations in the steroid field).

New areas for enzyme applications.

While enzymes for selective functional group transformations are becoming increasingly available to the synthetic organic chemist, more of these catalysts for the formation of C-C-conds are required. The use of aldolases for this purpose was described by G.M. Whitesides and C.-H. Wong (p. 193). The corresponding application of transaldolases and -ketolases may also be an attractive area of research.

The enantioselective addition of HCN to aldehydes, catalyzed by oxynitrilase (E.C. 4.1.2.10) has been known for many years (W. Becker, E. Pfeil, J. Am. Chem. Soc., **88** (1966) 4299 and earlier publications by E. Pfeil etal.).
The inexpensive and readily available enzyme from almonds adds the cyano group exclusively on the si-face of the carbonyl group in a broad range of structurally different aldohydes. Kilogram quantities of (R)-cyanohydrines have been prepared this way and converted into chiral synthons like α-hydroxyacids, aminoalcohols etc.

Particularly interesting from an organic synthetic point of view are enantioselective hydroxylations of non activated C-H-bonds and hydroxylations of aromatic compounds. These reactions cannot be carried out chemically and are so far only effected by whole cell systems. The possibilities of using isolated enzymes should be explored carefully.

So far an enzyme catalyzed reaction which corresponds to the Wittigreaction

$$ {}^{1}R_2C = O + {}^{2}R_2\overset{-}{C}\text{-}\overset{+}{X} \longrightarrow {}^{1}R_2C = C^{2}R_2 + X = O $$

does not seem to exist. The enzymatic preparation of an aldol-type product, followed by the application of a dehydratase could, however, possibly mimic this transformation.

Artificial coenzymes, genetic and proteinengineering

The practical application of artificial coenzymes and enzymes is still far off but may be possible in the future. The same is true for semisynthetic enzymes and the synthetic usefulness of site-specific mutagenesis. The use of genetic engeneering ist at the verge of general applicability and seems at present the most interesting route to follow.

Requests from synthetic organic chemists:

Organic chemists would like to have more coenzymes and enzymes be made commercially available, in particular: molecular weight increased nucleotides, more oxidoreductases, hydrolases for epoxides, nitriles, esters and lipids. More enzymes from thermophilic organisms should be available (e. g. from *Bacillus stereothermophilus*, *Thermoanaerobium brockii* etc.).

Final Remarks

Enzymatic methods should be seen as being complementary to chemical procedures rather than as a competition. Synthetic organic chemistry has developped over many years and a gradual "evolution" of chemoselectivity $\longrightarrow$ regioselectivity $\longrightarrow$ diastereoselectivity $\longrightarrow$ enantioselectivity has taken place over many decades. Enzymatic reactions can fit into all these categories and basic, systematic studies are needed here much like with the chemical methods in the past.

The collection of well written (and independently checked!) experimental procedures will be as helpful to promote an increased use of enzymes in organic synthesis as intensive educational efforts in this area both in universities and industry.

LIST OF CONTRIBUTORS

Adachi, K., Faculty of Pharmaceutical Sciences, University of Tokyo, 7-3-1 Hongo, Bunkyo-ku, Tokyo 113, Japan
Andresen, O., Novo Industri A/S, Novo Alle, DK-2880 Bagsvaerd, Denmark
Archelas, A., Laboratory of Organic an Bioorganic Chemistry, Faculté des Sciences de Luminy, Case 901, 70 route Léon Lachamp, F-13288 Marseille Cedex 9, France
van Balken, J. A. M., DSM, Research and Patents, P. O. Box 18, NL- 6160 MD Geleen, The Netherlands
Banfi, L., Institute of Organic Chemistry, University of Genova, Centro C.N.R., Palazzo delle Scienze, Corso Europa, I-16132 Genova, Italy
Benner, S. A., Department of Chemistry, Harvard University, 12 Oxford Street, Cambridge, Massachusetts 02138, U.S.A.
Berti, G., Institute of Organic Chemistry, Faculty of Pharmacy, University of Pisa, Via Bonanno 6, I-56100 Pisa, Italy
Boesten, W. H. J., DSM, Research and Patents, P. O. Box 18, NL-6160 MD Geleen, The Netherlands
de Bont, J. A. M., Department of Microbiology, Agricultural University, De Dreijen 12, NL-6703 BC Wageningen, The Netherlands
Brooks, D. W., Abbott Laboratories, Pharmaceutical Products Division, Abbott Park, Illinois 60064, U.S.A.
Brümmer, W. K., E. Merck Darmstadt, Frankfurter Str. 250, D-6100 Darmstadt, FRG
Cals, M. J. H., DSM, Research and Patents, P.O. Box 18, NL-6160 MD Geleen, The Netherlands
Dassen, B. H. N., DSM, Research and Patents, P. O. Box 18, NL-6160 MD Geleen, The Netherlands
Duarte, J. M. C., Department of Chemical Industries LNETI, 2745 Queluz, Portugal
Fersht, A. R., Department of Chemistry, Imperial College of Science & Technology, London SW 7 2AY, U. K.
Fourneron, J. D., Laboratory of Organic and Bioorganic Chemistry, Faculté des Sciences de Luminy, Case 901, 70 route Léon Lachamp, F-13288 Marseille Cedex 9, France

Fuganti, C., Dipartimento di Chimica del Politecnico, Centro del CNR per la Chimica delle Sostanze Organiche Naturali, Piazza Leonardo da Vinci 32, I-20133 Milano, Italy

Furstoss, R., Laboratory of Organic and Bioorganic Chemistry, Faculté des Sciences de Luminy, Case 901, 70 route Léon Lachamp, F-13288 Marseille Cedex 9, France

Gais, H.-J., Institut für Organische Chemie und Biochemie der Technischen Hochschule, Petersenstr. 22, D-6100 Darmstadt, FRG

Ghisalba, O., Central Research Laboratories of Ciba-Geigy Ltd., CH-4002 Basel, Switzerland

Godtfredsen, S. E., Novo Industri A/S, Novo Alle, DK-2880 Bagsvaerd, Denmark

Guanti, G., Institute of Organic Chemistry, University of Genova, Centro C.N.R., Palazzo delle Scienze, Corso Europa, I-16132 Genova, Italy

Guaragna, A., Institute of Organic Chemistry, University of Genova, Centro C.N.R., Palazzo delle Scienze, Corso Europa, I-16132 Genova, Italy

Günther, H., Institute for Organic Chemistry, Technical University Munich, Lichtenbergstr. 4, D-8046 Garching, FRG

Hilvert, D., Laboratory of Bioorganic Chemistry and Biochemistry, The Rockefeller University, 1230 York Avenue, New York, New York 10021, USA

Hoppe, D., Department of Organic Chemistry, University of Göttingen, Tammannstr. 2, D-3400 Göttingen, FRG

Ingvorsen, K., Novo Industri A/S, Novo Alle, DK-2880 Bagsvaerd, Denmark

Kaiser, E. T., Laboratory of Bioorganic Chemistry and Biochemistry, The Rockefeller University, 1230 York Avenue, New York, New York 10021, USA

Kerkhoffs, P. L., DSM, Research and Patents, P.O. Box 18, NL-6160 MD Geleen, The Netherlands

Kobayashi, S., Faculty of Pharmaceutical Sciences, University of Tokyo, 7-3-1 Hongo, Bunkyo-ku, Tokyo 113, Japan

Laane, C., Department of Biochemistry, Agricultural University, De Dreijen 12, NL-6703 BC Wageningen, The Netherlands

Leatherbarrow, R. J., Department of Chemistry, Imperial College of Science & Technology, London SW 7 2AY, U. K.

Lemière, G. L., Laboratory of Organic Chemistry, University of Antwerp (R.U.C.A.), Groenenborgerlaan 171, B-2020 Antwerp, Belgium

Meijer, E. M., DSM Research and Patents, P.O. Box 18, NL-6160 MD Geleen, The Netherlands

Nambiar, K. P., Department of Chemistry, Harvard University, 12 Oxford Street, Cambridge, Massachusetts 02138, USA

Narisano, E., Institute of Organic Chemistry, University of Genova, Centro C.N.R., Palazzo delle Scienze, Corso Europa, I-16132 Genova, Italy

Ohno, M., Faculty of Pharmaceutical Sciences, University of Tokyo, 7-3-1 Hongo, Bunkyo-ku, Tokyo 113, Japan
Peters, P.J.H., DSM, Research and Patents, P.O. Box 18, NL-6160 Geleen, The Netherlands
Presnell, S. R., Department of Chemistry, Harvard University, 12 Oxford Street, Cambridge, Massachusetts 02138, USA
Pugh, S. Y. R., Biotechnology Centre, Cranfield Institute of Technology, Cranfield, Bedfordshire MK 43, OAL, U.K.
Ramos Tombo, G.M., Central Research Laboratories of Ciba-Geigy Ltd., CH-4002 Basel, Switzerland
Reimerdes, E. H., Food Chemistry and Biotechnology, Bergische Universität-GH-Wuppertal, D-5600 Wuppertal 1, FRG
Riefling, B. F., E. Merck Darmstadt, Frankfurter Str. 250, D-6100 Darmstadt, FRG
Roberts, M. J. A., DSM, Research and Patents, P.O. Box 18, NL-6160 MD Geleen, The Netherlands
Roberts, S. M., Department of Microbiological Chemistry, Glaxo Group Research Ltd., Greenford, Middlesex UB6 OHE, U.K.
Schär, H. - P., Central Research Laboratories of Ciba-Geigy Ltd., CH-4002 Basel, Switzerland
Schoemaker, H. E., DSM, Research and Patents, P.O. Box 18, NL-6160 MD Geleen, The Netherlands
Simon, H., Institute for Organic Chemistry, Technical University Munich, Lichtenbergstr. 4, D-8046 Garching, FRG
Sinay, P., Laboratoire de Biochimie Structurale, U.A. 499, U.E.R. de Sciences Fondamentales et Appliquées, F-45046 Orléans, France
Stackhouse, J., Department of Chemistry, Harvard University, 12 Oxford Street, Cambridge, Massachusetts 02138, USA
Thanos, J., Institute for Organic Chemistry, Technical University Munich, Lichtenbergstr. 4, D-8046 Garching, FRG
Tramper, J., Department of Food Science, Food and Bioengineering Group, Agricultural University, De Dreijen 12, NL-6703 BC Wageningen, The Netherlands
Vigne, B., Laboratory of Organic and Bioorganic Chemistry, Faculté des Sciences de Luminy, Case 901, 70 route Léon Lachamp, F-13288 Marseille Cedex 9, France
Wandrey, C., Nuclear Research Center, Institute of Biotechnology, P.O. Box 1913, D-5170 Jülich, FRG
Wells, T.N.C., Department of Chemistry, Imperial College of Science & Technology, London SW 7 2AY, U.K.
Winterfeldt, E., Institut für Organische Chemie, Universtität Hannover Schneiderberg 1 B, D-3000 Hannover, FRG
Wong, C. - H., Department of Chemistry, Texas A & M University, College Station, Texas 77843, USA
Yde, B., Novo Industri A/S, Novo Alle, DK-2880 Bagsvaerd, Denmark
Young, D. W., School of Chemistry and Molecular Sciences, University of Sussex, Falmer, Brighton BN1 9QJ, U.K.

Subject index